W9-AKS-087

MONOGRAPHS ON THE
PHYSICS AND CHEMISTRY
OF MATERIALS

MONOGRAPHS ON THE PHYSICS AND CHEMISTRY OF MATERIALS

Strong solids (*Third edition*) A. Kelly and N. H. Macmillan
Optical spectroscopy of inorganic solids B. Henderson and G. F. Imbusch
Chemistry of the metal–gas interface M. N. Roberts and C. S. McKee
Quantum theory of collective phenomena G. L. Sewell
Experimental high-resolution electron microscopy (*Second edition*) J. C. H. Spence
Experimental techniques in low-temperature physics Guy K. White
Principles of dielectrics B. K. P. Scaife
Basic theory of surface states Sydney G. Davison and Maria Stęślicka
Acoustic microscopy Andrew Briggs

Acoustic Microscopy

ANDREW BRIGGS
Department of Materials, University of Oxford

CLARENDON PRESS · OXFORD
1992

Oxford University Press, Walton Street, Oxford OX2 6DP
Oxford New York Toronto
Delhi Bombay Calcutta Madras Karachi
Petaling Jaya Singapore Hong Kong Tokyo
Nairobi Dar es Salaam Cape Town
Melbourne Auckland
and associated companies in
Berlin Ibadan

Oxford is a trade mark of Oxford University Press

Published in the United States
by Oxford University Press, New York

A catalogue record for this book is available from the British Library

Library of Congress Cataloging in Publication Data
Briggs, Andrew.
Acoustic microscopy / Andrew Briggs.
(Monographs on the physics and chemistry of materials; 46)
Includes bibliographical references and index.
1. Materials—Microscopy. 2. Acoustic microscopes. I. Title.
II. Series.
TA417.23.B74 1992 620.1'1299—dc20 91-18378
ISBN 0-19-851377-1

Typeset by The Universities Press (Belfast) Ltd
Printed in Great Britain by
Bookcraft (Bath) Ltd
Midsomer Norton, Avon

For

Diana, Felicity, and Elizabeth

Most happy she that most assured doth rest,
But he most happy who such one loves best.

Spenser

Foreword

Professor Sir Peter Hirsch FRS

The properties of materials, whether natural or man-made, depend on microstructure. In order to understand the behaviour of materials, it is therefore essential to observe and characterize the often complex microstructures. There is now a wealth of powerful techniques available for this purpose, including optical microscopy, electron microscopy (both by transmission of thin specimens and by scanning of surfaces of bulk material), scanning tunnelling and atomic force microscopy, field ion microscopy, scanning proton microscopy, X-ray topography, NMR imaging, etc. Each of these techniques depends on a different contrast mechanism to reveal the structure, and different techniques can be used to identify different aspects of microstructure (for example changes in composition, or defects in crystals, or local changes in electronic states), and on different scales; each has its own particular range of applications for which it is particularly suitable.

Acoustic microscopy is an important addition to this powerful armoury. It depends on the elastic response of the material to acoustic waves, and therefore provides information on local changes in elastic properties; thus, for example, it is particularly sensitive to fine cracks (which might not be observable by other techniques). It has already been applied to a wide range of materials, including biological specimens, minerals, semiconductor devices, composites, ceramics, etc. As is the case for all other techniques, it is essential to have a clear understanding of the contrast mechanisms, so that the observations can be interpreted with confidence. This technique has now been developed to a stage at which potential users, such as materials scientists and biologists, need a comprehensive account of the basic technique, of the contrast mechanisms, and of the way the technique can be applied to obtain information on microstructure in different types of specimen. This book aims to achieve this objective, and is written by an author with much experience in this field with the users' needs in mind. I am sure that it will help to spread the further development and range of applications of acoustic microscopy, and lead to scientific advances in a variety of fields.

Acknowledgements

Sebastian. My kind Antonio,
I can no other answer make, but thanks,
And thanks, and ever thanks; and oft good turns
Are shuffled off with such uncurrent pay.
William Shakespeare, Twelfth Night III, iii (Theobald's emendation).

The three people in Oxford who contributed at the most personal level to the production of this book are Sue Spencer, Chris Goringe, and my wife Diana. Sue wrote the programs that allowed all the computer plotted figures to be produced, some adapting programs developed in research, others written entirely by her. Chris spent a long vacation putting together figures, tables, and references, and also suggesting numerous improvements to the book. To both of them I am deeply grateful. Diana helped in countless ways, from sorting out details of the text to simply making each day more worthwhile. I thank her.

But of course everyone else who has worked in the lab in Oxford has contributed too, both through the direct results described here, and through the development and refinement of ideas together. Sir Peter Hirsch started the whole venture, and has maintained constant support, encouragement, and guidance ever since. Yin Qin-rui got pictures of lead lanthanum zirconate titanate, and produced the first paper from the instrument at Oxford. Rolf Weglein taught us about the importance of defocusing. Chris Ilett discovered cracks in titanium nitride coatings on cutting tools, and set us on a course that we are still pursuing. Paul Warren found out how difficult it is to exploit the piezoelectric properties of a specimen. Mike Somekh contributed to almost every problem going on in the lab, and his lasting theoretical contributions include an understanding of the contrast from grain structure and the contrast from surface cracks and interfaces. John Weaver, too, contributed to the research of everyone around him, and he developed two electronics systems for quantitative work which have been in constant use ever since. Simon Peck made some wonderful studies of human teeth. John Rowe showed us how to analyse the quantitative data, and set up a splendid set of software for doing just that. Richard Tew showed us what a difference professional mathematics made. Bill O'Brien opened the way to biological applications. Jun-ichi Kushibiki established a quantitative facility here, and has given incalculable support in making lenses and in countless other ways. Chris Daft established a time-resolved method for measuring the properties of biological tissue. Bruce Thompson got us started on measuring short cracks. Andy Fagan got fascinating pictures of hard metals and a range of other materials. Sue Spencer implemented the quantitative facilities in a way that everyone in the lab could use. Chris Scruby brought enormous expertise of materials and non-destructive testing to the lab, and all of us have come to value his wise judgement and leadership. Duncan Fatkin demonstrated the usefulness of acoustic microscopy for studying advanced low-ductility materials. Charles Lawrence followed that up for a wide range of ceramic fibre composites. Modesto Montoto, Angel Rodriguez-Rey, and Tom Field applied the technique to the study of rocks. Alastair Sinton introduced professional computer programming to us, and applied it to measurements of protective polymer coatings. Trevor Gardner worked out how to relate the properties of teeth and bones to those of their constituent crystals. Roger Grundle and Jun Wang are extending Chris Daft's techniques to human bone-derived cells. Zenon Sklar, Paolo Mutti, and Neil Stoodley are applying the quantitative techniques to thin industrial coatings and semiconductor wafers. Tong-Guang Zhai, David Bennink, and Dieter Knauß are developing time-resolved measurements of cracks. This book is full of ideas that flowed from the creative activity of all these people, and many of the pictures and graphs have come directly from their work.

There are others too, who although they have never worked in Oxford, have nevertheless strongly influenced these pages. Cal Quate, Gordon Kino, Noriyoshi Chubachi, Henry Bertoni, Jan Achenbach, Martin Hoppe, Klaus Krämer, Yusuke Tsukahara, Jürgen Bereiter-Hahn, Andrew Kulik, Roman Maev, Isao Ishikawa, Steven Meekes, Michael Muha, and Dan Rugar have all contributed indirectly and directly. Sections 12.3 and 12.4 come largely from private letters from Henry Bertoni and Jan Achenbach. I am grateful to everyone who has provided figures (where they have been previously published this is acknowledged by a reference in the caption). Yusuke Tsukahara, Henry Bertoni, and Gordon Kino gave very helpful detailed comments on drafts of the book, saving it from numerous errors and showing how to make it much clearer. My parents and David Bowler corrected countless mistakes at the proof stage.

It is a pleasure to acknowledge the contribution of all these people, and to express my deep thanks to them.

Oxford Andrew Briggs
December 1991

Contents

List of symbols

These are the principal symbols for constants and variables, together with the parts of the text in which they appear, usually described by equation numbers, but sometimes by section where they are discussed in prose. The tables in which values are listed are also given. A few key units are also included.

a_0	radius of lens aperture	(2.7), §3.1
a_C	minimum geometrical aberration	(4.5)
a_D	radius of first minimum in Airy disc	(4.6)
a_G	transverse geometrical aberration	(4.4)
a_{ij}	multilayer matrix elements	(10.5)–(10.8); Table 10.1
a_K	periodicity	§3.5
a_l	radius of lateral longitudinal wave focal ring	§7.2.3
a_R	radius of Rayleigh wave focal ring	§7.2.2, (7.44)
a_{tot}	radius of focal spot	(4.7)
a_T	radius of transducer	§4.2; (7.37)–(7.46)
A_{an}	anisotropy factor, $2c_{44}/(c_{11}-c_{12})$	§11.2–(11.16)
A_{ij}	product of layer matrixes	(10.8)–(10.11)
$A(k_x)$	zeroth order coefficient in expansion of $R(k_x)$ near θ_l	(7.48)
A_s	size asymmetry	(10.3)
A_t	time asymmetry	(10.2)
$A(z)$	signal of transmission microscope, as function of focal separation	(7.21)
B	bulk modulus	6.8 (6.9); Table 6.1
B'	local substitution in cepstral analysis	(8.68)–(8.69)
B/A	second-order non-linearity parameter	§3.4, §9.3
$B(k_x)$	first order coefficient in expansion of $R(k_x)$ near θ_l	(7.48)–(7.52)
B_W	bandwidth	§5.3
c_{ijkl}	stiffness tensor element	(6.24)
c_{IJ}	stiffness constant in abbreviated matrix notation	(6.28)–(11.16); Tables 6.1, 11.1–11.3
C'	local substitution in cepstral analysis	(8.68)–(8.69)
d	sample thickness	(7.22)–(7.23); (9.1)
dB	unit of attenuation, as $10\times$ the logarithm to base 10 of the ratio of power	§3.1.2–§6.1; (9.8)
D	transducer–lens separation	§4.3
E	Young modulus	Table 6.1
E_l	energy flux normal to interface in transmitted longitudinal wave	(6.96)

Symbol	Meaning	Reference
E_r	energy flux normal to interface in reflected wave	(6.95)
E_s	energy flux normal to interface in transmitted shear wave	(6.97)
f	frequency	(3.4)–(8.68); §10.2
F	Fresnel distance	§4.2
F_A	figure of merit in pattern matching algorithm	(10.4)
$g(t, z)$	waveform distortion function	(8.51)–(8.65)
$G(x, x')$	Green function, relating scattered field $p_R(x)$ to incident field $p_i(x')$	(12.9)–(12.12)
GHz	gigahertz, 10^9 Hz	*passim*
h	distance of a ray from the axis of the lens at a given plane	(2.4)–(2.5); (4.1)–(4.9)
J_m	Bessel function, first kind, order m	(3.2), (4.13)
k	wavenumber, $2\pi/\lambda$	(6.4)–(12.42)
$\mathbf{k}$	wavevector	(11.2)
k_B	Boltzmann constant, 1.380658×10^{-23} J K^{-1}	(3.5)–(3.6)
k_i	polynomial terms for velocity in water at range of T	(3.11); Table 3.3
k_p	complex wavenumber k that gives a pole in $R(k_x)$	(7.25)–(12.45)
k_x	x-component of wavevector	(7.24)–(12.45)
k_y	y-component of wavevector	(12.16)–(12.29)
k_0	complex wavenumber k that gives a zero in $R(k_x)$	(7.24)–(12.34)
K	spatial frequency	§3.5
	local variable in Rayleigh ray contrast	(7.43)–(7.52)
l_i	direction cosine	(11.2)–(11.7); Table 11.1
L	linearity constant	(3.13)–(3.14)
L	distance from transducer	(4.14)
L_1	lens function for outgoing wave	(7.10)–(7.11); (12.2)–(12.29)
L_2	lens function for incoming wave	(7.10)–(7.11); (12.2)–(12.29)
M	user-chosen real number in denominator in Wiener filter	(8.63)
MHz	megahertz, 10^6 Hz	*passim*
M_{jk}	local matrix in multilayer reflection	(10.10), (10.12)
M_l	local variable in lateral longitudinal reflectance function	(7.50)
n	ratio of acoustic velocity in fluid to acoustic velocity in solid	(2.2)–§10.2
n_i	refractive index of medium i	(2.1)–(2.4)
ns	nanosecond, 10^{-9} s	*passim*
N.A.	numerical aperture of lens, sin θ_0, $1/(2 \times$ f-number)	§2.1–§4.2
Np	Naper, unit of attenuation as logarithm to base e of ratio of amplitudes	§3.2, §9.2–9.3

p	pressure	(6.8)–(6.88)
p_G	geometrically reflected field at a surface	(7.28)–(7.34)
p_{inc}	field incident on a surface	(7.31)–(7.35); (12.4)–(12.9)
$P_{inc}(k_x)$	Fourier transform of p_{inc}	(7.29)–(7.30)
p_{ref}	reflected field at a surface	(7.28)
p_R	Rayleigh reflected field at a surface	(7.28)–(12.27)
P	power	(6.73)–(6.74)
P_0	power at which fractional depletion s occurs	(3.13)
$\langle P \rangle$	mean power	(6.72)–(6.74)
$\langle P_n \rangle$	mean noise power	(3.5)
$P(\theta)$	lens pupil function	(7.11)–(12.36)
q	focal length of lens	(2.2)–(7.46)
$Q(t)$	local function in Fourier $V(z)$, $P(t)R(t)t$	(7.17)–(7.18); (8.8)–(8.14)
$QQ(t)$	autocorrelation function of $Q(t)$, local function in Fourier $V(z)$	(8.14)–(8.15)
r_0	radius of curvature of lens	(2.2)–(4.11)
R	reflection coefficient	(6.67)–(11.15)
R_c	resolution coefficient	(3.10); Table 3.1
R_R	Rayleigh wave stress amplitude reflection coefficient	(12.5)–(12.46)
s_i	distance from a surface, along the lens axis	(2.1)–(2,5)
$s_{a,b}$	depth below a surface	(4.1)–(4.9)
$s_{a,b,c}$	size of turning points in pattern matching algorithm	(10.3)–(10.4)
s_d	fractional depletion of acoustic beam due to non-linear harmonic generation	(3.13)
$\bar{s}(f)$	two-way lens response in frequency domain	(8.61)–(8.64)
s_F	dimensionless distance for aperture, $\lambda L/a_T^2$	(4.14)
s_{ijkl}	stiffness tensor element	(6.25)
s_{IJ}	stiffness constant in abbreviated matrix notation	(6.28)
$\mathrm{sinc}(x)$	$\sin(x)/x$	§3.1.1–§5.1; (10.1)
$s(t)$	two-way lens response in time domain	(8.51)–(8.54)
$s(k_x, k_x')$	2-D scattering function from a crack	(12.2)–(12.45)
$s(k_x, k_x', k_y)$	3-D scattering function from a crack	(12.28)–(12.29)
$\bar{S}_0(f)$	reference signal in frequency domain	(8.61)–(8.68)
$\bar{S}(f)$	signal in frequency domain	§8.3; (8.51)–(8.68)
$S_0(t)$	reference signal in time domain	(8.51)–(8.55)
$S(t)$	signal in time domain	(8.52)–(8.55)
t	local variable in Fourier $V(z)$, $(1/\pi)\cos\theta$	(7.15)–(7.19); (8.2)–(8.14)
	time	(3.7)–(10.2)
t_0	time interval between echoes from lens and from specimen	(3.7)–(3.9)
$t_{a,b,c}$	time of turning points in pattern matching algorithm	(10.2)
T	pressure or stress amplitude transmission coefficient	(4.15); (6.67)–(6.97)

Symbol	Meaning	Reference
T	temperature	(3.5)–(3.12)
T_R	Rayleigh wave stress amplitude transmission coefficient	(12.5)–(12.46)
u	local variable in Fourier $V(z)$, kz	(7.15)–(7.19), (8.2)–(8.15)
	particle displacement	(6.11)–(6.79), (11.1)–(11.6)
$U(r)$	amplitude distribution in focal plane	(3.2)
v	velocity	(3.4)–(11.12)
v_0	velocity in coupling fluid	(3.4)–(9.7); Table 10.1; §10.2–§12.2
v_l	longitudinal wave velocity in a solid	(6.7)–(6.89); Table 6.3; §7.2.3–§11.2
v_R	Rayleigh wave velocity	(4.12); §6.3; Table 6.3; (7.8)–(8.50)
v_s	shear wave velocity in a solid	(6.6)–(6.89); Table 6.3; §8.1–§11.2
V	video signal (envelope detected)	(7.10)–(7.13)
$V(z)$	video signal as an explicit function of defocus	(7.14)–(12.36)
w	resolution	(3.1)–(3.3)
W	phase aberration	(2.6)
X	local variable in Rayleigh wave equation, $(v_R/v_s)^2$	(6.55)–(6.57); Table 6.2
Y	local variable in Rayleigh wave equation, $(v_s/v_l)^2$	(6.55)–(6.57); Table 6.2
z	defocus, displacement of specimen surface relative to focal plane of lens (towards lens is negative)	Chapter 7; (7.1)–(12.36)
Z	impedance	(4.15), (6.65)–(6.96), (8.58)–(12.46)
α	attenuation	§3.1.2, §6.1
α_0	attenuation/f^2	§3.2, Table 3.1
	normal component of wavevector in fluid	(6.77)–(6.88)
	attenuation in coupling fluid	(7.9)–(7.52), (8.18)–(8.37)
α_{acc}	acceptable round trip attenuation in the coupling fluid	(3.8)
α_{He}	attenuation in superfluid helium	(3.12)
α_l	longitudinal wavevector normal component into solid	(6.42)–(6.88); §10.2
α_R	attenuation of Rayleigh wave	(7.9)–(7.34); (12.4)–(12.28)
α_s	shear wavevector normal component into solid	(6.43)–(6.88); §10.2
β	wavevector component along surface	(6.42)–(6.87)
β_L	non-linear coupling constant	(3.14)
γ_s, γ_t	pragmatic constants in pattern matching algorithm	(10.4)
Γ_{ij}	Christoffel matrix element	(11.1)–(11.6); Table 11.1

ε_{ij}	strain tensor element	(6.23)–(6.31)
ε_{I}	strain tensor element in abbreviated matrix notation	(6.27)–(6.28)
$\eta_{\mathrm{R,T}}$	local variables in Rayleigh ray contrast	(7.45)–(7.46)
η	shear viscosity	(6.20)–(6.22); (12.41)
θ	zenithal angle	(7.10)–(12.36)
θ_0	semi-angle of lens subtended at focus	(2.3)
θ_m	complex angle of refraction	(10.6)–(10.7)
θ_{opt}	optimum lens angle for subsurface imaging	(4.9)
θ_r	semi-angle of spherical surface subtended at centre	(2.3)–(2.6)
θ_{R}	Rayleigh angle, $\sin^{-1}(v_0/v_{\mathrm{R}})$	(7.2)–(9.9)
Θ	dilation	(6.31)–(6.47)
λ	Lamé elastic constant	(6.2)–(6.6); Table 6.1
	wavelength	(4.6)–(4.14), (12.1)–(12.44)
λ_0	wavelength in coupling fluid	(3.1)–§11.5
λ_{R}	Rayleigh wavelength	Table 6.2; §6.3–(12.44)
μ	Lamé elastic constant, shear modulus	(6.6); Table 6.1
μm	micrometre, 10^{-6} m	*passim*
ξ	transverse displacement	(6.1)–(6.4)
ρ	density,	(6.1)–(12.41)
ρ_1	density of solid	(6.82)–(7.53)
ρ_0	density of fluid	(6.82)–(12.41)
σ	Poisson ratio	Table 6.1; (6.56)–(6.58)
σ_{ij}	stress tensor element	(6.24)–(6.32), (10.5)–(10.9)
σ_{I}	stress tensor element in abbreviated notation	(6.26)–(6.28)
σ_{s}	shear stress	(6.1)–(6.2)
$\boldsymbol{\sigma}_{\mathbf{T}}$	traction	(6.65)–(6.72)
τ	time resolution	(5.3)
	relaxation time	(6.16)–(6.21)
ϕ	azimuthal angle	(7.10)–(7.12); (11.11)–(12.35)
	phase	(7.1)–(7.5); (8.21)–(8.50)
	scalar potential	(6.36)–(6.83)
χ	local substitution in layered media calculation	(10.6)–(10.7)
$\boldsymbol{\psi}$	vector potential	(6.36)–(6.82)
ω	angular frequency, $2\pi/f$	(6.4)–(6.22)

1
Son et lumière

Οὐ γνώσεσθε; οὐκ ἀκούσεσθε; (Isaiah 40:21 (LXX))

1.1 Composites

One of the scientists in our laboratory, Charles Lawrence, has a knack of having a beautiful picture on the screen of one of our acoustic microscopes whenever anyone walks past. So it always seems, anyway. He has studied quite a number of ceramics and composites in the acoustic microscope (Scruby *et al.* 1989; Lawrence *et al.* 1989, 1990). His favourite picture is shown in Fig. 1.1(a). It is a section through a glass matrix composite. The matrix is a borosilicate glass, similar to glass better known under the trade name Pyrex. The role of the matrix is largely to hold the fibres together (preferably without touching), and to protect their surfaces mechanically and chemically so that they can support loads at elevated temperatures in aggressive environments. The composition of the glass can be tailored for optimum compatibility with the fibres, and the relatively low viscosity at high temperature enables almost complete penetration of the matrix into the fibre bundles during fabrication. The commercial name of the fibres is Nicalon; they are composed mainly of silicon carbide. The composite has a flexural strength of 1.25 GPa and a work of fracture of 40 kJ m^{-2}. For comparison, an optical picture of another area of the specimen is shown in Fig. 1.1(b). Newcomers to acoustic microscopy find optical comparisons reassuring, though as confidence is gained with the acoustic microscope such comparisons become less and less necessary. Crucial questions for the development of a composite like this one are: are its properties, such as its strength and toughness at high temperatures, the best that can ever be obtained? If so, how do you make sure that those properties are consistently achieved? If not, how do you improve them still further?

Great care must be taken not to introduce damage in the preparation of sections of ceramic composites for microscopy; otherwise it will be hard to distinguish what was intrinsic to the specimen from what was introduced during the preparation of the section. This specimen was first cut off perpendicular to the direction of the fibres using a slow annular diamond wheel (100–150 r.p.m.), with a flow of water to lubricate the cutting process and to keep it cool. The cut off piece was mounted in a cold-cure epoxy resin, with bubbles removed to prevent them from becoming traps for diamond paste in subsequent polishing. When the

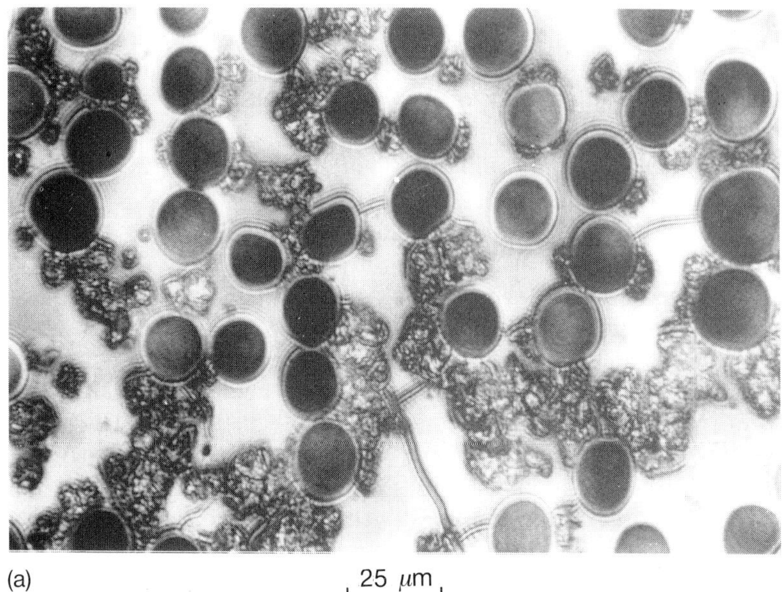

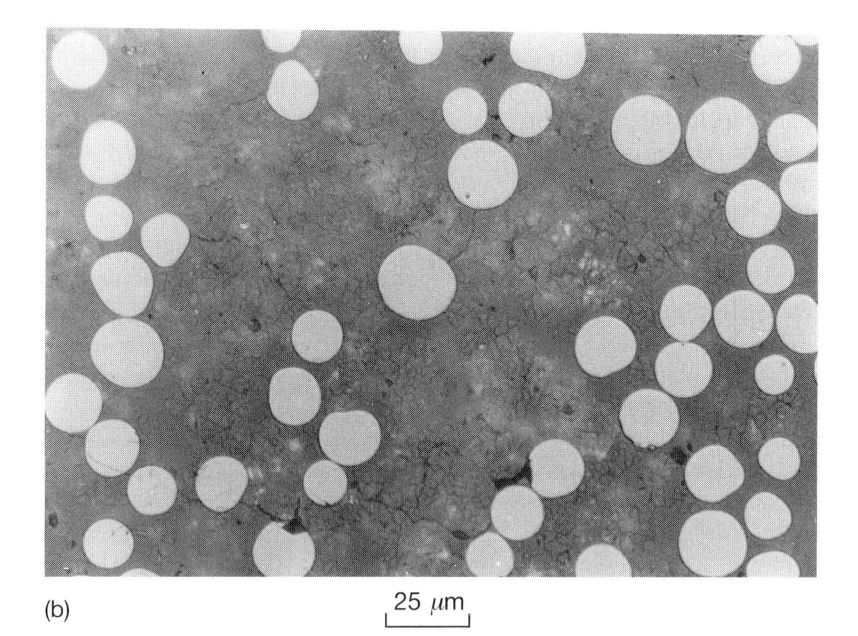

FIG. 1.1. A section through a glass matrix composite: (a) Acoustic micrograph, 1.9 GHz, $z = -2.8\,\mu$m; (b) optical micrograph. The matrix is borosilicate glass, Corning code 7740: composition by weight, SiO_2, 81 per cent; Al_2O_3, 2 per cent;

resin had set, the edges were rounded off, and the specimen was mounted on a precision polishing jig. The surface was lapped and polished, with a relatively low load, using oil-soluble 6 μm diamond paste on a scrolled cast iron platten, and then in turn 1, 0.25, and 0.1 μm paste on a scrolled brass platten. The specimen was thoroughly cleaned in an ultrasonic bath between every stage. It was given a final polish with recirculating Syton-W30 (0.125 μm colloidal particles of silica suspended in water) using a thin coating of polyurethane foam on the platten. In the acoustic microscope you do not want any surface relief, so this stage had to be finely judged—not too short or the damage from the 0.1 μm paste would not be removed, not too long or softer material would be polished away too much. The specimen was then examined in the acoustic microscope.

Figure 1.1(a) shows a wealth of detail. First, and most obvious, the fibres have a different contrast from the matrix. In the imaging conditions here, the fibres are darker than the matrix. By simply counting the fibres and measuring their diameters, the mean SiC fibre diameter is 16 μm, and the area fraction is 0.49. 'Well,' you say, 'that's easy, any old microscope would have told you that.' But the matrix appears to be composed of two phases. About 70 per cent of the matrix area appears homogeneous, while the remaining 30 per cent appears granular, with a grain size less than 10 μm. The bright, homogeneous region in Fig. 1.1(a) is borosilicate glass, behaving itself perfectly well. The crystalline phase was subsequently confirmed, using both energy-dispersive and wavelength-dispersive spectroscopy on an electron microprobe and also X-ray diffraction, to be crystobalite. Crystobalite is an allotrope of silica with a cubic crystal structure, which is stable at high temperature (1470–1713°C) under atmospheric pressure. The precipitation of crystobalite, known as devitrification, is likely to occur when borosilicate glass is held for a prolonged time between the glass softening temperature and the crystallization temperature. Unfortunately, the fabrication procedure for 7740/SiC composite involves exactly these conditions. An acoustic micrograph of a section through a monofilament specimen parallel to the fibre axis is shown in Fig. 1.2(a). The crystobalite seems to grow to form a layer 3–5 μm thick around the fibre, but it also seems to nucleate away from the fibre too.

FIG. 1.1. legend (*continued*)

B_2O_3, 13 per cent. The fibres are silicon carbide, manufactured by Nippon Carbon under the name Nicalon (Yajima *et al.* 1978): nominal diameter, 15 μm; tensile strength, 2.75 GPa; Young modulus, 196 GPa; Poisson ratio, 0.15; density, 2.55 Mg m^{-3}. The composite has fibre volume fraction 0.49; flexural strength, 1.25 GPa; Young modulus, 120 GPa; work of fracture, 40 kJ m^{-2} (Lawrence 1990).

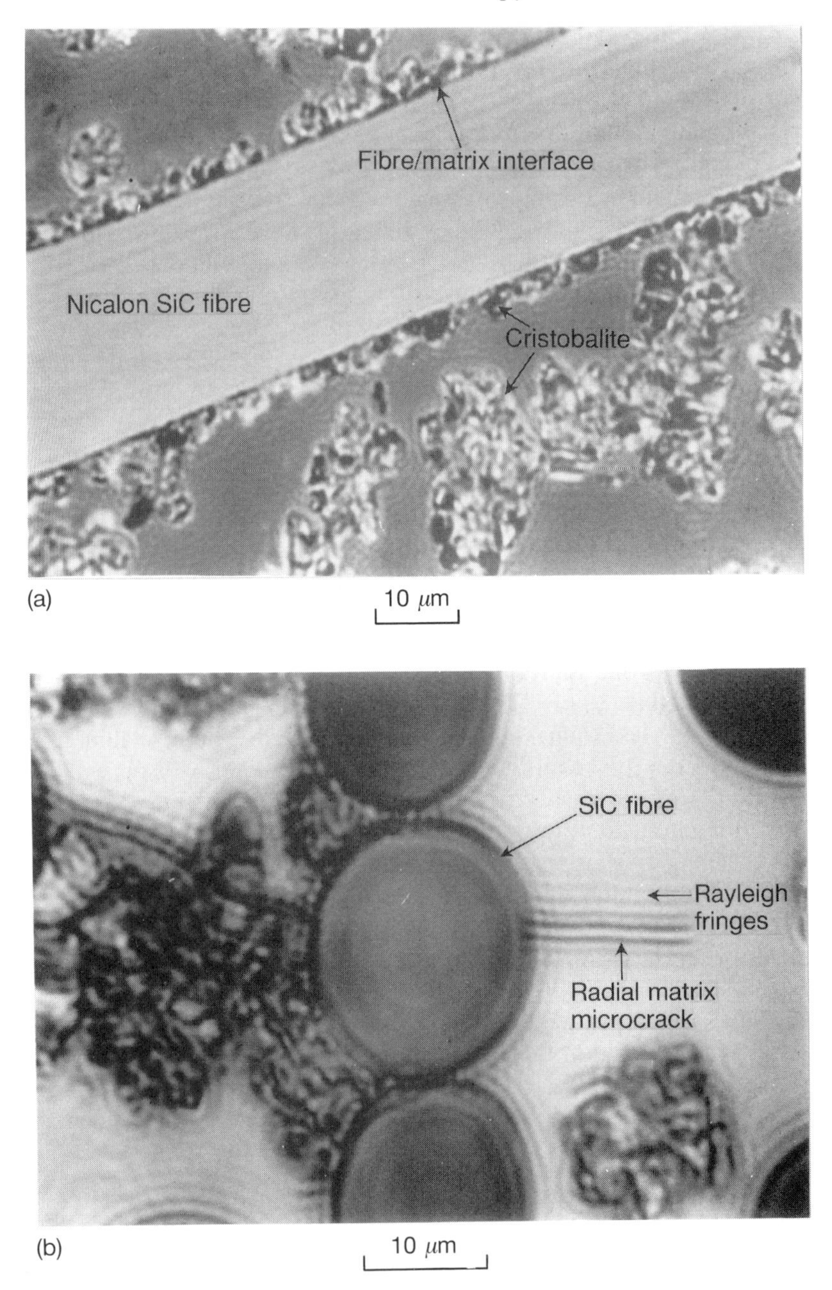

FIG. 1.2. Acoustic images of the same 7740/SiC glass matrix composite as in Fig. 1.1. (a) Section parallel to the axis of a fibre; (b) section perpendicular to the fibre axes; 1.9 GHz (Lawrence 1990).

There is a phenomenon in the acoustic images that is not seen by any other kind of microscopy. In all of the acoustic micrographs of these specimens there are fringes, both in the glass and in the silicon carbide. In the silicon carbide they are more widely spaced than in the glass, and they could be mistaken for growth rings, perhaps associated with the processing of the fibres, such as are found in a tree trunk. But they are not growth rings; they are an interference effect associated with waves that can be excited in the surface of solids. These surface waves have longitudinal and shear components that decay exponentially with depth: they are called Rayleigh waves after the Cambridge scientist Lord Rayleigh, who discovered them theoretically in 1885. In an acoustic microscope surface waves can be excited rather strongly in a solid, and interference with specularly reflected waves gives rise to the interference fringes observed. The spacing of the interference fringes is half the Rayleigh wavelength. The Rayleigh wavelength in the fibres must be greater than it is in the matrix, because the separation of the fringes is smaller in the matrix than it is in the fibres. In the matrix the fringes are associated with at least three kinds of features. First, there are fringes around the fibres, which are loci of constant distance from the interface between the fibre and the matrix, just as the fringes inside the fibres are. Then there are fringes, too, that seem to run around areas of crystobalite, or perhaps at constant distance from interfaces between crystobalite and glass. Finally, especially in Fig. 1.2(b) there are fringes that seem to run alongside cracks in the glass. The acoustic microscope has a greatly enhanced sensitivity to cracks or boundaries that run up to the surface, and is able to give strong contrast from them even when they are much finer than the wavelength being used for imaging. This sensitivity is associated with the excitation of Rayleigh waves in the surface, and their scattering by cracks or change of velocity as they pass from one material to another. The fringes are a manifestation of Rayleigh wave activity. The contrast from surface cracks and boundaries will be discussed in detail in Chapter 12, but there is a great deal of ground to be covered before then.

1.2 Rocks

Two geologists from Oviedo University in Spain, Modesto Montoto and Angel Rodriguez-Rey, heard about what acoustic microscopy could do. They came to our laboratory and spent considerable periods of time learning about the technique and using it to study their samples (Rodriguez-Rey *et al.* 1990). A pair of pictures of an area of a polished 30 μm thick specimen of granodiorite, a type of granite rock, are shown

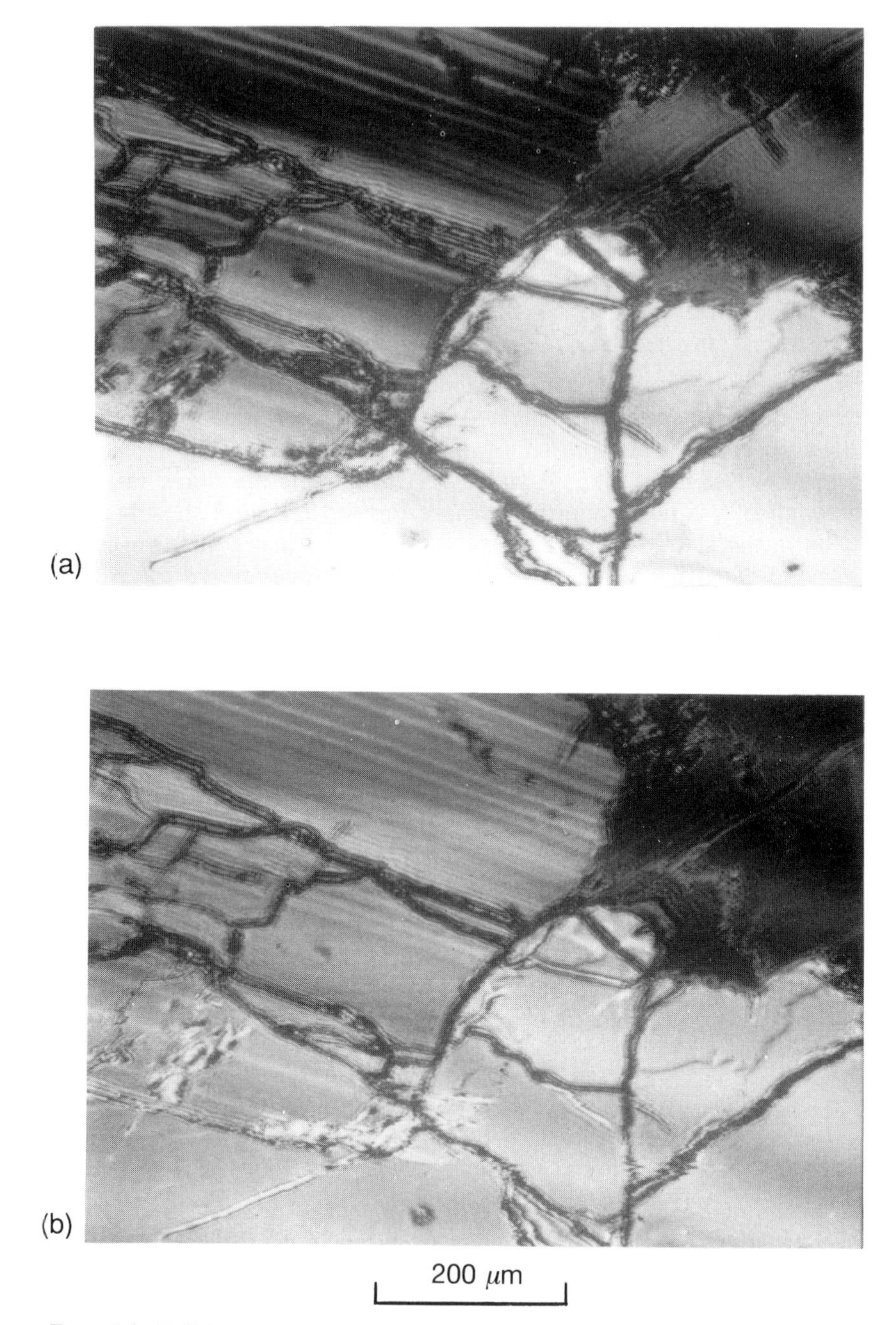

FIG. 1.3. Polished granodiorite section containing plagioclase, biotite, and quartz; 400 MHz. (a) $z = -10\ \mu m$; (b) $z = -32\ \mu m$ (Rodriguez-Rey *et al.* 1990).

in Fig. 1.3. This specimen was taken from the quarries used in the sixteenth century to provide building stones for the monastery of San Lorenzo el Real de El Escorial, near Madrid in Spain. Once again great care was taken to obtain a flat surface with no relief polishing, though in this case it was possible to use standard techniques for preparation of such specimens for conventional petrographic study (except that only one surface is needed for acoustic microscopy in reflection). Three different minerals are present in this area of the specimen. In the top right corner is biotite, a highly anisotropic phyllosilicate. In the top left region is plagioclase, which appears in the rock as zoned crystals ranging in composition from oligoclase to andesine. Finally, the mineral in the lower region of the picture is quartz.

The difference between the pictures is due to the different distance from the lens at which each was taken. No-one can operate an acoustic microscope for very long without discovering that the contrast can vary dramatically as the specimen is moved up and down relative to the lens. Indeed, it is quite possible for contrast reversals to occur between, for example, one grain in a specimen and another. In Fig. 1.3 the contrast of the biotite varies quite markedly between (a) and (b), relative to the plagioclase and the quartz. Plagioclase and quartz give rather similar contrast to each other in both pictures; they are distinguished by the twinning microstructure in the plagioclase. The variation of the signal from any given point on the specimen with the distance between the lens and the specimen is so important that an entire chapter (Chapter 7) is devoted to it; indeed, it is also the main subject of the concluding chapter (Chapter 13).

In the disposal of radioactive waste, one potential pathway of radionuclides back to the environment is through cracks in the rock in which it is stored. Of course, this would only be relevant if a container failed and released waste into the surrounding rock, but it is important to understand what would happen in that case. Figure 1.4 is a series of images of another section of granodiorite, which might be a candidate type of rock for the disposal of nuclear waste. Figure 1.4(a) is an acoustic image. The specimen had previously been studied by light microscopy and electron microscopy. By impregnating it with fluoresceine, and then examining it in a fluorescence microscope, it is possible to make the cracks show up as bright against an otherwise dark background; this is shown in Fig. 1.4(b). In Fig. 1.4(a) the same cracks show up with a characteristic fringe pattern like the ones in the acoustic pictures of the composite in §1.1. Comparison of Fig. 1.4(a) and (b) suggests that both techniques are rather effective for revealing cracks. Rock in sections 30 μm thick can also be examined in a transmitted-light polarizing microscope; Fig. 1.4(c) illustrates how the grain structure can be revealed in this way, because of the crystal optical anisotropy. On the left side of the picture there is a grain which looks uniformly bright, and which can be interpreted as a single quartz grain with only a few cracks in it. The other half of the picture has a grain with a different colour which has rather more detail in it. If you look very closely, it is possible to discern rather fuzzy borders delineating different levels of brightness within that grain. But it is not easy. These borders correspond to the two rather wiggly lines in the right-hand part of Fig. 1.4(a). They are boundaries where small misorientation within a grain has occurred by the formation of dislocations in order to accommodate strain during the geological processes of deformation of the rock (cf. the quartz in Fig. 1.3). They are blurred in the transmission polarized light micrograph because the picture is averaged over the whole 30 μm thickness of the sample. The contrast in the acoustic microscope comes from the slight change in crystallographic orientation across the misorientation boundary. The contrast in images of anisotropic specimens, together with quantitative measurements, will be described in Chapter 11.

There is a particularly interesting crack indicated by * in the acoustic micrograph of this specimen. It is detected and identified as a crack in the acoustic microscope by the characteristic fringes that run either side of it. This crack had not been detected in any other microscope before it was seen in the acoustic microscope. The specimen was subsequently examined again in a scanning electron microscope (s.e.m.). By working first at low magnification, in order to identify the area corresponding to Fig. 1.4(a), and then progressively increasing the magnification, it was possible to find the crack. It occupies most of the diagonal in the s.e.m. picture in Fig. 1.4(d). Even so, it is not easy to see, and the s.e.m. picture emphasizes the astonishing enhancement of the contrast from surface cracks that is available in the acoustic microscope.

1.3 Biological matrix

For a long time there has been a desire to understand better the relationship between the biomechanics and the biochemistry of bone development. The scientific motive is to understand how the mechanical properties develop on a microscopic scale as a function of time, and also in response to chemical and mechanical stimuli. The medical motive is to understand the interaction with prosthetic materials and devices, with a view to developing better materials and procedures for operations such as joint replacements. Hip and knee replacements for people over the retiring age have for some time been two of the most successful operations in the more developed world; it would be very nice to be able

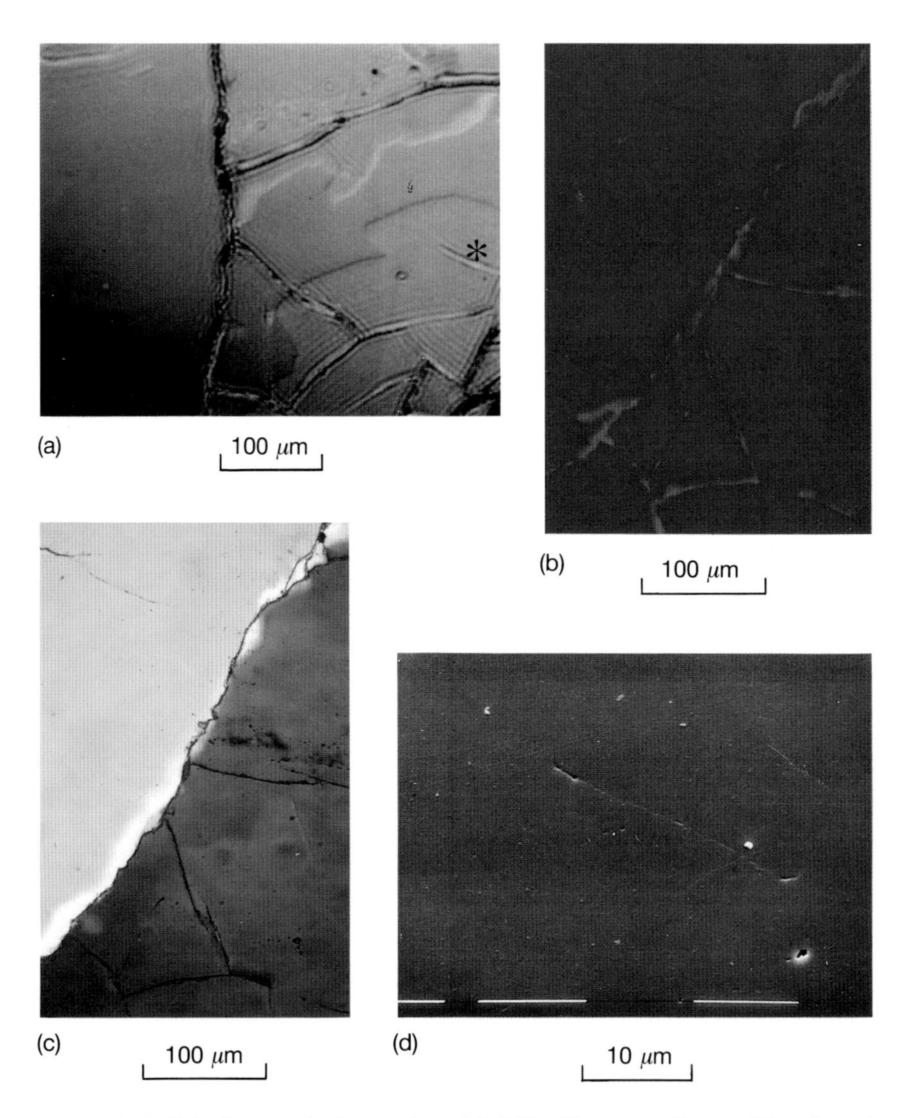

FIG. 1.4. Polished granodiorite section: (a) 350 MHz, $z = -10\ \mu m$; (b) reflected fluorescence microscopy; (c) transmitted polarized light; (d) s.e.m. (Rodriguez-Rey *et al.* 1990).

to extend them to younger age groups, and to other joints (Bulstrode 1987). Surgeons at the Nuffield Orthopaedic Centre in Oxford are using acoustic microscopy to study the development of the mechanical properties of the matrix secreted by bone-derived cells. This is the material that eventually achieves the mechanical properties associated with bone. Pictures of two specimens, taken by Roger Grundle, are shown in Fig. 1.5.

The preparation of human bone-derived cells for such a study is a major undertaking in itself. The details are somewhat technical but very important (Beresford *et al.* 1983*a,b*); the cells are described as bone-derived because they are a heterogeneous population that includes osteoblasts. During orthopaedic surgery, pieces of cancellous bone, the spongy part, are obtained sterile so that they are free from micro-organisms. The pieces are washed three times in phosphate-buffered sodium chloride to remove blood and marrow debris. They are then placed in special Nunc tissue culture flasks that have a flat bottom of area 25 cm^2. Each flask contains 10 ml of culture fluid, consisting of Dulbecco's modification of Eagle's medium, a mixture of nutrients including salts and amino acids, plus 10 per cent by volume fetal calf serum (the fluid part of blood) supplemented with 50 units ml^{-1} penicillin and 15 $\mu g\, ml^{-1}$ streptomycin. The purpose of the antibiotics is to kill off any bacterial growth that may get into the flask by mistake. The bone explants are kept at 37°C in an incubator with a mixture of 95 per cent air and 5 per cent carbon dioxide. The medium is changed once after the first 24 hours, and three times a week after that. The cells start to grow out of the explant within 7 days. After 3 to 4 weeks the whole area of the bottom of the flask is completely covered. At that point the cells are treated with trypsin, an enzyme which releases them from the surface of the flask so that they float in the medium. They are then given fresh medium, separated out in a centrifuge, and given yet more fresh medium. A sample of the cells is treated with trypan blue dye, which distinguishes between dead and living cells in a light microscope, so that you can see what proportion have survived all this tender loving care, and another sample is counted in a haemocytometer. Twenty-five thousand cells are then placed on to a 30 mm Falcon tissue culture dish, which has surface charge characteristics appropriate for bone-derived cells. The cells adhere well to the surface of the material that these dishes are made of, and the dishes fit directly on the stage of an acoustic microscope. The material also has acoustic properties that are suitable for the kinds of experiments that will be described in Chapter 9. The tissue culture dish is put inside a standard 10 cm Petri dish to keep out further unwanted micro-organisms, and the whole is placed in the incubator as before. The medium continues to be changed three times a week, and you can see how the

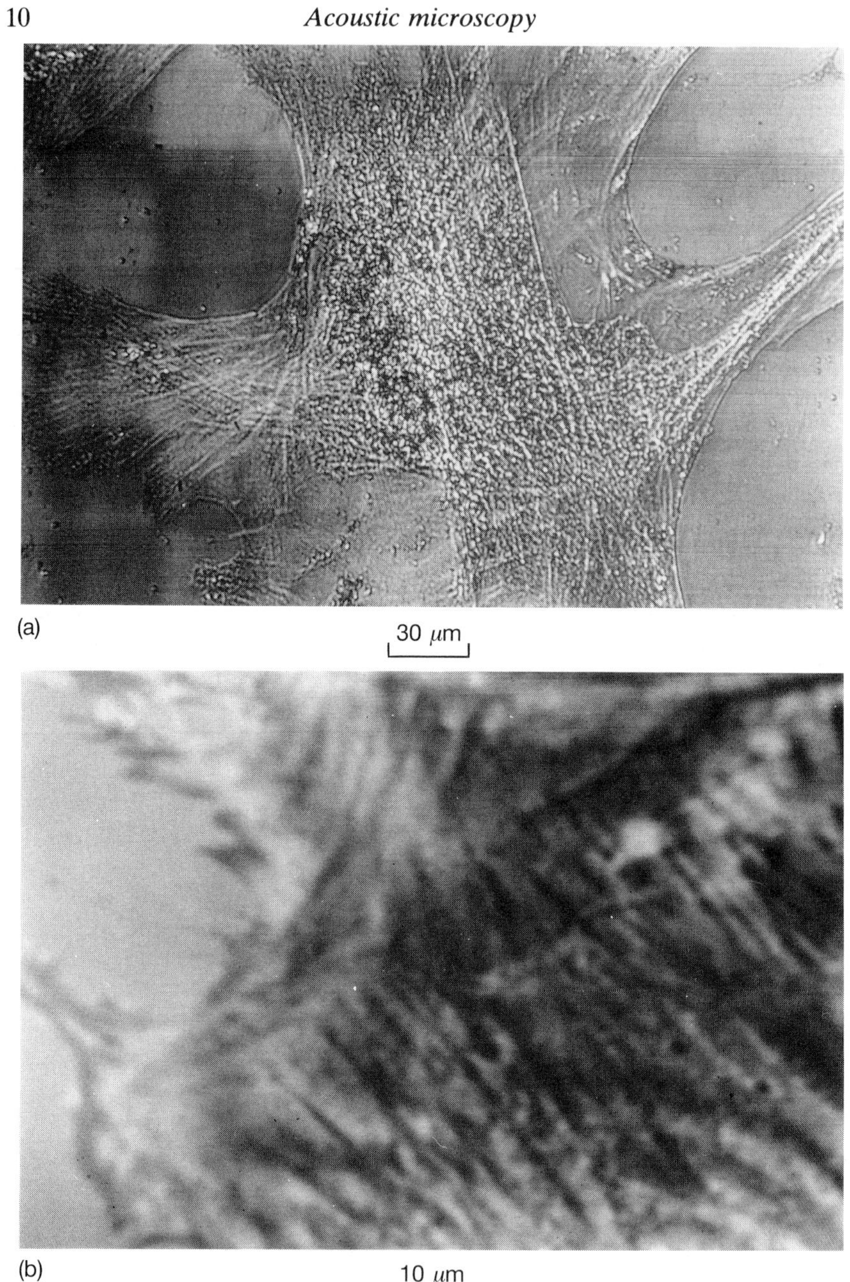

FIG. 1.5. Matrix from human bone-derived cells: (a) fixed, after 5 weeks, 1.5 GHz; (b) unfixed, after 1 week, 1.3 GHz. Both pictures were taken with the coupling fluid at 37°C and with the substrate at the focus of the lens (photographs courtesy of Roger Gundle).

cells are getting on by looking at them in a phase contrast microscope with green light. The rate of growth of cells, and especially how much matrix they produce, how fast they produce it, and the rate at which it mineralizes, can be controlled by varying the concentration of fetal calf serum in the medium, and also the concentrations of additives such as ascorbate, beta-glycerophosphate, and dexamethasone.

The acoustic micrograph in Fig. 1.5(a) came from a 5-week-old preparation. It was fixed in alcohol, and stained for alkaline phosphatase and, with von Kossa stain, for biomineral material. The biomineral material of interest here is hydroxyapatite, the principal crystalline mineral constituent of bone. The ordered structure visible within the matrix is not seen with either the light or electron microscopes. But the acoustic microscope can also work perfectly well with unfixed, unstained specimens. Figure 1.5(b) is an acoustic micrograph of matrix and cells from a 17-year-old male. In addition to the standard ingredients of culture medium, these cells were specifically stimulated with beta-glycerolphosphate and a vitamin C preparation. Because the acoustic microscope can look at living cells, the same culture could be examined again and again over an extended period of time. Figure 1.5(b) came from an examination in the acoustic microscope after 1 week. One of the difficulties with these cells was that they wanted to climb on top of one another as they grew. Nevertheless, there is still a fascinating amount of detail about the matrix structure.

The significance of acoustic micrographs like this one does not come solely, or even primarily, from the fact that the cells are living. It comes also from the fact that what you are seeing in Fig. 1.5 is contrast from the mechanical structure of the specimen. In cases where this is the property of primary interest, acoustic microscopy offers the possibility of seeing it directly with submicron resolution.

1.4 What else?

Applications of acoustic microscopy are growing so fast that it is impossible to keep up with them all, still less to include them all in a book like this. The references in this book contain no less than 79 important publications from 1989–90, and no doubt there are many others too. Reviews over the last decade have included: applications to the human retina, Chinese hamster ovary cells, steel, integrated circuits, and the interior of opaque objects (Wickramasinghe 1983); printed circuit boards and a section of a cat's ear (Attal 1983); grain structure in a transparent ferroelectric ceramic, quartzite, a eutectic CaF_2–MgF_2

synthetic crystal, defects in a $AgGaSe_2$ single crystal and in an epi-silicon layer, glass fibre reinforced polycarbonate, triballoy powder metal, a delaminated permalloy film on crown glass, photoresist inclusions in an integrated circuit and pinhole defects in the metallization of another integrated circuit, and endothelial cells from *Xenopus laevis* tadpole hearts (Hoppe and Bereiter-Hahn 1985); teeth, cracks, and semiconductor materials (Briggs 1988); alumina, silicon carbide, sialon, reaction-bonded silicon nitride, hot-pressed silicon nitride, zirconia, and ceramic–matrix composites (Fatkin *et al.* 1989); granodiorite, caries lesions in human teeth, a histological section of a mouse muscle, creep crack growth in a cermet, non-destructive testing of silicon nitride ball bearings, the fibre–matrix interface in a ceramic fibre composite (Briggs *et al.* 1989); adhesion and disbonding, surface layers on GaAs and silicon, cracks in cermet coatings on hardmetals, and partially stabilized zirconia (Briggs 1990*a*); and semiconductor packaging, magnetic recording media, paint, and photoresist (Briggs and Hoppe 1991). The purpose of what follows is to introduce you to the design and operation of an acoustic microscope in Chapters 2–5; to provide some background acoustics and use it to explain the concepts of the contrast and how to make measurements in Chapters 6–8; and then to apply all this to four important classes of specimens in Chapters 9–12. This may save some people from wasting time trying to do what acoustic microscopy is not suitable for. Much more important, I hope it will show you what you can do with acoustic microscopy, and how to do it, and then how to understand what you have done.

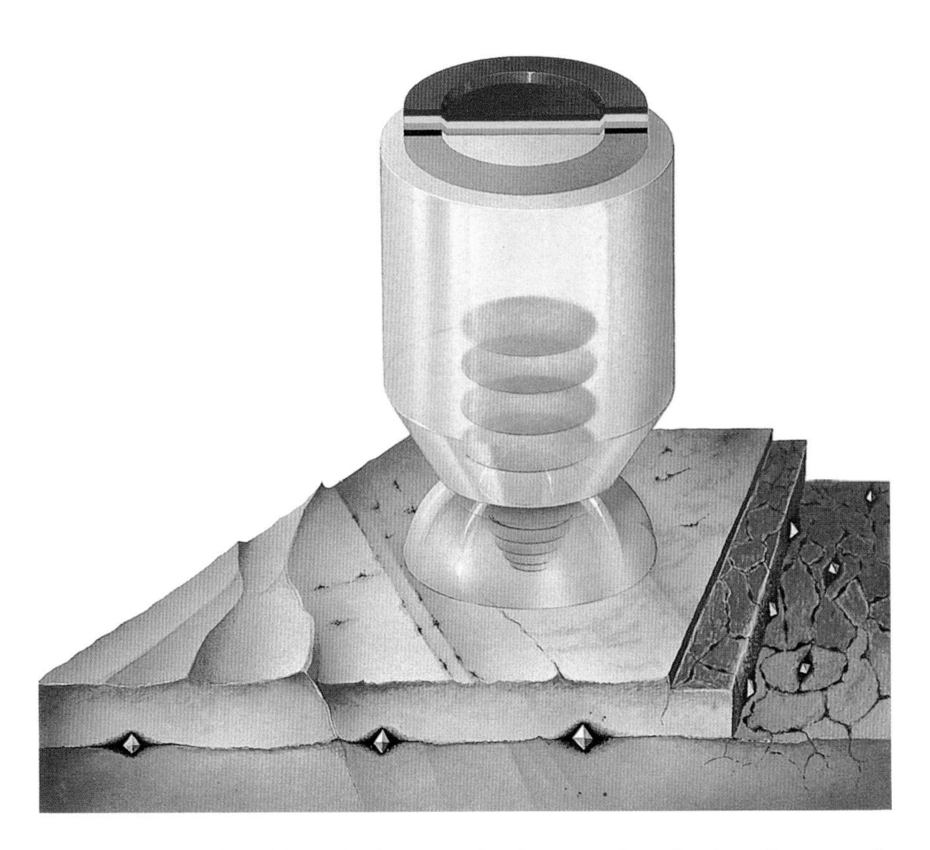

FIG. 2.1. A lens for high-resolution acoustic microscopy in reflection. The central transparent part is a single crystal of sapphire, with its c-axis accurately parallel to the axis of the cylinder. The sandwich structure at the top is the transducer, with the yellow representing an epitaxially grown layer of zinc oxide between two gold electrodes. The pink shaded areas within the sapphire represent the plane-wavefronts of an acoustic pulse; they are refracted at the lens cavity so that they become spherical in the coupling fluid. A lens for use at 2 GHz would have a cavity of radius 40 μm (courtesy of Leica, Wetzlar)

2
Focusing and scanning

If you have a particle or a field of some kind whose interaction with material objects varies with their constitution, the chances are that you will be able to fashion a microscope from the phenomenon. (John Maddox 1985)

He murmured almost to himself, 'sixteen surfaces, all opaque, all misaligned.' (Quate 1985)

2.1 Focused acoustic beams

Successful development of the scanning acoustic microscope in its present form began with the realization that it is not possible to make a high-resolution acoustic lens that can image more than one point of an object at a time, but it is possible to make an acoustic lens that has excellent focusing properties on its axis. The design that has been universally adopted for high-resolution work is shown in Fig. 2.1. The lens basically consists of a disc of sapphire with the axis of the disc aligned accurately parallel to the crystallographic *c*-axis of the sapphire. In the centre of one face of the disc a concave spherical surface is ground. This surface provides the focusing action and, to optimize transmission, it is coated with a quarter-wavelength thick matching layer. On the other face a thin film of gold is deposited to form a ground electrode. A transducer is then placed on this face, usually by epitaxially growing zinc oxide (ZnO) using vacuum sputtering. Finally, a small dot of gold is deposited to define an active area of the transducer that is accurately opposite the focusing surface. When the transducer is energized, plane acoustic waves are generated which travel through the disc. In use, the lens is placed in contact with a coupling fluid (usually water), and when the waves cross the spherical interface between the lens and the fluid they are refracted towards a focus on the axis of the lens. The very high refractive index encountered when acoustic waves pass from a solid to a liquid enables the waves to be brought to a good focus with a single lens surface in this way, even when the numerical aperture is large. The analogous situation in optics, where refractive indices are generally much closer to unity, would give a smaller numerical aperture, and severe geometrical aberrations.

In geometrical optics the numerical aperture of a lens is given by first-order theory and spherical aberrations are given by third-order theory (Jenkins and White 1976; Hecht 1987). In first-order theory the approximation is made that $\sin\theta$ may be replaced by θ. An object in a medium of refractive index n_1 at a distance s_1 from a spherical surface of radius r_0 concave towards a second medium of refractive index n_2 forms an object at a distance s_2, and these quantities are related by

$$\frac{n_1}{s_1} + \frac{n_2}{s_2} = \frac{n_2 - n_1}{r_0}. \tag{2.1}$$

The particular case of parallel rays from infinity incident on the spherical surface is illustrated in Fig. 2.2. In this case $1/s_1 = 0$, and s_2 becomes equal to the focal length

$$q = \frac{r_0}{1 - n} \tag{2.2}$$

where $n = n_1/n_2$. The notation will be used throughout that n is the ratio of the velocity in the coupling fluid to the velocity in the solid, whether

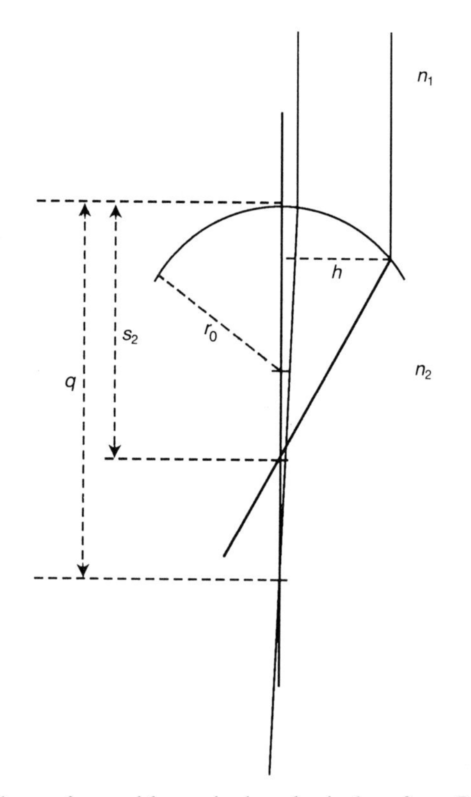

FIG. 2.2. Parallel rays focused by a single spherical surface. In the paraxial limit $h \to 0$, and $s_2 \to q$ (eqns (2.1) and (2.2)).

the solid be the lens or the specimen. In an acoustic system, n can be quite small, so that q is only slightly greater than r_0. This means that θ_0, the semi-aperture angle of the lens subtended at the focus, will be only slightly less than θ_r, the semi-angle subtended by the spherical surface at its centre of curvature. The relationship between these two is given by simple trigonometry

$$\tan \theta_0 = \frac{\sin \theta_r}{\cos \theta_r + n/(1-n)}. \tag{2.3}$$

The resolution of a microscope is inversely proportional to the numerical aperture of the lens, which is defined as $N.A. = \sin \theta_0$. The highest numerical aperture theoretically possible is therefore $N.A. = 1$. In optics the conventional definition has an additional factor of the refractive index of the fluid between the lens and the object in order to relate the resolving power to wavelengths in air, so that numerical apertures greater than unity are possible. The definition here is more appropriate to an acoustic microscope lens, where the wavelength in the coupling fluid is a more useful reference. The f-number, familiar from photography and equal to the aperture diameter divided by the focal length, is related to the numerical aperture by $\mathrm{f} = 1/(2 \times N.A.)$. A small refractive index makes it possible to have a big numerical aperture using only a single refracting surface.

The aberrations may be calculated using third-order theory, in which $\sin \theta$ is approximated by $\theta - \theta^3/3!$. Rays from an object on the axis in medium 1 are refracted to cross the axis again in medium 2 at a distance that now depends on the distance h from the axis at which they pass through the lens surface (Hecht 1987) such that

$$\frac{n_1}{s_1} + \frac{n_2}{s_2} = \frac{n_2 - n_1}{r_0} + h^2 \left\{ \frac{n_1}{2s_1} \left(\frac{1}{s_1} + \frac{1}{r_0} \right)^2 + \frac{n_2}{2s_2} \left(\frac{1}{s_2} - \frac{1}{r_0} \right)^2 \right\}. \tag{2.4}$$

The term in curly brackets describes the deviation from the first-order theory. For paraxial rays $h = 0$ and that term vanishes; this is the first-order result. But for other values of h the rays do not cross the axis at the same point as the paraxial rays, and this causes aberration. When parallel rays are incident from infinity as illustrated in Fig. 2.2, $1/s_1 = 0$. If the aberration is not too big, then in the term in the curly brackets the approximation $s_2 \approx q$ can be made. With $n = n_1/n_2$, (2.4) then reduces to

$$\frac{1}{s_2} = \frac{1}{q} + \frac{n^2 h^2}{2 q r_0^2}. \tag{2.5}$$

The magnitude of the aberration is proportional to h^2, but it decreases as

n^2, the square of the relative refractive index. Once again, it is apparent that a small refractive index is beneficial.

To illustrate how these results apply to the lens of a scanning acoustic microscope, the refractive index for acoustic waves passing from sapphire to water is $n = 0.135$, giving $q = 1.16r$. Thus, if the refracting surface were to be a complete hemisphere, θ_0 would be 81° and the numerical aperture would be 0.988. More realistically, if the spherical surface subtended a semi-angle of 60° at its centre of curvature, then the semi-aperture angle of the lens at the focus would be $\theta_0 \approx 53°$, giving a numerical aperture of 0.8. It would not be possible to achieve such a high numerical aperture with a single surface if the refractive index were smaller. For glass with $1/n = 1.5$, the focal length would be $3r$, and the numerical aperture for $\theta_r = 60°$ would be 0.33. In this and the following comparisons with light, the geometry has to be one of parallel rays *in vacuo* or in air incident on a convex glass surface and being brought to a focus inside the glass: that is why the slightly unusual reciprocal refractive index (less than one) is used.

Because of the n^2 dependence, the effect of the small refractive index is even more remarkable in reducing aberrations. A lens for use at 1 GHz would typically be ground with $r = 0.1$ mm. If the refractive index were 0.67, then the longitudinal spherical aberration, $f_2 - s_2$, would be 43 μm by third-order theory (an exact calculation yields 66 μm, which is even worse). But with $n = 0.135$ the longitudinal spherical aberration reduces to 0.79 μm (the exact calculation yields 1.02 μm). The transverse spherical aberration is found by multiplying the longitudinal aberration by $\tan \theta_0$; for $n = 0.135$ this yields 1.3 μm. This is the diameter of the pattern formed by the outermost rays at the paraxial focus. The circle of least confusion in geometrical optics is found at $s_2 + \frac{1}{4}(f_2 - s_2)$ and has diameter equal to a quarter of the transverse aberration. This is considerably less than a wavelength at 1 GHz.

Since the aberrations of an acoustic lens can be much less than a wavelength, an alternative is not to think of them in geometrical terms, but rather to consider the phase aberrations introduced into the wavefront (Lemons and Quate 1979). In the absence of any aberration the wavefront after passing though the lens would be a sphere whose centre was at the focus. The difference between such a sphere and the actual wavefront is the aberration

$$W = r_0 \frac{1 - \{1 - 4n(1-n)\sin^2(\frac{1}{2}\theta_r)\}^{1/2} - 2n(1-n)\sin^2(\frac{1}{2}\theta_r)}{1-n}$$

$$\approx 2r_0 n^2(1-n)\{\sin^4(\tfrac{1}{2}\theta_r) + 2n\{1-n\}\sin^6(\tfrac{1}{2}\theta_r)\}. \tag{2.6}$$

When $n = 1$, corresponding to a lens with no refraction at all, $W = 0$. As n decreases, W passes through a maximum (which happens to be at $n \approx 0.67$ for $\sin \theta = 0.5$), and then W falls off again, eventually as n^2 for $n \ll 1$. In the example given above with $r = 0.1$ mm, $\theta_r = 50°$, and $n = 0.135$, the wavefront distortion would be $0.1\ \mu$m. This is less than a tenth of a wavelength in water at 1 GHz, and the wavefront aberrations are even smaller if the focus is taken as a point slightly closer to the lens than the paraxial focus. For light, with $1/n = 1.5$, the wavefront distortion would be $1\ \mu$m, which would be quite unacceptable in a good microscope.

Chromatic aberrations do not arise in the acoustic microscope because in its usual mode of operation it may be considered essentially monochromatic. Even when it is necessary to take the spread of frequencies in the acoustic pulses into account, the media through which the waves pass are essentially non-dispersive; in solids over the frequency range of interest the phonons are very near the centre of the first Brillouin zone where the dispersion relationship is linear, especially for sapphire.

In optics the skill of making compound objective lenses for microscopes with high numerical aperture that are accurately compensated for both geometrical and chromatic aberrations has been developed over a period of over 200 years. In acoustics this is neither possible nor necessary. The prospect of using conventional optical techniques to design and fabricate acoustic lenses with more than one element presents insuperable practical difficulties. But the very high refractive index means that, provided only focusing on the axis is required, then a single lens surface provides a high-aperture lens with aberrations considerably less than a wavelength. The size of the spot to which the waves are focused is about one wavelength; this will be analysed in more detail in Chapter 3. By scanning this spot over the area of a specimen, a scanned image can be obtained.

2.2 Scanning in transmission

The transducer–lens combination that has been described serves to illuminate (or insonify) a spot in the focal plane that is about a wavelength in diameter. After the acoustic waves have interacted with a specimen they must be detected in some way. It is not necessary in principle to have a focused detector for this purpose. In the scanning electron acoustic microscope (Davies 1983; Briggs 1984; Cargill 1988) the excitation of the acoustic waves in the specimen is focused, but the detection has no spatial discrimination. In this respect it is like conventional scanning electron microscopy but, because efficient detection of acoustic waves is generally coherent, there can be unwanted interference effects if the wavelength is small compared with the area being scanned (Cargill 1980); there is some evidence that scanning electron acoustic microscopy can image not only cracks with great sensitivity, as can the ordinary acoustic microscope (see Chapter 12), but also residual stress fields (Cantrell *et al.* 1990; Cantrell and Qian 1991). On the other hand, in the scanning laser acoustic microscope (Kessler and Yuhas 1979; Kessler 1985, 1988; Lin *et al.* 1985; Wade and Meyyappan 1987; Wey and Kessler 1989) the insonification is broadcast throughout the specimen, and the detection is by a focused optical probe that measures local surface tilt on the surface of the specimen. But in the scanning acoustic microscope both the illumination and the detection are performed by focusing elements and, since these are focused at the same point, the configuration may be described as confocal. The first confocal acoustic microscopes worked in transmission and, although this is now of mainly historical interest, the transmission arrangement will be described first because in some respects it is simpler and will serve to introduce some principles.

An early high-resolution transmission scanning acoustic microscope is illustrated in Fig. 2.3. Two acoustic lenses are mounted facing each other on a common axis, and with a common focal plane. A continuous radio-frequency (r.f.) signal is fed to the transducer on one of the lenses, which converts this to an acoustic wave which is then brought to a focus. The waves from the focal region are refracted by the second lens so as to arrive as approximately plane wavefronts at the receiving transducer, which converts them back into an r.f. electrical signal. This signal is amplified, and then passes to a diode detector that gives an output approximately proportional to the r.f. power. This signal, sometimes called the video signal, is amplified by a low-frequency amplifier and used to control the brightness of the spot in a cathode ray tube. Thus the brightness of this spot can be related to the amount of acoustic power that is transmitted through the specimen.

In order to build up an image, the specimen is scanned in the common focal plane of the lenses. The scan is performed in a raster pattern; 'raster' is simply the German word for scan, but in English it has come to mean, in the context of microscopy, a scan in a series of lines (by convention horizontal) covering a rectangular area. The specimen is mounted on a support that can be vibrated fairly rapidly in one direction (typically at the frequency of the mains electricity supply), and then more slowly in the other direction. For example, motion in the faster direction may be achieved by supporting the specimen on a pair of leaf springs and using an electromagnetic driver similar to that in an audio loudspeaker;

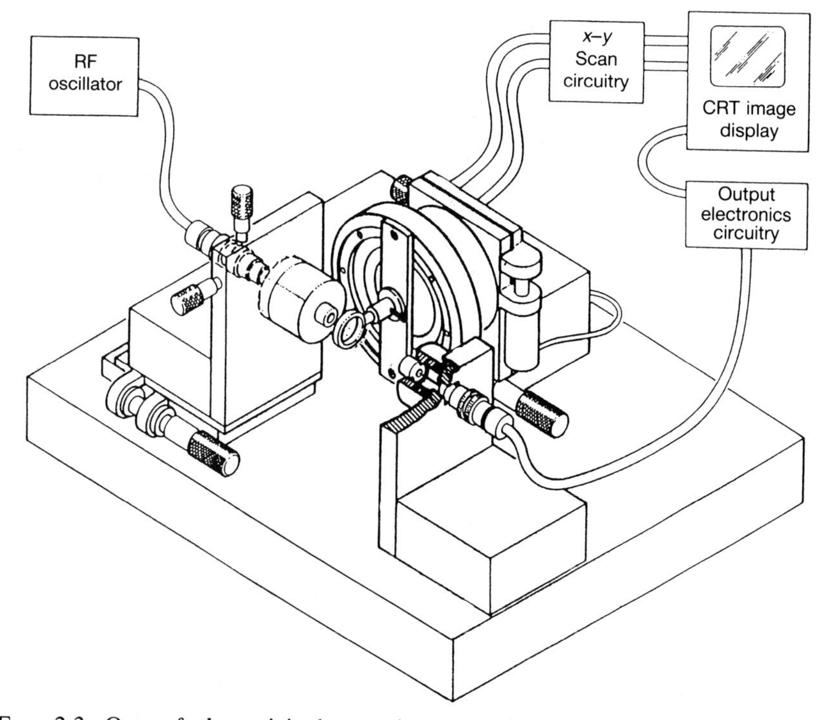

FIG. 2.3. One of the original scanning acoustic microscopes. It worked in transmission, so that it was not necessary to use pulsed waves, and the detected transmitted signal could be used directly to modulate the beam in a cathode ray tube. The slow scan was provided by a small motor driving a lead screw, and the fast scan by a modified loudspeaker coil (Lemons and Quate 1974, 1979).

motion in the slower direction may be achieved by a screw drive powered by an electric motor moving the specimen holder in a precision linear translation stage, or by another system similar to the first scan direction. The time taken to obtain an image of, say, 500 lines at 50 Hz would be 10 seconds.

To display this image, the scan of the electron beam in the cathode ray tube is synchronized with the mechanical scan of the specimen, either in real time with a long-persistence phosphor, or using a framestore and scan rate converter. The magnification is the ratio of the size of the displayed image to the extent of the image scan. For a fixed display size the magnification is increased by reducing the amplitude of the specimen scan. Since this can be reduced almost indefinitely, there is scarcely any limit to the magnification that can be achieved. Magnification is not everything, however, and the limit to the detail that can be seen lies in the resolution of the instrument, which is determined by the lens and its frequency of operation, and not by the area of the specimen scanned. Once the amplitude of the specimen scan has been reduced so that the display size is less than about a hundred resolution diameters across, then there is little benefit in further magnification; indeed there is even a loss because there is no additional detail but only a smaller area of specimen is seen. At the other end, the lowest useful magnification has no theoretical limit; the limit is rather a practical one imposed by the largest amplitude of which the mechanical scanning arrangement is capable.

In a conventional optical microscope, provided the lenses have been ground with spherical symmetry, the magnification is inevitably the same in both horizontal and vertical directions of the image. In a scanning microscope this cannot be assumed. Therefore, the magnifications in the two directions must be calibrated independently. This is achieved by choosing a test target and adjusting the range of the scan in the two directions. If the specimen has a grid pattern (a 400 mesh electron microscopy specimen grid is excellent for this purpose), then the magnification may be adjusted until the mesh in the image is square, with right angles at the corners and sides of equal lengths in the two directions. If this has been carried out with the directions of the grid parallel to the two scan directions, then a particularly severe test of the scan is to repeat the test with the grid direction diagonal to the scan. Any non-uniformity in the scan will then appear as deformations of the grid lines or the angle between them.

The operation of a transmission scanning acoustic microscope requires the lenses to be set up so that they are accurately confocal. This requires holders that can be moved relative to each other along three axes, with rather fine adjustment, and that are rigid to better than a wavelength even when a specimen is vibrating between them. The separation must first be set. If r_0 is the radius of curvature of each lens, a_0 the aperture radius, and n the refractive index, then the focal planes of the two lenses will coincide when the separation between their front surfaces is

$$s = 2r_0\left\{\frac{n}{1-n} + \left(1 - \frac{a_0^2}{r_0^2}\right)^{1/2}\right\}. \tag{2.7}$$

Aligning the axes of the two lenses is more difficult. A medium-power stereo microscope may be used, arranged so that each eye views one of the lenses obliquely, in order to align the lens approximately. A check should be made that no signal due to electromagnetic leakage is obtained when the lenses are separated by air, and then water may be introduced between them. The alignment and focus must then be adjusted iteratively to find the maximum transmitted signal, bearing in mind that because of

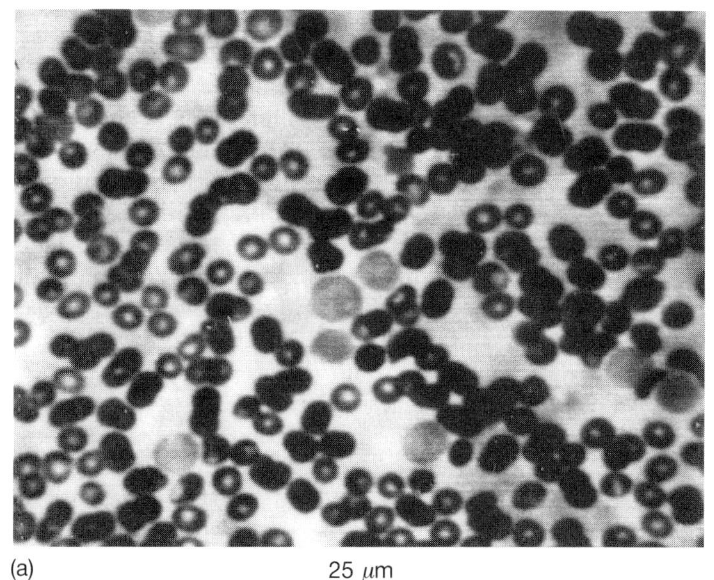

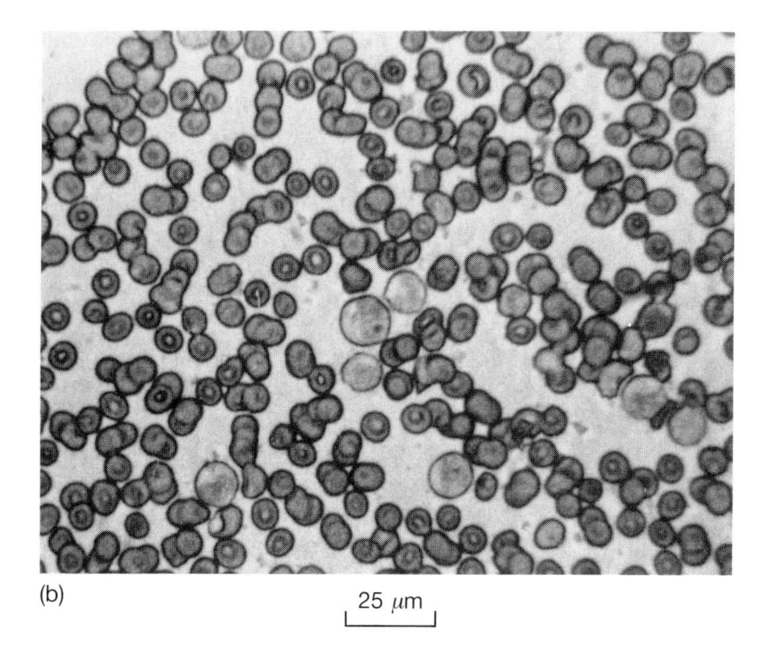

FIG. 2.4. Human bone marrow smear, stained with Wright–Giemsa's stain: (a) transmission acoustic, 900 MHz; (b) transmitted light (Lemons and Quate 1979).

the diffraction pattern of each lens there may be small subsidiary maxima in rings around the much larger central maximum of the axis of a lens.

Because of attenuation, transmission acoustic microscopy is not possible at high resolution through thick specimens. It is therefore better suited to applications in which the specimen is available in a thin section or smear. Figure 2.4 shows acoustic (a) and optical (b) images of a smear of human bone marrow (Lemons and Quate 1979). This was prepared by using a microscope slide to smear fresh bone marrow onto a mylar film 2 μm thick supported in tension. The smear was fixed in formalin to stabilize the cells and to make them adhere to the membrane when they were scanned in the acoustic microscope. To enhance the contrast of the optical image Wright–Giemsa's stain was applied. This made little qualitative difference to the appearance of the acoustic images; indeed, a potential advantage of acoustic miscroscopy for biological studies, expecially of living cells, is that staining is quite unnecessary. Two kinds of cells are visible in both acoustic and optical images. The most numerous are erythrocytes (red blood corpuscles), which are circular with a bi-concave cross-section, and contain neither nucleus nor organelles. In the acoustic image they appear dark because the acoustic waves are both reflected from their surface and attenuated in passing through them; this attenuation is less near the centre where they are thinner. The larger cells that can be seen are granulocytes. These allow more acoustic energy to pass through them, and some internal structure can be discerned. The erythrocytes have an average diameter of 7.5 μm, and in the optical image they are seen with a resolution of about 0.5 μm. The acoustic image was obtained at a frequency of 0.9 GHz, giving a wavelength in water $\lambda_0 \approx 1.7$ μm, though the resolution may be a little better than that for reasons to be discussed in §3.1.1.

Images have also been obtained of solid materials, in transmission, for example, diffusion bonds in steel (Derby *et al.* 1983). While some features of interest can be discerned in the pictures, they serve chiefly to emphasize the severe difficulty of working in transmission. Grain scattering in the specimen can cause such severe attenuation that it is necessary to use a relatively low acoustic frequency with correspondingly poor resolution; even so, the multiple refraction at grain boundaries has an effect similar to that of bathroom glass. Even when thin specimens without these problems can be used, the problem of alignment of the lenses for transmission becomes increasingly severe with high frequencies and shorter wavelengths. By working in reflection the problem of alignment is obviated. There are also other more fundamental advantages in working in reflection, most notably when surface waves can be excited in the specimen.

2.3 Reflection acoustic microscopy

Any lens is inevitably confocal with itself. If the transmission arrangement is conceptually folded over, so that the same lens is used for both transmitting and receiving, the need to adjust the lenses to be confocal is obviated. This is achieved in reflection acoustic microscopy. The fundamental principles of reflection acoustic microscopy remain the same as those of transmission. Mechanical scanning is again used to build up an image, despite countless suggestions to try electronic scanning, though usually the lens is scanned over a stationary specimen. The chief difference is that, in order to use the same lens both for transmitting and for receiving, pulsed signals are used. The very first reflection scanning acoustic microscope did use continuous waves (c.w.) at 600 MHz, and was able to resolve electrodes of a transistor that were 2 μm wide, and even the structure of a smaller transistor with electrodes 2.5 μm apart and 1 μm wide (Lemons and Quate 1974). The contrast in the images was phase contrast, due to variations in the phase of the reflected acoustic wave as it interfered with the generating signal. This principle has been refined using a spectrum analyser to detect the c.w. signal (Kulik *et al.* 1989; Gremaud *et al.* 1990). By slowly sweeping the frequency and subsequently Fourier-transforming the data, reflected pulses can be synthesized. If the frequency is swept more rapidly, there will be a constant difference in frequency between the outgoing signal and the returning signal that depends on the distance of the reflector (Faridian and Somekh 1986); detection can then be performed at any chosen frequency difference. But in reflection microscopy short pulses are generally used in order to separate the reflected signal from the transmitted signal. The electronic circuitry for generating and for detecting the acoustic waves is correspondingly more sophisticated than was necessary for c.w. transmission microscopy.

A schematic block diagram of the radio-frequency (r.f.) chain of the reflection scanning acoustic microscope is shown in Fig. 2.5 (Atalar and Hoppe 1986). The whole system is controlled by a small computer. The r.f. signal is generated by a synthesizer set at the required frequency. This signal is gated by a very fast PIN switch; as will be seen in §3.2, the speed and isolation of this switch is a determining factor in the performance of the whole microscope. This and the other fast switches in the system are controlled from a central pulse timing unit. The pulse is amplified, and fed to one port of a microwave single-pole-double-throw (s.p.d.t.) switch. In the position shown this connects the pulse to the lens, via a network that matches the largely capacitive transducer to the 50 Ω electrical impedance generally used in microwave electronics;

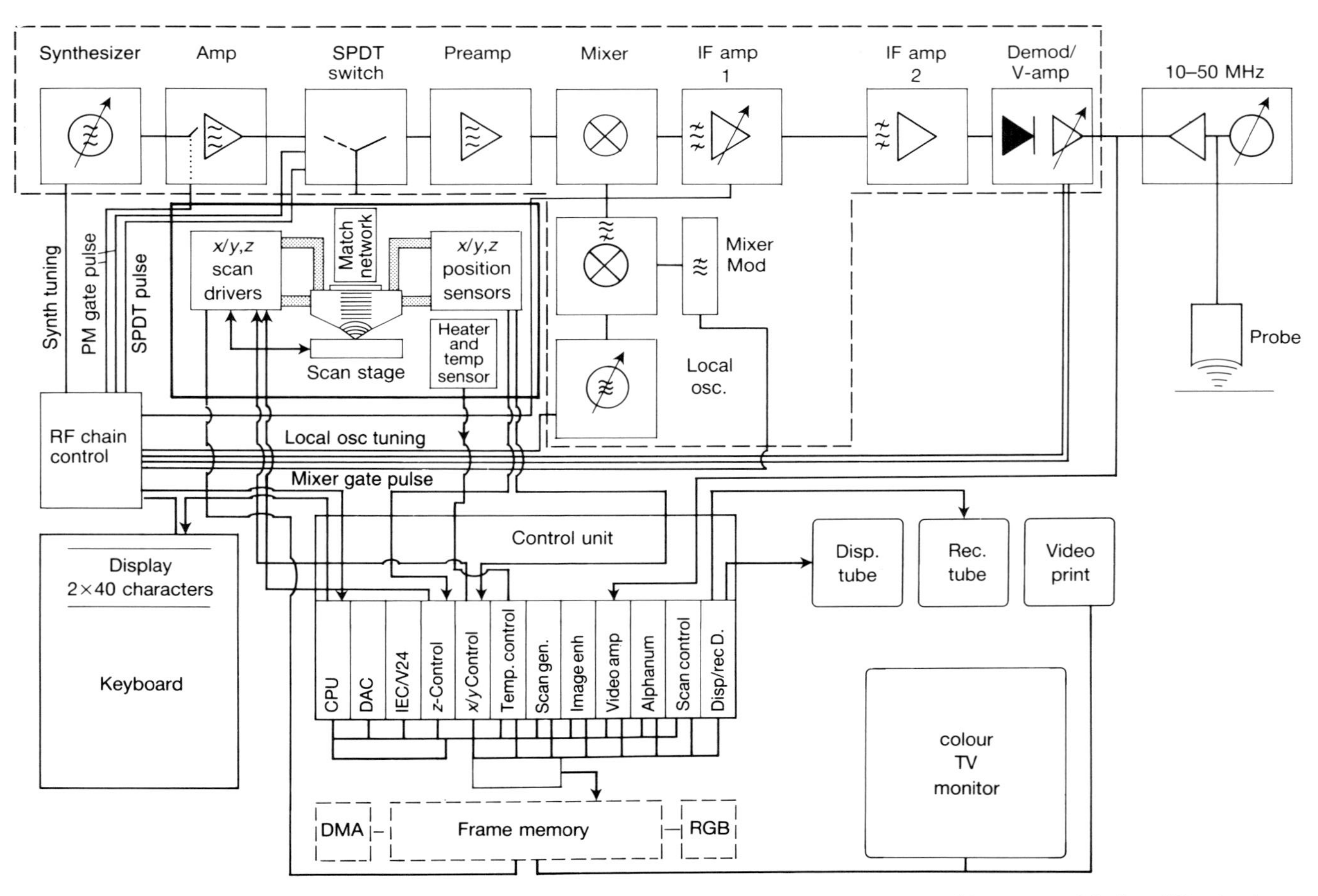

FIG. 2.5. Schematic block diagram of electronics for a reflection acoustic microscope. (Courtesy of Leica, Wetzlar.)

ideally the matching network should make the transducer look like a pure 50 Ω radiation resistance. The pulse generates an acoustic wave that propagates through the lens, is reflected by the specimen, and returns back through the lens to generate an electrical pulse at the transducer. By this time the s.p.d.t. switch has been changed over to send the signal to the input of a low-noise preamplifier. A vital feature of this amplifier is that it should have fast recovery characteristics, so that it can recover quickly, both from any transmitted signal that breaks through the s.p.d.t. switch due to imperfect isolation and from any transient switching spikes that may come from the s.p.d.t. switch itself. From the preamplifier the signal passes to a superheterodyne receiver. There are additional microwave switches gating the local oscillator and the second stage of the intermediate frequency (i.f.) amplifier, in order to give protection against unwanted signals and to select by time-gating only the signal from the lens that corresponds to a reflection from the specimen. The timing of the various pulses is microprocessor-controlled by digital delay techniques. After further amplification with various controls to set the gain and offset, the i.f. signal is rectified by a diode-detector; the rectified signal in

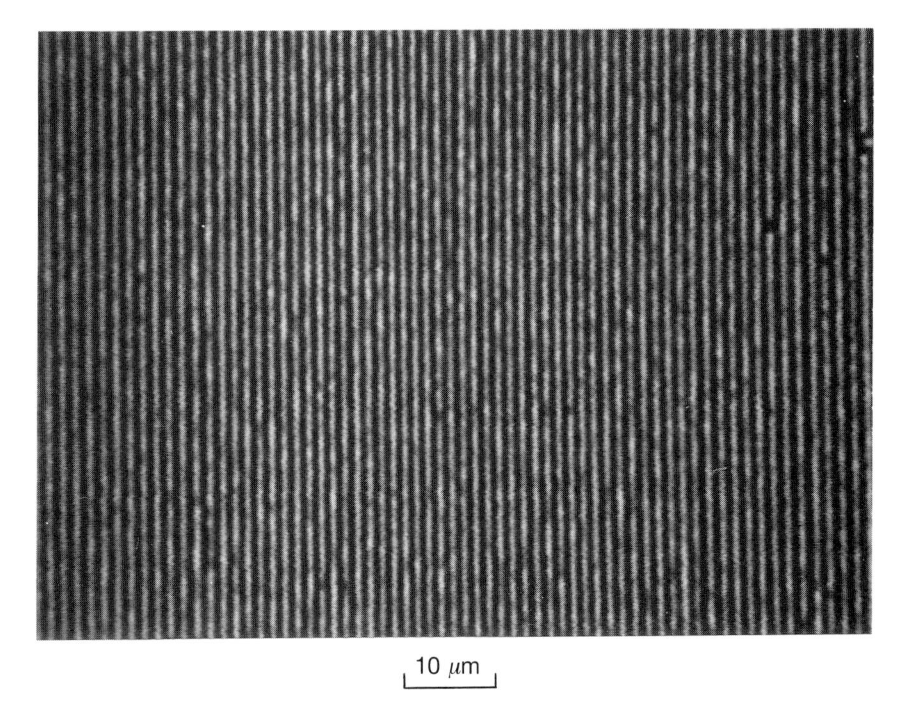

FIG. 2.6. A grating of 0.8 μm period imaged at 2.0 GHz, at 60°C. (Courtesy of Leica, Wetzlar.)

turn passes to a circuit that measures the magnitude of the gated pulse. Additional electronic circuits control the mechanical scan of the lens. The video signal determines the brightness of the image at each point. The image can be displayed either in real-time on a long-persistence phosphor screen, or on a standard monitor via a framestore. A digital framestore has the advantage that digital image handling techniques can then be used.

The resolution of reflection instruments such as the one described here may be tested by imaging a specimen with a fine grating ruled on it. Figure 2.6 shows an image of a grating with a period of 0.8 μm at 2.0 GHz. At lower frequencies the pattern was not resolved at all (cf. Hoppe and Bereiter-Hahn 1985), but at 1.7 GHz, and above it can be seen quite well. The enormous amount of creative research that has gone into making acoustic microscopy with this kind of resolution routinely possible should not be underestimated (Jipson and Quate 1978). Chapter 3 considers the factors that determine and limit that resolution.

3
Resolution

A born provincial man who has a grain of public spirit as well as a few ideas, should do what he can to resist the rush of everything that is a little better than common towards London. . . . And he counted on quiet intervals to be watchfully seized, for taking up the threads of investigation—on the many hints to be won from diligent application, not only of the scalpel, but of the microscope, which research had begun to use again with new enthusiasm of reliance. '. . . and then at home you always want to pore over your microscope and phials. Confess you like those things better than me.' (*Middlemarch*, George Eliot 1871–2)

This rule is convenient on account of its simplicity; and it is sufficiently accurate in view of the necessary uncertainty as to what exactly is meant by resolution. Perhaps in practice somewhat more favourable conditions are necessary to secure a resolution that would be thought satisfactory. (Rayleigh 1879)

3.1 Diffraction and noise

3.1.1 *Diffraction*

It is well known that the resolution of a conventional microscope with a lens of numerical aperture $N.A.$, using light of wavelength λ_0, is

$$w = 0.61\lambda_0/N.A. \tag{3.1}$$

If the aperture is a lens, and if the wavelength is taken to be the wavelength in the medium between the lens and the image or object plane so that the refractive index of the medium is by definition $n = 1$, then the numerical aperture is equal to the aperture radius a_0 divided by the focal length q, i.e. $N.A. = a_0/q$. If the semi-angle subtended at the focus by the circumference of the lens aperture is θ_0, then, equivalently, $N.A. = \sin\theta_0$. The basis of equation (3.1) is that the focused Fraunhofer diffraction pattern from a circular aperture illuminated by plane waves has a first zero at that distance from the axis in the focal plane. This criterion is known as the Rayleigh criterion, although the reduction of the size of the diffraction pattern of stars with increase of the size of a telescope objective had already been studied by Herschel, Fraunhofer, Dawes, and Airy. Sir George Airy (1801–1892) showed that the amplitude distribution in the focal plane, with r as the radial coordinate, is (Wilson and Sheppard 1984; Hecht 1987)

$$U(r)=\frac{\exp(-\mathrm{i}2\pi q/\lambda_0)}{\mathrm{i}2\pi q/\lambda_0}\exp\left(\frac{\mathrm{i}\pi r^2}{q\lambda_0}\right)\pi a_0^2\left\{\frac{2\mathrm{J}_1(2\pi r a_0/q\lambda_0)}{2\pi r a_0/q\lambda_0}\right\}. \qquad (3.2)$$

This is called a point-spread function, because it describes how what should be a point focus by geometrical optics is spread out by diffraction. The expression in the curly brackets is the one that is of interest. The other terms are phase and overall amplitude terms, as are usual with Fraunhofer diffraction expressions. The function J_1 is a Bessel function of the first kind of order one, whose values can be looked up in mathematical tables. $2\mathrm{J}_1(x)/x$, the function in the curly brackets, is known as jinc(x). It is the axially symmetric equivalent of the more familiar $\mathrm{sinc}(x)\equiv\sin(x)/x$ (Hecht 1987), the diffraction pattern of a single slit, usually plotted in its squared form to represent intensity. Just as sinc(x) has a large central maximum, and then a series of zeros, so does jinc(x). $\mathrm{J}_1(x)=0$, but by L'Hospital's rule the value of $\mathrm{J}_1(x)/x$ is then the ratio of the gradients, and $\mathrm{jinc}(0)=1$. The next zero in $\mathrm{J}_1(x)$ occurs when $x=3.832$, and so that gives the first zero in jinc(x). This occurs at $r=(3.832/\pi)\times(q/2a)\lambda_0$ in (3.2), which is the origin of the numerical factor in (3.1).

As Lord Rayleigh (alias John Strutt, 1824–1919) himself confessed, there is an element of arbitrariness in his resolution criterion. In a scanning acoustic microscope, as in a scanning optical microscope, there are at least two significant points that have arisen since Lord Rayleigh's day. First, because the brightness of each point is determined electronically, any dip in the signal between two points can be amplified, with a suitable offset, indefinitely, subject only to noise limitations. The Sparrow criterion (Hecht 1987) explicitly recognizes this, and says that two points are resolved if there is a dip in the signal, however small, between their two images. Second, because both the acoustic wave arriving at a point on the specimen and the reflected wave returning to the transducer are focused, there are actually two point-spread functions, and they must be multiplied together. In a reflection acoustic microscope, the same lens does both jobs; thus, $U(r)$ for each point is squared before the point-spread functions of adjacement images are added (this is quite different from squaring the total signal after they have been added; Wilson and Shepard 1984). The upshot of all this is that the resolution is a bit better than the Rayleigh criterion would suggest, and is (Kino 1980, 1987)

$$\boxed{w=0.51\lambda_0/N.A.}\,. \qquad (3.3)$$

This is the same as (3.1) except that the numerical constant is smaller.

In practice, in an acoustic microscope the ideal of a numerical aperture is never even approximately realized. It is one of the experimental frustrations of acoustic microscopy that no satisfactory method exists for measuring the pupil function of the lens (i.e. the amplitude of the waves as a function of radial position on the lens surface) but, whatever else it is, it is certainly not uniform everywhere on the lens surface and zero elsewhere.

Does this mean that the resolution of an acoustic microscope is undefined? No, because the Sparrow criterion gives a perfectly good experimental measure; indeed, detailed studies have been made of the transfer function and the spatial frequencies present in a line scan (Block *et al.* 1989; Heygster *et al.* 1990). What is clear is that, whatever the exact numerical constant in (3.3), the resolution is proportional to the wavelength in the fluid λ_0. The way to improve the resolution is therefore to make the wavelength smaller. The wavelength depends on the velocity of sound in the fluid v_0 and the frequency f such that

$$\boxed{\lambda_0 = v_0/f}. \tag{3.4}$$

Why can the frequency not be increased indefinitely, thus making unlimited resolution possible?

3.1.2 *Noise*

If the input of an amplifier is connected to a device with a correctly matched electrical impedance, then even in the absence of any useful signal from the device there will be noise in the output of the amplifier equivalent to a noise power input (Robinson 1974)

$$\langle P_\mathrm{n} \rangle = 4k_\mathrm{B}T\,\Delta f \tag{3.5}$$

where k_B is Boltzmann's constant (1.38×10^{-23} J K^{-1}), T is the absolute temperature, and Δf is the bandwidth of the amplifier (combined with any other bandwidth-limiting filters or elements in the circuit). At 293 K this yields 4×10^{-12} mW per megahertz of bandwidth.

For a broadband r.f. amplifier of bandwidth Δf_1 sending a signal to a square-law diode detector and thence to a low-frequency video amplifier of bandwidth Δf_2 the noise power is (Dicke 1946; Robinson 1974)

$$\langle P_\mathrm{n} \rangle = k_\mathrm{B}T(2\Delta f_1\,\Delta f_2)^{1/2}. \tag{3.6}$$

This would give the noise power for a continuous wave (c.w.) microscope such as the transmission microscope described in §2.2. However, for a pulsed instrument with heterodyne detection the bandwidth is defined in the intermediate frequency (i.f.) stage, and the i.f. bandwidth Δf may be used in (3.5).

Ratios of power in electronic circuits are measured in decibels. A hypothetical unit of a Bel would describe the logarithm to base 10 of the ratio of two power levels; the decibel (dB) is a unit of a tenth of a Bel. Hence the ratio of two power levels P_1 and P_2 may be described as $\alpha = \{10 \times \log_{10}(P_1/P_2)\}$ dB. Conversely, $P_1 = P_2 \times 10^{\alpha/10}$. The decibel is dimensionless and, since it is a logarithmic unit, amplifications and attenuations expressed in decibels can be added and subtracted. A useful unit of absolute power is the decibel milliwatt (dBm): a power expressed in dBm is a power relative to 1mW with the ratio measured in dB. In these units the thermal noise at 293 K is -114 dBm MHz^{-1}.

The performance of the acoustic microscope can be described in these terms (Bray 1981). Equation (3.5) gives the noise output power for an ideal amplifier, but in practice a real amplifier will give a noise output greater than this by an amount known as its noise figure, N. For a reasonable low-noise amplifier, $N \approx 6$ dB. At room temperature with a bandwidth of 100 MHz, the noise level would be -82 dBm. The losses in the acoustic path include the following. In the conversion of the electrical signal to an acoustic signal and vice versa by the transducer, there may be a two-way loss of 20 dB. There may be a two-way illumination loss at the lens surface, because some of the acoustic wave falls outside the lens, of 10 dB, and a two-way transmission loss through the lens, due to partial reflection at the interface between the lens and the coupling fluid, of perhaps 6 dB. It is often preferable, for reasons that will gradually become clear in chapters 9 onwards, to operate the microscope at a certain amount of defocus, which may cost a further 10 dB. The peak power to the transducer might be 25 dBm, so that before allowing for the attenuation in the fluid the signal from a perfect reflector would have a strength of -21 dBm, and a signal-to-noise ratio of 67 dB. Thus an attenuation of 35 dB in the coupling fluid would leave the instrument with a dynamic range of 32 dBm for the pictures. Although the figures would be different for different instruments, they illustrate the kind of thinking that goes into the design of a system and its lenses. For a given minimum pulse length, it is the attenuation in the fluid that determines the highest frequency and the shortest wavelength that can be used, and therefore the best resolution that can be obtained.

3.2 The coupling fluid

There is, alas, no analogy in acoustics to a vacuum in optics. Acoustic waves need a medium to support their propagation. At the frequencies of interest for acoustic microscopy there are some solids through which acoustic waves can propagate with relatively small loss; single crystal sapphire is such a material and this is one reason for choosing it for the lens. Between the lens and the specimen the medium must be a fluid, in order to permit scanning, and at the frequencies of interest almost all fluids exhibit an attenuation that is proportional to the square of the frequency (a theoretical introduction to this will be given in §6.1.2).

The echo from the surface of the lens can never be entirely eliminated, and because the reflection from the specimen must pass through the coupling fluid it will suffer attenuation that may well make it smaller than the lens surface echo. Therefore these two reflections must be separated in time. Although it would be inconceivable to achieve this in a light microscope, the relatively slow velocity of acoustic waves makes it possible in an acoustic microscope. There is a limit to the shortest r.f. pulse that can be achieved with adequate isolation. The shortest focal length lens ever used in reflection required a delay between the lens echo and the specimen reflection of 17 ns (17×10^{-9} s) (Hadimioglu and Quate 1983); a more conservative design criterion, including allowance for a certain amount of defocus, would be 60 ns (Atalar and Hoppe 1986).

A resolution coefficient may be defined to compare the shortest wavelength that can be used in different coupling fluids (Attal and Quate 1976; Wickramasinghe and Petts 1980). For a given pulse length, the minimum focal length q is proportional to the velocity v_0. If the time interval between echoes is required to be t_0, then

$$q = v_0 t_0/2. \tag{3.7}$$

The maximum frequency f that can be used is determined by the attenuation, which is proportional to the square of the frequency. If the attenuation per unit distance travelled is written as $\alpha = \alpha_0 f^2$, and the acceptable attenuation within the considerations of §3.1 is α_{acc}, then the constraint on the frequency is

$$f \leq \surd(\alpha_{acc}/2\alpha_0 q). \tag{3.8}$$

The shortest wavelength λ_{min} that can be used is therefore

$$\lambda_{min} \equiv v_0/f_{max} = \surd(v_0^3 \alpha_0 t_0/\alpha_{acc}). \tag{3.9}$$

Thus the resolution coefficient R_c may be defined as

$$R_c \equiv \surd(v_0^3 \alpha_0). \tag{3.10}$$

since this gives the quantity $\lambda_{min}\surd(\alpha_{acc}/t_0)$, and therefore enables the resolution available from different fluids to be compared directly. This coefficient applies when a lens is to be made specifically for that fluid; if a lens were already available the relevant quantity would be $v_0 \surd \alpha_0$. The resolution coefficients for various fluids are listed in Table 3.1. In this

Table 3.1
Acoustic parameters of various fluids

Fluid	Temperature T(K)	Velocity, v_0 (μm ns^{-1})	Impedance, Z (Mrayl)	Attenuation, α_0 (dB μm^{-1} GHz^{-2})	Resolution coefficient, R_c (μm dB$^{1/2}$ ns$^{-1/2}$)
Water, H_2O	298	1.495	1.49	0.191	0.799
Water, H_2O	333	1.551	1.525	0.086	0.566
Methanol, CH_3OH	303	1.088	0.866	0.262	0.581
Ethanol, C_2H_5OH	303	1.127	0.890	0.421	0.776
Acetone, $(CH_3)_2CO$	303	1.158	0.916	0.469	0.853
Carbon tetrachloride, CCl_4	298	0.930	1.482	4.67	1.94
Hydrogen peroxide, H_2O_2	298	1.545	2.26	0.087	0.566
Carbon disulphide, CS_2	298	1.310	1.65	0.087*	0.442
Mercury, Hg	297	1.449	19.7	0.050	0.391
Gallium, Ga	303	2.87	17.5	0.0137	0.570
Air (dry)	273	0.33145	0.4286×10^{-3}	—	—
Air (dry)	293	0.34337	0.4137×10^{-3}	1.6×10^5	80
Air (dry)	373	0.386	—	—	—
Argon, Ar (4 MPa)	293	0.323	0.023	3.58	0.347
Argon, Ar (25 MPa)	293	0.323	0.145	0.721	0.156
Xenon, Xe (4 MPa)	293	0.178	0.042	8.28	0.216
Oxygen, O_2	90	0.900	1.0	0.086	0.250
Nitrogen, N_2	77	0.850	0.68	0.120	0.271
Hydrogen, H_2	20	1.19	0.08	0.049	0.287
Xenon, Xe	166	0.63	1.8	0.191	0.219
Argon, Ar	87	0.84	1.2	0.132	0.280
Neon, Ne	27	0.60	0.72	0.201	0.208
Helium, He	4.2	0.183	0.023	1.966	0.110
Helium, He	1.95	0.227	0.033	0.610	0.084
Helium, He	0.4	0.238	0.035	0.015*	0.014
Helium, He	0.1	0.238	0.0345	4×10^{-5}*	7×10^{-4}

* For these two fluids the attenuations do not follow a simple f^2 law, and the values given correspond to measurements at 3 GHz for carbon disulphide and at 1 GHz for helium at 0.4 K and at 0.1 K. The data in this table were taken from Lemons and Quate (1979), Kaye and Laby (1986), Selfridge (1985), Wickramasinghe and Petts (1980), Heiserman *et al.* (1980), and Foster and Rugar (1985).

table, the units for attenuation and resolution coefficient contain the dimensionless dB. In many references attenuation is expressed in the alternative dimensionless unit of the neper (Np), which is the unit of the natural logarithmic ratio of two amplitudes. To convert an attenuation in nepers to the same quantity in dB, multiply by $20 \log_{10}(e) \approx 8.686$. For each fluid the acoustic impedance is also given. Acoustic impedance is the ratio of minus the traction or stress component acting on a surface to the particle displacement velocity (§6.4.1); it is equal to the product of velocity and density, and the units are megarayls, where $1 \text{ rayl} \equiv 1 \text{ kg m}^{-2} \text{ s}^{-1}$. Knowledge of the acoustic impedances enables transmission and reflection at a boundary to be calculated in a way closely analogous to electromagnetic theory (§6.4.2).

Table 3.1 enables the resolution available using different coupling fluids to be compared. The smaller the value of the resolution coefficient in the final column, the shorter the wavelength and therefore the finer the resolution available for given values of α_{acc} and t_0. Down to gallium, all the fluids are liquids at or close to room temperature (values are quoted at 303 K for California). Of these, only mercury ($R_c \approx 0.4\ \mu\text{m dB}^{1/2}\ \text{ns}^{-1/2}$) has a resolution coefficient that is more than a factor of two smaller than the value for cold water ($R_c \approx 0.8\ \mu\text{m dB}^{1/2}\ \text{ns}^{-1/2}$). Compared with hot water ($R_c \approx 0.57\ \mu\text{m dB}^{1/2}\ \text{ns}^{-1/2}$) the advantages are less, none of the resolution coefficients being even as much as 35 per cent smaller. Of the organic solvents, methanol alone compares favourably with water; it is occasionally used to excite surface waves in slow materials, and also for imaging corrosive materials. The other three that are listed have values of R_c greater than $0.6\ \mu\text{m dB}^{1/2}\ \text{ns}^{-1/2}$, and all other organic solvents have high values of α_0 (Kaye and Laby 1986). Hydrogen peroxide and carbon disulphide have resolution coefficients that are slightly smaller than the coefficient for hot water, but the small advantage is more than outweighed by the nasty properties of these two fluids. The two liquid metals also have attractive resolution coefficients, and indeed high-resolution images have been obtained at up to 1 GHz with both gallium (Jipson 1979) and mercury (Attal 1980, 1983; Attal *et al.* 1989). The chief advantage for which both these liquid metals were exploited in those experiments was not, however, the resolution coefficient. Rather it was the better match of acoustic impedance to the solid being examined in order to facilitate interior imaging. Gallium also has a better velocity match (particularly to shear waves), and so, in terms of the acoustic properties alone, it would be the best fluid for interior imaging of high-velocity materials such as semiconductors. Gallium oxidizes strongly in air, however, so it is extremely difficult to obtain a clean surface of the liquid to form good contact with the lens; indeed the experiment with gallium has never been repeated. Mercury is slightly easier to use. A small ball the same size as the spherical cavity in the lens is placed there before being brought into contact with the specimen. In this way contact can be established with surfaces of both the lens and the specimen, even though mercury wets neither. But neither of these liquid metals is suitable for general use; they are too difficult to work with and they can destroy both the specimen and the microscopist. It is apparent from the figures that air is not a strong candidate.

The remainder of Table 3.1 after air consists of elements that are gases at room temperature. The first three figures are for room temperature (in England) with the gases under compression. Although the attenuation at a given frequency is much higher than it is in water, it decreases with pressure until, at high enough pressures, it may be sufficiently reduced to enable the relatively low velocity to bring the value of the resolution coefficient below that for water. A high-pressure acoustic microscope has been demonstrated with argon at 6 MPa, and at 160 MHz a resolution of 1.5 μm was achieved (Wickramasinghe and Petts 1980). However, because of the great impedance mismatch at the specimen surface, the contrast was almost entirely topographical. As a route to higher resolution of topography, cryogenic fluids are much better. All the cryogenic fluids listed offer a better resolution coefficient than water. Acoustic micrographs have been obtained at 1.8 GHz in liquid argon, and up to 2.8 GHz in liquid nitrogen, giving a wavelength of 300 nm. But it is immediately apparent from Table 3.1 that the best candidate is liquid helium, both because it has by far the lowest velocity, and even more importantly because the attenuation drops so dramatically with temperature. If resolution is the ultimate goal, then the coupling fluid to use is liquid helium. But for acoustic microscopy outside a cryostat, the perhaps surprising result from all these comparisons is that almost the only fluid significantly better than water is hot water. It turns out that water is also the best choice for another completely different reason that will be explained in Chapter 7, namely the excellent coupling to Rayleigh waves in many solids.

Values of the acoustic properties of water at different temperatures are given in Table 3.2 (Kaye and Laby 1986). The velocity may be described by a polynomial expansion in temperature; five terms are sufficient to achieve a fit that is considerably better than the accuracy of the best available measurements (Del Grosso and Mader 1972). Thus the polynomial expression for the acoustic velocity in water is

$$v_0 = \sum_{i=0}^{5} k_i T^i. \tag{3.11}$$

The coefficients k_i are given in Table 3.3; with T in °C. The polynomial expansion is useful in analysis programs to calculate the velocity from the

Table 3.2
Acoustic properties of pure water

T (°C)	v_0 (μm ns^{-1})	Z (Mrayl)	α_0 (dB μm^{-1} GHz^{-2})
0	1.40239	1.402	0.469
10	1.44727	1.447	0.321
20	1.48234	1.480	0.226
30	1.50913	1.503	0.165
40	1.52886	1.517	0.130
50	1.54255	1.524	0.104
60	1.55099	1.525	0.086
70	1.55480	1.520	0.075
80	1.55449	1.511	0.076
90	1.55048	1.497	0.063
100	1.54311	1.479	0.062

The velocity has a maximum value $v = 1555.147$ m s^{-1} which occurs at 74.2°C (Kaye and Laby 1986).

temperature. The attenuation in water decreases with temperature, but the rate of decrease is less at higher temperatures. The attenuation and velocity are plotted against temperature in Fig. 3.1. For practical operation of an acoustic microscope with water as the coupling fluid, the smallest lens radius for routine operation is 40 μm (focal length 46 μm). With water at 60°C a lens of this radius can be operated comfortably up

Table 3.3
Polynomial coefficients* for the acoustic velocity in water

i	k_i (m s^{-1} K^{-i})
0	$0.140238754 \times 10^{4}$
1	$0.503711129 \times 10^{1}$
2	$-0.580852166 \times 10^{-1}$
3	$0.334198834 \times 10^{-3}$
4	$-0.147800417 \times 10^{-5}$
5	$0.314643091 \times 10^{-8}$

* When these coefficients are used in eqn (3.11), the standard deviation of the fit to the best experimental measurements is 0.003 m s^{-1}. The measurements were made at a frequency of 5 MHz, and the accuracy was estimated to be 0.015 m s^{-1} (Del Grosso and Mader 1972).

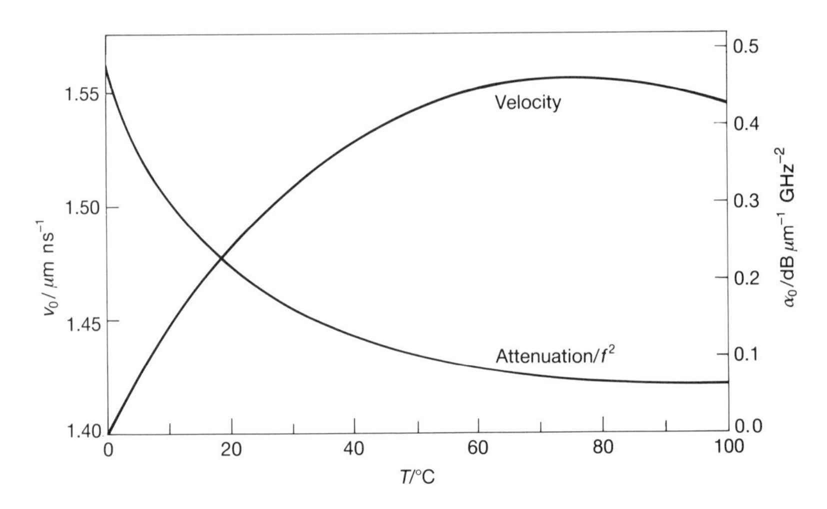

FIG. 3.1. The variation of acoustic velocity and attenuation in water as a function of temperature (Lemons and Quate 1979; cf. Table 3.2).

to 2 GHz; at that frequency the round-trip attenuation in the water would be 35 dB. For lower temperatures or larger focal lengths, the maximum frequency must be reduced accordingly.

3.3 Cryogenic microscopy

Although almost all practical acoustic microscopy is done with water, it is instructive to see what has been done in cryogenic acoustic microscopy, partly because it represents a series of formidable instrumental achievements, partly because it gives an appreciation of the limits to resolution, and partly because the images are largely free from the very important but perhaps unfamiliar interactions of acoustic waves in water with a solid. The only fluid that can approach negligible attenuation for acoustic microscopy at high frequencies is superfluid liquid helium (Heiserman 1980; Heiserman *et al.* 1980; Foster and Rugar 1985). Below the lambda point at 2.17 K, the attenuation is controlled less and less by viscosity and thermal conductivity, until below 0.7 K these processes become insignificant and attenuation occurs only by much weaker phonon scattering processes. At frequencies below 20 GHz the dominant scattering process involves the collision of a phonon of the acoustic wave with a thermal phonon to form a new single phonon of higher energy. This leads to an attenuation that is proportional to the frequency and to the fourth power of the temperture. The attenuation also depends on the pressure; at the saturated vapour pressure it is

$$\alpha_{\mathrm{He}} = 0.488 f T^4 \,\mathrm{dB}\,\mu\mathrm{m}^{-1}\,\mathrm{GHz}^{-1}\,\mathrm{K}^{-4}. \tag{3.12}$$

The very strong temperature dependence means that, if the temperature can be reduced sufficiently, much higher frequencies become possible than in any other fluid. The low acoustic velocity in helium gives further improvement to the resolution available. Temperatures below 0.2 K are necessary to realize these benefits, and to obtain these a dilution refrigerator is used. A special scanner has been designed to operate in the confined space of a crystostat, and this is schematically illustrated in Fig. 3.2. The lens is mounted on a length of semi-rigid coaxial line that serves both as the electrical connection to the transducer and as a flexible support for the mechanical scan. The scan is driven in each direction by an electromagnetic coil and motion is sensed by an identical coil placed opposite; integration of the velocity signal gives the position. The scanner assembly is mounted on the mixing chamber of the refrigerator. The specimen is mounted on the end of a top-loading rod, which enables specimens to be changed relatively rapidly (in an hour or

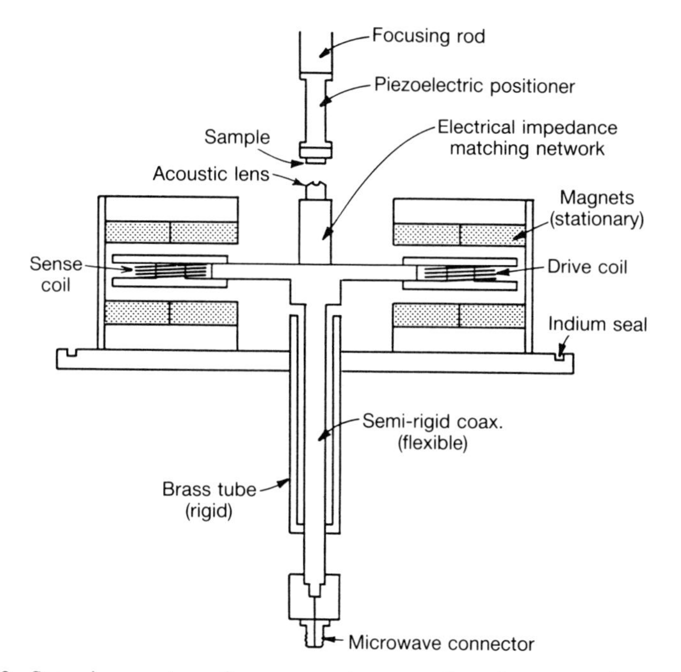

FIG. 3.2. Scanning system for cryogenic acoustic microscopy in a dilution refrigerator (Rugar *et al.* 1982; Foster and Rugar 1985).

two) compared with the much longer time of several days that would be required for the chamber to cool. Coarse focusing is achieved by an external micrometer connected to this rod, and a piezoelectric positioner at the bottom is used for fine focusing. If pure ^{4}He were used in the refrigerator there would be superflow up the top-loading access tube, with evaporation at the warmer end and condensation again on the liquid surface. This would constitute a substantial heat leak. To prevent this, 2 per cent ^{3}He is mixed with the ^{4}He to add viscosity to the superfluid, substantially reducing the superflow. It also contributes an additional attenuation mechanism in the helium, but this has not been found to be a limiting factor; and at 0.1 K, even at 8 GHz with a 200 μm lens, the attenuation in the fluid is less than 3 dB (Hadimioglu and Foster 1984).

The experimental problems that had to be solved in order to develop high-resolution cryogenic acoustic microscopy should not be underestimated. Because the impedance of helium is so very different from that of sapphire, the matching layer on the lens surface becomes particularly critical; by using a layer of amorphous carbon a quarter wavelength thick it is possible to achieve 8 per cent transmission of the acoustic power. As the frequency is increased the problem of matching the transducer impedance to a 50 Ω microstrip transmission line becomes increasingly severe. The thickness and radius of the top gold electrode must be carefully chosen, and a parallel stub used on the microstrip line. A further problem that increases with frequency, and is particularly acute in liquid helium, is that, as the power is increased, non-linear generation of harmonics depletes the intensity of the beam at the fundamental frequency (Foster and Putterman 1985). Provided that the depletion remains small, then the power P_0 at which a fractional depletion s_d occurs from an acoustic wave at a frequency f focused by a lens of numerical aperture $N.A.$ is (Foster and Rugar 1985)

$$P_0 = \frac{4 s_d L (N.A.)^2}{f^2} \tag{3.13}$$

where L is a constant characterizing the acoustic linearity of the fluid. For a fluid of density ρ_0, acoustic velocity v_0, and a non-linear coupling constant $\beta_L = 1 + (\rho_0/v_0)(\partial v_0/\partial \rho_0)$, i.e. one plus the Gruneisen constant, then

$$L = \frac{\rho_0 v_0^5}{16\pi^3 \beta_L^2}. \tag{3.14}$$

The smaller the value of L the lower the power at which non-linear depletion will set in. Helium has both a low density and a low velocity; the low velocity is particularly devastating because this appears to the

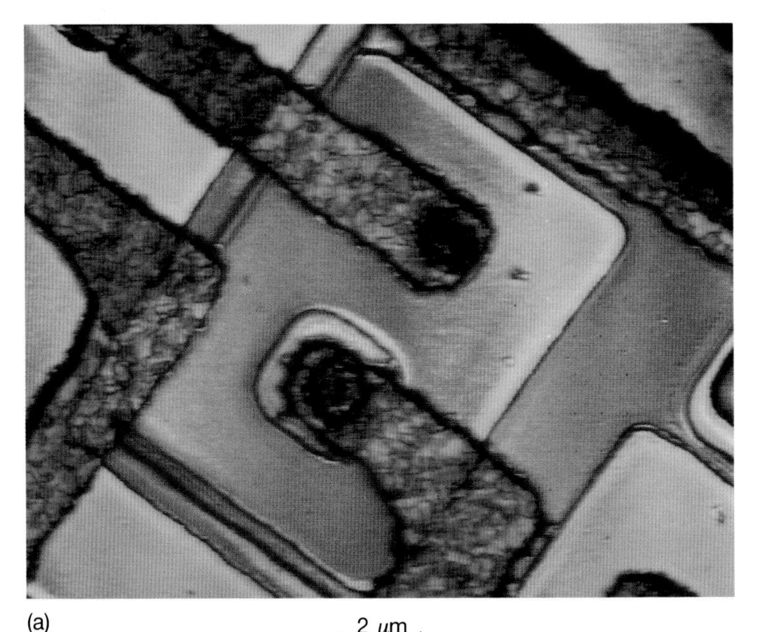

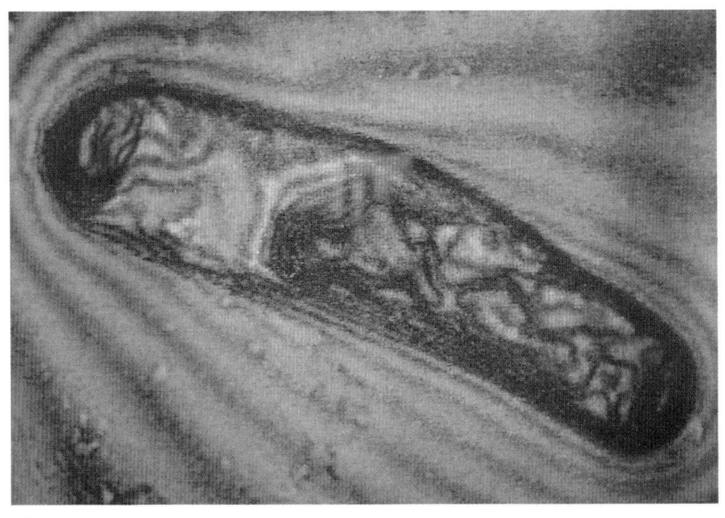

FIG. 3.3. Acoustic micrographs taken in superfluid helium at 0.2 K. (a) Bipolar transistor on a silicon integrated circuit. The aluminium lines making connections to the base and the emitter are $2\,\mu m$ wide and $0.5\,\mu$m thick. Three images were taken at different heights, and superimposed with colour coding. The lens had a numerical aperture $N.A. = 0.625$ and a depth of focus less than 150 nm, $f = 4.2$ GHz. (b) Myxobacterium, with different planes similarly colour-coded and superimposed, $f = 8$ GHz (Foster and Rugar 1985).

fifth power in L. The result is that helium at 0.1 K has a value of $L = 1.51 \times 10^{-5}$ mW GHz^2, which is more than three orders of magnitude smaller than any other liquid, and nearly five orders of magnitude smaller than water ($L = 1.2$ mW GHz^2). Moreover, the power for a given level of non-linear depletion decreases as the square of the frequency. The limit to the acoustic power level imposed by non-linearity in liquid helium is therefore particularly severe, and becomes more so as higher resolution is approached. At 8 GHz a lens with $N.A. = 0.5$ will give 10 per cent depletion ($s_d = 0.1$) when the transmitted power is $P_0 = 2.4 \times 10^{-8}$ mW (−76 dBm). Even at complete saturation the power transmitted through the focus in the helium, and therefore reflected back to the lens, will not be much more than this, and the input power to the preamplifier will be even less than that because of imperfect acoustic and electrical matching at the lens and the transducer. It is therefore essential to minimize the thermal noise by using a superconducting coaxial line to connect to the transducer, and a low-noise FET preamplifier. All these are immersed in helium to maintain a temperature of 4.2 K. If the bandwidth is 20 MHz and a noise temperature of 20 K can be achieved, then from (3.5) the noise power is −107 dBm.

In view of the formidable technical difficulties, the results that have been achieved are all the more impressive. Figure 3.3(a) is an acoustic image of a bipolar transistor on a silicon integrated circuit taken at 4.2 GHz. There is a very severe acoustic impedance mismatch between helium and a material such as silicon. Even for normally incident waves 99 per cent of the power is reflected straight off the surface, and for waves incident at an angle greater than 3° all the power is reflected. For a lens of $N.A. = 0.5$, only about 0.001 per cent of the power penetrates the specimen and the reflected wave contains very little information about the elastic properties (Christie and Wyatt 1982). Therefore, the cryogenic reflection images contain information that is almost entirely topographic. Because the microscope is a confocal system, the signal from a reflecting surface that is not in focus is small (Wilson and Sheppard 1984) and, with a large numerical aperture, a significant signal is obtained only from planes that are very close to focus. As the distance between the reflecting surface and the focal plane changes, so the contrast level changes. This effect can be seen towards the left of the central rectangle in Fig. 3.3(a), where the intensity changes from right to left, probably because the specimen was slightly tilted in the microscope. In a specimen with significant topography, such as an integrated circuit in which each layer is typically a micron thick, a different focus may be required to image areas that have different heights. The transistor in Fig. 3.3(a) was imaged at three different values of focus, and the picture is a colour composite of these three images, with a different colour allocated

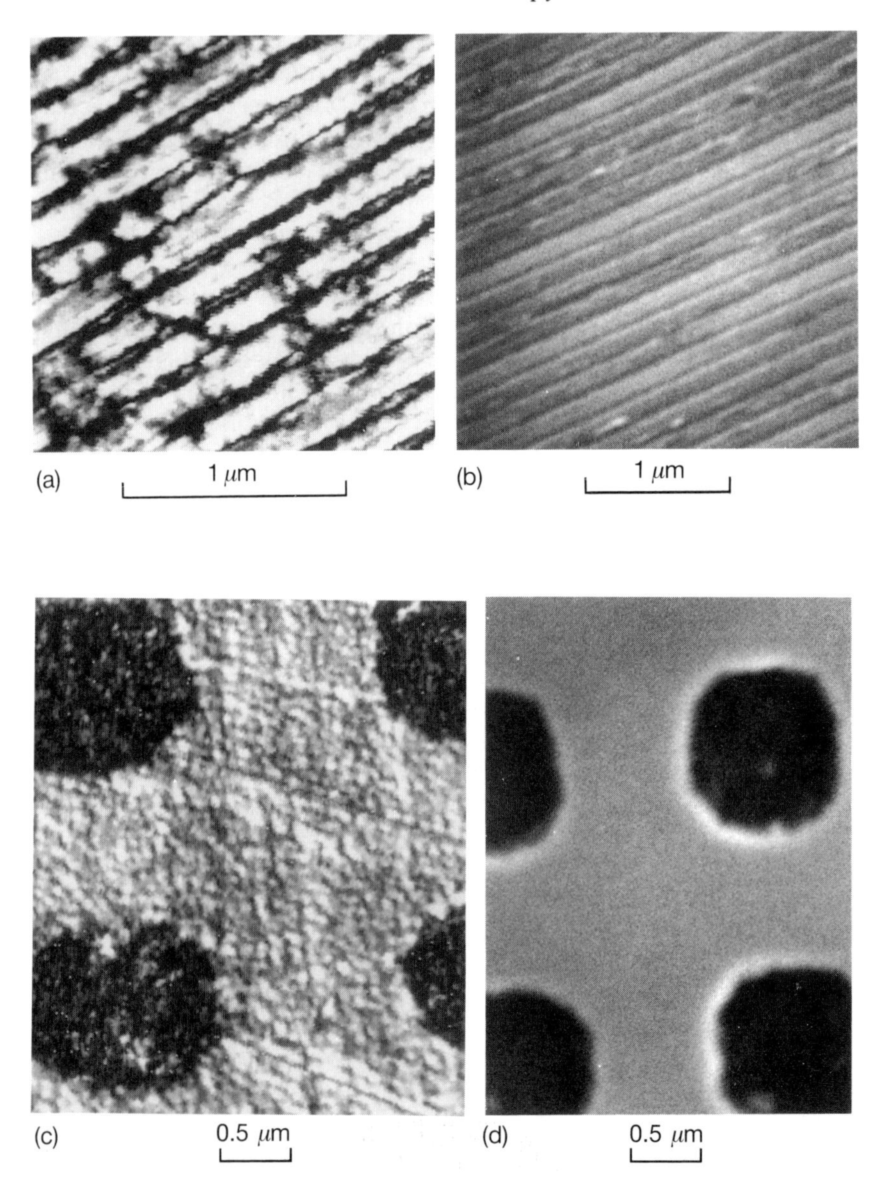

FIG. 3.4. The highest resolution ever achieved by acoustic microscopy, using pressurized superfluid helium (at 0.4 K and 2.14 MPa, at 15.4 GHz) compared with scanning electron microscopy of the same specimens coated with 10 nm carbon, at 5 keV. (a) Acoustic and (b) s.e.m. pictures of a 200 nm period titanium grating on a silicon substrate; (c) acoustic and (d) s.e.m. pictures of an array of 1 μm diameter holes with 2 μm spacing in a thin film of chromium on glass. (Muha *et al.* 1990).

to each focal plane (Hammer and Hollis 1982); this is one of the few cases in which colour significantly enhances the information in an acoustic micrograph. A biological example of the power of the cryogenic acoustic microscope is given in Fig. 3.3(b), which is an image at 8 GHz of a myxobacterium. The detail at each end of the bacterium is again topographical. The depth of focus was about 100 nm; like Fig. 3.3(a), Fig. 3.3(b) is a composite of three images taken at different focal planes.

Two of the highest-resolution acoustic pictures ever taken are shown in Fig. 3.4. The specimens are a titanium grating of 200 nm period on a silicon substrate and a two-dimensional array of 1 μm diameter holes in a thin chrome film on a glass substrate (Muha *et al.* 1990). Figure 3.4(a) and (c) are acoustic micrographs and, for comparison, s.e.m. images are shown in Fig. 3.4(b) and (d). The acoustic microscope used here exploited the fact that, when superfluid helium is pressurized above 2 MPa ($\approx$20 atmospheres), depletion of energy from acoustic waves by phonon scattering and harmonic generation are either eliminated or at least drastically reduced; indeed, the microscope can even be operated at higher temperature than was used for Fig. 3.3(a) and (b). Because of the high resistivity of the films, for the s.e.m. the specimens were coated with 10 nm of carbon, and imaged at 5 keV, to reduce charge accumulation. The acoustic images were obtained at a frequency of 15.3 GHz, at a temperature of 0.4 K, and a pressure of 2.14 MPa. From the 3 dB width of some of the features it was estimated that the resolution was 15 nm.

In order to image the interaction of phonons with internal structure in a specimen, scattering by thermal phonons can be used to attenuate the acoustic beam in a cryogenic microscope (Foster 1984). Experiments to test this confirmed that the cryogenic microscope was indeed able to detect on one surface of a sapphire disc the arrival of thermal phonons generated by a heater on the other side. The disc was 2 mm thick, and a thin-film chromium film was deposited on the back face. Short pulses of current were applied to the film in order to generate pulses of thermal phonons in the sapphire. The microscope was focused on the other face, with an acoustic frequency of 4.2 GHz, and the arrival of thermal pulses caused sufficient attenuation of the reflected acoustic waves (about 10 per cent for a temperature pulse of 15 K) to enable them to be detected. This depended on making the coherent acoustic pulse from the microscope arrive in the focal region at the same time as the phonons from the thermal pulse. By varying the relative timing of these two it was even possible to distinguish between the arrival of phonons that had propagated longitudinally and transversely through the solid. To illustrate how the attenuation of the acoustic beam by scattering from thermal phonons may be used to form images in a cryogenic microscope, Fig. 3.5 shows

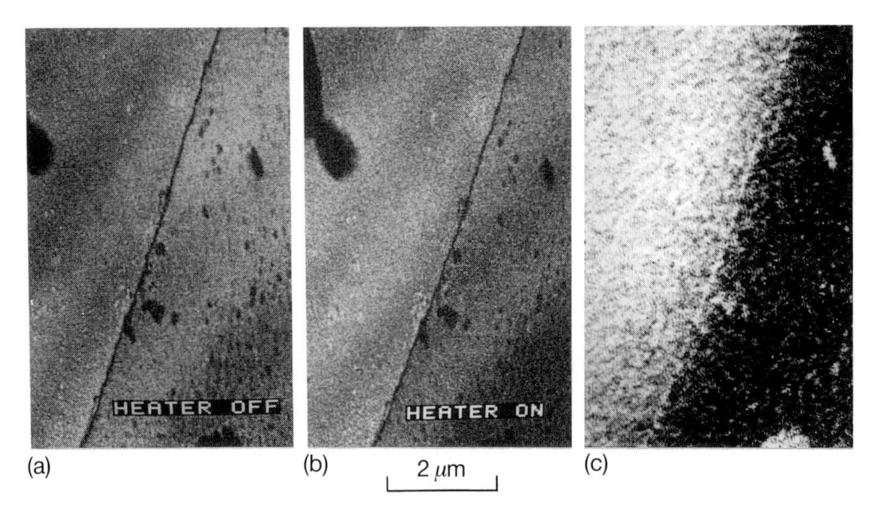

FIG. 3.5. Thermal phonon imaging, 4.2 GHz. (a) Direct acoustic micrograph; (b) heater pulsed synchronously with acoustic microscope; (c) the second picture divided by the first (Foster and Rugar 1985).

three images of a sapphire surface with a thin film chromium heater on the right (Foster and Rugar 1985). In this case the heater is on the same side of the specimen as the surface being imaged. Figure 3.5(a) was taken with the heater off. The dark spots are due to particles of dirt. Figure 3.5(b) was taken with the heater pulsed at a time that ensured that thermal phonons were radiated into the helium at the same time as the arrival of the microscope pulse. The slight darkening that this causes is not easy to distinguish from other contrast in the picture, and so an image was formed by dividing Fig. 3.5(b) by Fig. 3.5(a); the result is shown in Fig. 3.5(c). This emphasizes the contrast due to the attenuation from phonon scattering, so that the region with the heated chromium film appears dark. The contrast from dirt particles on the sapphire surface on the left is eliminated by the division, but some particles on the chromium film appear bright in Fig. 3.5(c), because of the reduced phonon emission.

3.4 Non-linear enhancement of resolution

Non-linear effects in the propagation of acoustic waves in a fluid occur when the velocity of the motion of the particles becomes a significant fraction of the phase velocity of the acoustic wave. When this happens, harmonic generation takes place, so that power is transferred from the fundamental frequency of the wave to higher harmonic frequencies. In superfluid helium this leads to the limit on the power that can be propagated through the focal region, but seems to have little effect on the resolution; in other fluids it has been discovered that the harmonic generation can lead to enhanced resolution (Rugar 1984). A simple explanation of this is as follows.

When the changes in pressure associated with the propagation of a wave become sufficiently large, they lead to corresponding changes in the density and bulk modulus over a wavelength. The wave travels faster at the peaks in the pressure in each cycle and slower at the troughs. The wave becomes distorted into a sawtooth waveform, with the pressure rising gradually in the direction of propagation over a wavelength, and then dropping rapidly before the next wavelength begins (Bjørnø and Lewin 1986). This involves the introduction of higher harmonic components into the wave. In the acoustic microscope the intensity is greatest in the focal region, and so this is where, as the power is increased, harmonic generation first occurs. The higher harmonics have shorter wavelengths associated with them than the fundamental, so the diffraction limit to the resolution should be correspondingly better for the harmonics. It is perhaps somewhat less obvious that there should also be an improvement in resolution when the fundamental frequency is detected at the transducer. However, when any wave passes through a focal region there is a phase change of π (Born and Wolf 1980). If there are no other phase changes associated with the reflection from the specimen, the shape of the waveform returning to the transducer from the focal region will be reversed. The faster speed of the pressure peaks, and the slower speed of the pressure troughs, will now lead to a distortion of the sawtooth waveform back to a more sinusoidal waveform. This implies a conversion of energy from the higher harmonics, which contain information with spatial resolution better than the diffraction limit associated with the fundamental frequency, back to the fundamental frequency at which they can propagate through the coupling fluid and be detected at the transducer. A theoretical analysis (Rugar 1984) indicates that both the harmonic generation and the reconversion to the fundamental frequency occur within a few wavelengths of the focus. Perhaps the most astonishing aspect is that the information obtained at the higher frequency is converted back to the lower frequency so that it can be carried back to the lens. At the harmonic frequency, the attenuation would be too high to allow an adequate signal to propagate that far. The resolution associated with the second harmonic is better by a factor of $\sqrt{2}$ compared with the resolution of the fundamental beam. The reason for this can be visualized from two extreme cases. Suppose that the profile of the beam is Gaussian. If all the second harmonic were excited in the focal plane, then the amplitude profile of the harmonic would follow the intensity profile of the fundamental, which in turn would have a width (however defined) $1/\sqrt{2}$ of the width of the amplitude profile of the fundamental. But of course the enhanced resolution cannot come from excitation in the focal plane, otherwise the phase change of π that is necessary for subsequent down-conversion could not occur. So the other extreme is that the harmonic is excited well away from the region of the focal phase change. Now the wavelength is half the fundamental wavelength, but then the size of the aperture of excitation is again $1/\sqrt{2}$ of the fundamental beam width at any point, by an argument similar to that in the focal plane. So once again the diffraction spot size is $1/\sqrt{2}$ smaller. The reality must lie somewhere between these two cases.

The improvement of resolution as a result of non-linear effects can be demonstrated by changing the power level to the transducer. Figure 3.6 shows a series of pictures of a grating of gold lines on a silicon substrate. The grating has a periodicity of 290 nm; the gold lines were approximately 40 nm wide and 50 nm thick. The images were obtained at 2 GHz with liquid nitrogen as the coupling fluid and a lens of numerical aperture $N.A. = 0.35$. The smallest periodicity that should be detected at the fundamental frequency is 0.7λ, or about 300 nm. This is just too big to enable the structure of the grating to be resolved. For nitrogen, $L = 3.9 \times 10^{-5}$ W GHz^2 and so, from (3.13), 10 per cent harmonic generation would occur at a power of 2×10^{-6} W. The transducer and lens had a two-way insertion loss of 72 dB, and so this would correspond to an input power to the transducer of 8 mW. The four images of the grating were taken with input power levels of (a) 10, (b) 32, (c) 100, and (d) 320 mW. In the first image the grating is not resolved, but it gradually becomes more visible as the power is increased.

Attempts to image non-linear properties of a specimen have been frustrated by the non-linear properties of the coupling fluid. Because the onset of non-linear effects occurs in the fluid at much lower powers in any solid, lens configurations have been invented to try to make sure that focused beams of different frequencies overlap only in the specimen, so that any sum or difference frequencies that are detected must have been generated there. The most elegant such arrangement used three zinc oxide transducers that could operate up to nearly 1 GHz mounted on the curved back surface of a fused quartz lens, with the geometry designed so that the three transducers were confocal in the water (Tan *et al.* 1985). Beams of different frequencies could be generated by two of the transducers, and reflected mixing products such as $2f_1 - f_2$ could be

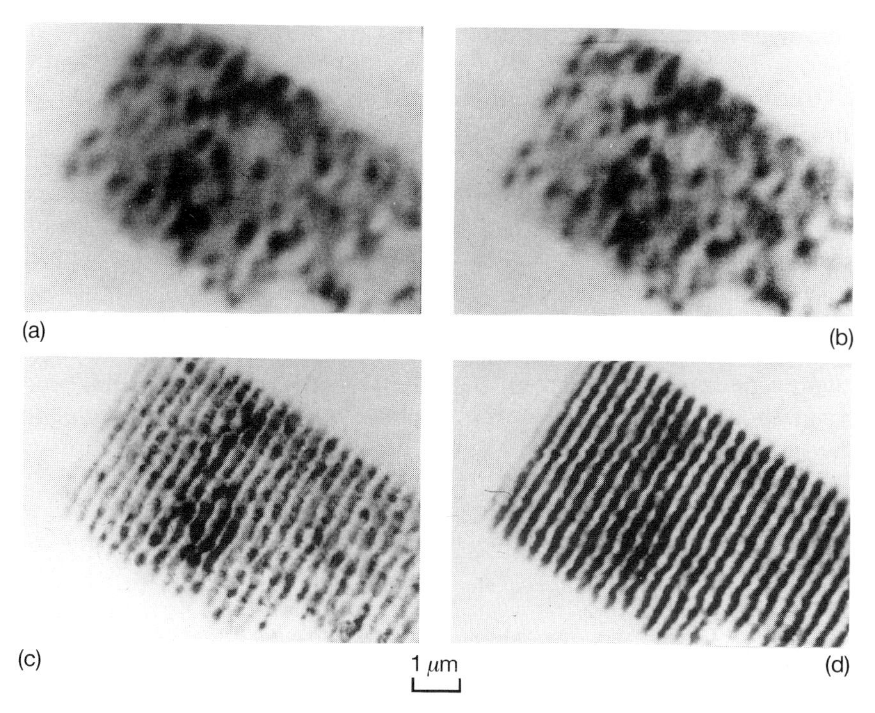

FIG. 3.6. A 290 nm gold grating on silicon in liquid nitrogen, with $\lambda_0 = 430$ nm at 2.0 GHz, at increasing levels of peak power to the transducer: (a) 10 mW; (b) 32 mW; (c) 100 mW; (d) 320 mW (Foster and Rugar 1985).

detected by the third. Alas, despite determined efforts, no non-linear effects were ever detected that could be unequivocally ascribed to the specimen rather than to the water. It is possible to measure a second-order non-linearity parameter B/A (Bjørnø and Lewin 1986) using a short quasi-c.w. toneburst (Nikoonahad and Liu 1990), and it would be very nice to be able to measure non-linearity in material and biological specimens in an acoustic microscope.

For many reasons most acoustic microscopy will continue to be performed with water as the coupling fluid, and the highest-resolution water-coupled images ever obtained have made use of nonlinear resolution enhancement (Hadimioglu and Quate 1983). Pulses 3 ns long were applied to a 4.4 GHz transducer. The lens had a focal length of 15 μm and an aperture diameter of 22 μm ($N.A. \approx 0.73$). This would give in the linear regime a wavelength of 370 nm and a resolution of 240 nm. Water has a non-linear parameter $\mathsf{L} = 1.2 \times 10^{-3}$ W GHz^2, so that higher input powers are needed to achieve non-linearity than for the cryogenic fluids.

At low power it proved impossible to resolve the spacing of a grating of periodicity 200 nm, but as the power was increased above 23 dBm (200 mW) the grating became resolved. This corresponds to an improvement in resolution of $\sqrt{2}$ to 170 nm. An example of images obtained at this resolution is given in Fig. 3.7. The specimen in this figure is an integrated circuit containing bipolar transistors; the aluminium lines are about 3 μm wide. Figure 3.7(a) is an acoustic image at 4.4 GHz with 25 dBm input power. An acoustic image of part of this area at higher magnification is shown in Fig. 3.7(b). The resolution can be compared with that achieved in a light microscope using a 100×, 1.25 numerical aperture oil-immersion objective (Fig. 3.7(c)). The scanning electron microscope image in Fig. 3.7(d) offers higher resolution still, and comparison with this indicates the extent to which features are visible in the acoustic image that are not simply topographical in origin.

Figure 3.7 demonstrates the resolution that can be achieved using water-coupled acoustic microscopy. It represents a remarkable achievement in the development of the instrument. Never before had it been

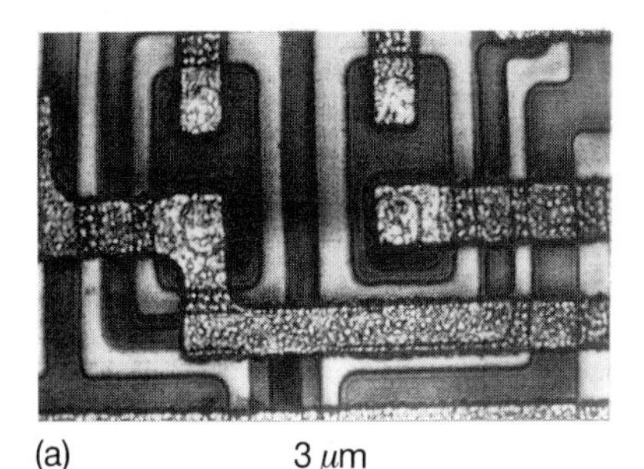

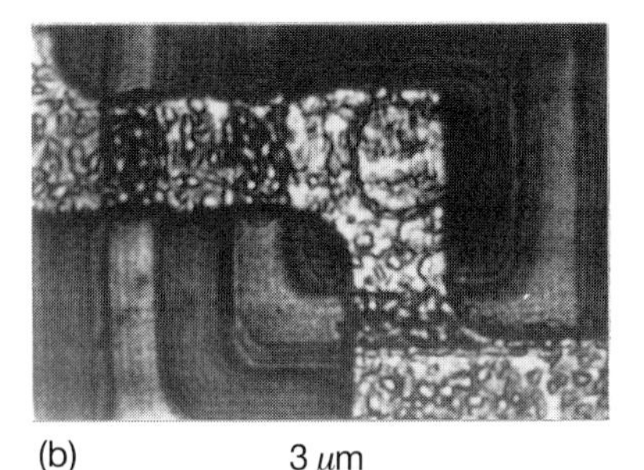

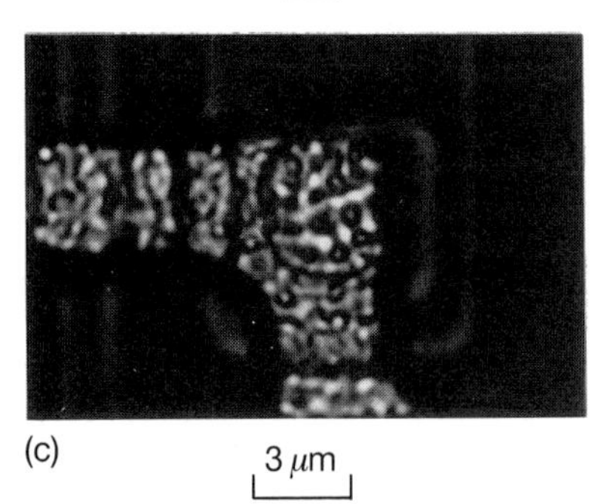

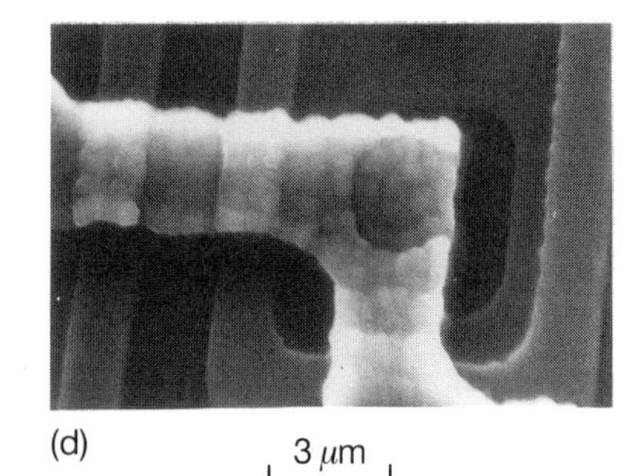

FIG. 3.7. An integrated circuit containing a bipolar transistor: (a) 4.4 GHz, 25 dBm; (b) the same but at higher magnification; (c) light microscope; (d) s.e.m. (Hadimioglu and Quate 1983).

possible to obtain an acoustic image through water with such high resolution. Nevertheless, comparison with the s.e.m. image shows that resolution alone cannot explain why the acoustic microscope has aroused such interest. But if it is possible to image how the acoustic waves interact with the specimen, and specifically to understand what they tell you about the elastic properties, then that is indeed significant.

3.5 Aliasing

More often than not a digital framestore is used to store and display the image. The area of the stored image corresponding to each data point is called a pixel. It can sometimes happen when a relatively large area is scanned at low magnification that the pixel size is large compared with the resolution of the lens. When this is the case curious images can occur when an object containing some degree of periodicity, such as a semiconductor memory chip, is being looked at. This phenomenon is known as aliasing.

The best way to think about this problem is in the spatial frequency domain. A spatial frequency axis is shown in Fig. 3.8. Along this axis, a spatial frequency K corresponds to $2\pi/a_K$, where a_K is the periodicity of interest. So a value of K near the origin corresponds to periodic features that are widely spaced, whereas a large value of K corresponds to features very close together. The concept of spatial frequency is very useful for developing detailed confocal optics (Wilson and Sheppard 1984). The highest spatial frequency that can be stored and displayed by a framestore corresponds to one pixel being bright, the next being dark, the next being bright, and so on. In other words the relevant spacing is $2a_P$, where a_P is the pixel spacing. The value π/a_P is called the Nyquist limit. It is not possible to distinguish between positive and negative

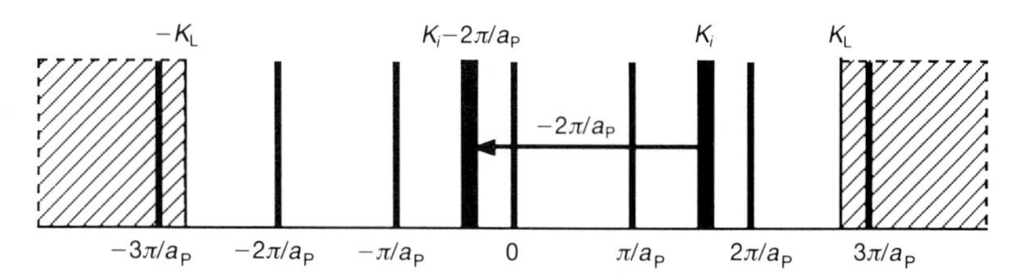

FIG. 3.8. A one-dimensional spatial frequency diagram. A spatial frequency component with periodicity a_{K_i} is represented by the point with coordinate $K_i = 2\pi/a_{K_i}$ on the K-axis. If a digital image recording system has a pixel spacing of a_P, then periodic structures whose spatial frequency lies outside the range $-\pi/a_P \leq K \leq +\pi/a_P$ will appear as structures with spatial frequency shifted by an integral multiple of $2\pi/a_P$ to lie within that range.

spatial frequencies in this context, and so the range of values that the framestore can cope with is $-\pi/a_P \le K \le +\pi/a_P$.

What happens if you try to image an object with periodicity outside this spatial frequency range? The spatial frequency will be displaced by an integral multiple of $2\pi/a_P$ so as to lie in the range that the framestore can handle. There is an almost exact analogy with phonon wavevectors in a crystal lattice or, if you prefer, with why stage-coach wheels appear to go backwards in movies. The effect is illustrated in Fig. 3.8 with a feature with a periodic structure of spatial frequency K_i. It will be stored in the framestore with a spatial frequency $\pm|K_i - 2\pi n_K/a_P|$, with $n_K = 1$ in the example here. This will appear as a periodic structure in the image that bears no apparent relationship to the object.

An upper limit to the spatial frequencies that can be passed to the framestore is imposed by the highest spatial frequencies to which the lens is sensitive, which is $K_L = 2\pi N.A./\lambda_0$. This is twice the spatial frequency that would be indicated by the discussion of point resolution in §3.1.1 (Wilson and Sheppard 1984). Frequencies outside the range of the lens, the shaded area beyond $\pm K_L$ in Fig. 3.8, do not enter the framestore. Aliasing arises when spatial frequencies are present that are resolved by the lens but lie outside the Nyquist limit of the framestore. The problem is well-known from digitizing signals in time, and a key point is that once a signal outside the Nyquist limit has been digitized it is impossible to sort out the mess. Frequencies outside the Nyquist limit must be removed from the analogue signal by an anti-aliasing filter before the signal is digitized. Because the x- and y-directions of a raster scan are not equivalent, it is not easy to introduce a sensible analogue filter into the video signal. Therefore the best filter is the point-spread function of the lens itself; but for that to be effective, the magnification must be reduced so that the Nyquist limit of the framestore lies outside the cut-off spatial frequency of the lens.

This problem need not cause too much alarm. Most specimens do not have extended periodic structures; usually a microscope is operated so that the framestore is not limiting the resolution, and the capacity of framestores is increasing, if not by the day, then at least by the year. But the problem can arise, particularly when large areas are being examined, and when it does the only remedies are to reduce the frequency or the numerical aperture of the lens, or else to reduce the area scanned.

3.6 Does defocusing degrade the resolution?

It will become very clear by the end of this book that in a great deal of acoustic microscopy of materials the contrast is dominated by Rayleigh waves excited in the surface of the specimen (Briggs 1985). A summary of the properties of Rayleigh waves will be given in Table 6.2, and their role in the contrast will be introduced in §7.2.1. What all that means in terms of acoustic pictures will be described at a more practical level using the example of caries lesions in tooth enamel in §9.4.1. For now it is enough to say this: when Rayleigh waves are responsible for the contrast, the best pictures are generally obtained when the lens is defocused some distance towards the specimen. But, it is natural to ask, does this not degrade the resolution?

The answer is 'no' (Smith *et al.* 1983). Rayleigh waves are excited by rays from the lens that are incident on the specimen at a particular angle, the Rayleigh angle $\theta_R = \sin^{-1}(v_0/v_R)$. Such rays will form a cone, with its apex at the focus, or virtual focus, of the lens. These rays will intersect the plane of the surface of the specimen in a circle, and as the negative defocus is increased, so the circle will get bigger. It might be thought that, if Rayleigh waves indeed dominate the contrast, then the resolution could not be better than the size of this circle, but that is not so. The focusing action does not end when the Rayleigh waves are excited at the circumference of this circle. Rather, they propagate as a converging circular wave, passing through a diffraction limited spot in the centre, before diverging again and exciting a cone of rays in the water which propagate back through the lens to the transducer. Of course, any feature within that circle will have some effect on the signal, but the overwhelming concentration of energy is at the centre. And thus it is that, even when quite severe defocus is used to enhance the contrast, the resolution can still be nearly as good. One of the experimental demonstrations of this happens also to have been on human tooth enamel (Peck and Briggs 1987).

4

Lens design and selection

You get used to looking through lenses; it is an acquired skill. . . . The microscope also teaches you to move your hands wrong, to shove the glass slide to the right if you are following a creature who is swimming off to the left—as if you were operating a tiller, or backing a trailer, or performing any other of those paradoxical manouvers which require either sure instincts or a grasp of elementary physics, neither of which I possess. . . . I had about five minutes to watch the members of a very dense population, excited by the heat, go about their business until—as I fancied sadly—they all caught on to their situation and started making out wills. (*Lenses,* Annie Dillard, in *Teaching a Stone to Talk* 1986)

The heart of the acoustic microscope system is its acoustic objective. (Atalar and Hoppe 1986)

4.1 Interior imaging

Acoustic waves can penetrate materials that are opaque to other kinds of radiation, such as light. However, the large velocity mismatches that make the aberrations of the lens so small when the acoustic waves are focused in the coupling fluid have the opposite effect when the focus is inside a solid. The aberrations are governed by the difference between the sine and the tangent of the angle of refraction in each case. When plane waves in a fast medium are refracted through a concave spherical surface into a slow medium, the aberrations are small because the refraction is towards the normal and therefore towards the centre of curvature of the surface. The smaller the refractive index (taken as the ratio of the velocity in the liquid to the velocity in the solid), the smaller is the angle of refraction and thus the difference between its sine and its tangent, and therefore the smaller are the aberrations. But if this focused beam is refracted at the plane surface of a solid in order to focus below the surface, the refraction is away from the normal, and the aberrations can be correspondingly large. The situation is analogous to the problem of the apparent depth of a swimming pool: in the paraxial limit when the observer is looking down in a direction close to the vertical, the apparent depth is approximately independent of the angle of observation, but if he looks in a more horizontal direction the pool appears to become shallower. In the acoustic case the velocity in the solid is faster than in the coupling medium, so an object would appear to be deeper than it really is, by an amount that can increase considerably even for modest angles of observation.

A simple geometrical analysis of this can be given in terms of third-order aberration theory for a single surface (Jenkins and White 1976; Hecht 1987). The situation is illustrated in Fig. 4.1. Rays propagating towards a virtual focus at a distance s below the surface of a solid are refracted so that a ray that passes through the surface at a distance h from the axis crosses the axis at a depth s_a, with the paraxial focus at s_b. The refractive index n is the ratio of the velocity in the liquid to the velocity in the solid; in acoustics this usually has a value smaller than

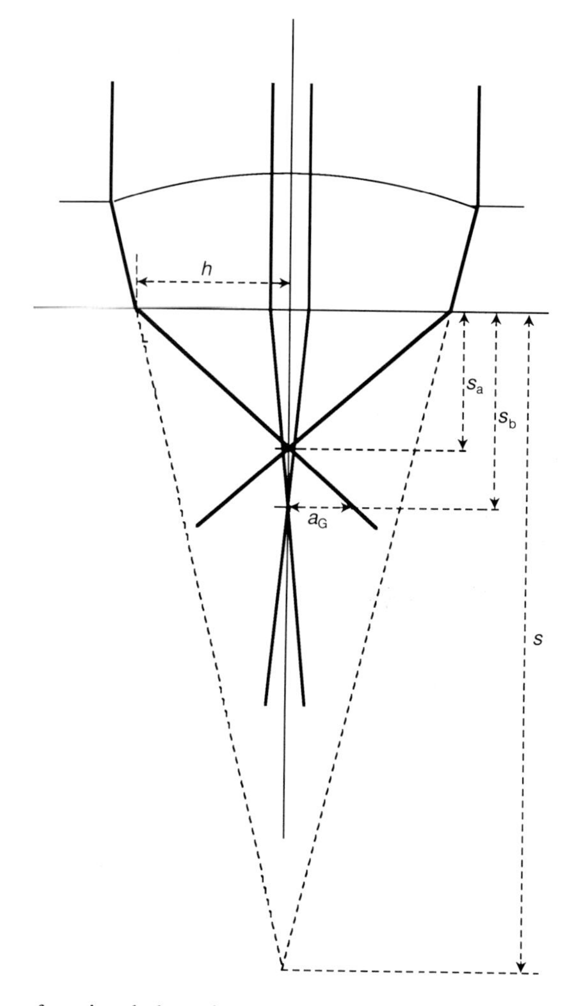

FIG. 4.1. Rays focusing below the surface of a solid, with aberrations arising from refraction of the interface (eqns (4.1)–(4.4)).

one. These quantities are related by

$$\frac{1}{s}+\frac{n}{s_a}=\frac{h^2(1-n)^2}{2}\frac{1}{s^3}. \tag{4.1}$$

But

$$\frac{1}{s}+\frac{n}{s_b}=0; \tag{4.2}$$

hence

$$n\left(\frac{1}{s_a}-\frac{1}{s_b}\right)=\frac{h^2(1-n)^2}{2}\frac{1}{s^3}. \tag{4.3}$$

The transverse aberration is

$$a_G \equiv \frac{h}{s_a}(s_b-s_a)=-\frac{h^3}{s_b^2}\frac{(1/n-1)^2}{2}. \tag{4.4}$$

This is the transverse aberration at the paraxial focus. In simple geometrical terms the circle of least confusion is at a quarter of the distance from where the outermost ray crosses the axis to the paraxial focus. But when diffraction effects are comparable with geometrical aberrations, the relevant focal plane is half way between the paraxial focus and where the outermost ray crosses. Thus the minimum geometrical aberration is

$$a_C=-\frac{h^3(1/n-1)^2}{4s_b^2}. \tag{4.5}$$

If the wavelength in the solid is λ_1, then the Airy disc due to diffraction has its first minimum at a distance from the axis (Hecht 1987)

$$a_D=\frac{1.22\lambda_1}{2h}s_b. \tag{4.6}$$

The total size of the focal spot due to these two effects, geometrical aberration and diffraction, may be written

$$a_{tot}=(a_C^2+a_D^2)^{1/2}.$$
$$=\left[\left\{\frac{(1/n-1)^2}{4s_b^2}h^3\right\}^2+(0.61\lambda_1 s_b^2 h^{-1})^2\right]^{1/2}. \tag{4.7}$$

The minimum in a_{tot} occurs when

$$\frac{h_{opt}}{s_b}=\frac{1.089}{(1-n)^{1/2}}\left(\frac{\lambda_1}{s_b}\right)^{1/4}. \tag{4.8}$$

The optimum lens angle (in the coupling fluid) would be

$$\theta_{opt}=\sin^{-1}[n\sin\{\tan^{-1}(h_{opt}/s_b)\}]. \tag{4.9}$$

As an example of the application of this result, the minimum aberration for acoustic imaging with a wavelength of 25 μm at a depth of 0.5 mm below the surface of a material of refractive index $n = 0.25$ would be given by $h/s_b = 0.7$, so that the optimum lens angle would be $\theta_{opt} = 8.3°$. This result depends rather weakly on s_b, so that halving the depth would increase the optimum angle by less than 1°.

The geometrical approach does not tell the whole story. The calculation of the transverse aberration assumes negligible wavelength, but this is self-evidently negated by the requirement that the total focal spot size due to the combination of aberration and diffraction be a minimum; indeed the criteria used give a geometrical aberration of exactly one-third of the size of the Airy disc. A full diffraction model would enable the acoustic field to be calculated at each point within the solid. This would take phase and amplitude relationships into account, and would enable the field to be visualised. The results of such a calculation are shown in Fig. 4.2. The calculation was performed by finding the field on the surface of the solid, and then using a Green function to calculate the contribution from each point on the surface to a given point in the solid, taking into account the dependence of the transmission coefficient on the angle of refraction. The variation of intensity along the axis of the lens is shown in Fig. 4.2(a). The ratio of the wavelength in the solid to the focal depth in the solid is $\lambda_1/s_b = 0.05$, as in the example above, and the semi-angle of the lens is $\theta_0 = 10°$. The focal intensity is concentrated in a region that extends about two wavelengths above and below the paraxial

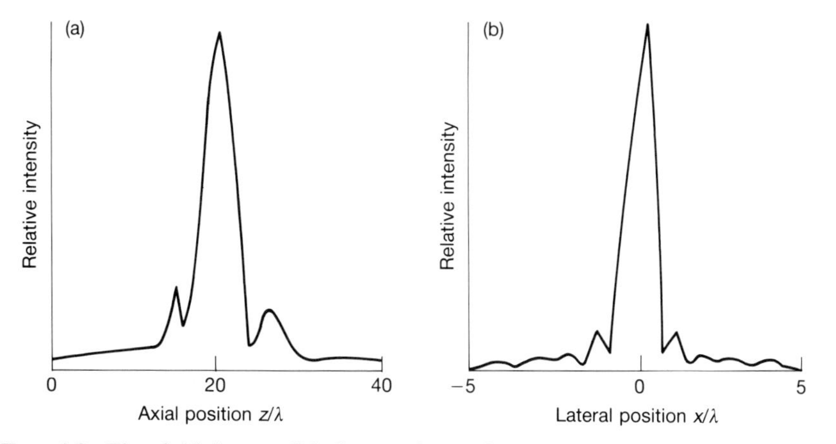

FIG. 4.2. The field in a solid due to focused acoustic waves incident on the surface. The lens semi-angle in the fluid was 10°, the ratio of acoustic velocities in the fluid and in the solid was 0.25, and the paraxial focus was at a depth of 20 wavelengths in the solid. (a) Intensity distribution along the axis; (b) lateral intensity distribution in the focal plane. (Courtesy of Bruce Thompson.)

focus, and has its maximum very close to the paraxial focus itself. The distribution is not symmetrical about the focal plane, but neither is it extensively smeared out towards the surface as the geometrical argument might suggest. The distribution in the focal plane is shown in Fig. 4.2(b). The first minimum occurs at a distance from the axis 35 per cent greater than the radius of the aberration-free Airy disc. This gives a measure of the degradation of the resolution by the aberration at the surface of the solid. A somewhat surprising result of a series of calculations of this type is that from the point of view of resolution there appears to be no optimum aperture. As the aperture is increased, the size of the focal spot, whether defined as the radius at half the maximum amplitude or as the radius of the first minima, seems to go on getting smaller for lens semi-angles right up to the longitudinal critical angle. This conclusion is different from that suggested by the geometrical approach, but it is not clear where the discrepancy arises. It may be because the outer rays play a less and less important role as the aperture is increased, leading to a degradation not of resolution but of the overall efficiency of the system. It is possible that the geometrical approach, with its advantage of a simple analytical formula, may give an indication of a useful compromise between optimum resolution and optimum efficiency.

An experimental test has been performed on a lens designed to give a focus at a depth of 10 wavelengths in brass ($n = 0.32$), with a converging semi-angle in the specimen of 30° (Pino *et al.* 1981). The field in the focal plane was found to be in reasonable agreement with diffraction calculations similar to those outlined above. It was also found that the quality of the focus was not severely degraded for a range of focusing depths from 2 to 20 wavelengths; over this range the radius of the first minimum in the diffraction pattern appeared to remain within about $1.4\lambda_1$ (at 30° the radius of the aberration-free Airy disc would be $1.22\lambda_1$). For comparison, eqns (4.8) and (4.9) would suggest a value for the semi-angle of convergence in the brass of 41.5° with $\theta_{opt} = 12°$, giving a focal radius with aberrations of $1.23\lambda_1$, slightly better than that found for 30° convergence in the solid.

Imaging below the surface of a solid is often performed at frequencies sufficiently low that the attenuation in the coupling fluid is not a limiting factor (though attenuation in the solid may be). For surface imaging, however, attenuation in the coupling fluid must be the first consideration, and this is the starting point for design and selection of high-resolution lenses.

4.2 Surface imaging

The starting point for the design of lens for surface imaging is the choice of the frequency for which it is to be used. This determines the attenuation per unit distance in the coupling fluid (Table 3.1). In water the attenuation is about 0.23 dB μm^{-1} GHz^{-2} at 20°C, falling to less than 0.09 dB μm^{-1} GHz^{-2} at 60°C (Table 3.2). The pulse length to be used determines the bandwidth of the receiving system, and thus the noise level. The dynamic range required must be specified. Allowing for other sources of noise and of losses, the maximum acceptable attenuation in the coupling fluid A_{max} can be specified. Knowing the specific attenuation per unit distance at the temperature and frequency, $\alpha_0 f^2$, the focal length is found from (3.8) to be

$$q = \frac{A_{max}}{2\alpha_0 f^2}. \tag{4.10}$$

Hence the radius of curvature of the lens surface can be found from the relationship

$$r_0 = (1-n)q \tag{4.11}$$

where n is the ratio of the velocity in the fluid to the velocity in the lens. The aperture of the lens must also be specified. For high-resolution work this should be as large as possible. Aberrations can almost be neglected, so these do not generally pose a limitation. The limit to the aperture is rather a practical one, namely, that if the aperture is made too large the working distance of the lens (the distance from the end of the lens to the focal plane) becomes very small ($0.15r_0$ for $n = 0.134$ in the limit of a hemispherical lens surface). The benefits diminish as the semi-angle becomes large both because the numerical aperture, which varies as $\sin\theta$, increases at a decreasing rate, and also because the transmitted intensity through the lens surface becomes small at high angles. Because of the diminishing returns it is not usually worth having a semi-angle greater than 60°. There is, however, a very important minimum angle for surface imaging of solids. For many applications the most interesting contrast arises from the excitation of Rayleigh waves. These couple to waves in the fluid at a Rayleigh angle given by Snell's law as

$$\sin\theta_R = v_0/v_R, \tag{4.12}$$

where v_0 is the velocity in the fluid and v_R is the Rayleigh velocity. The sine of the angle of refraction of the Rayleigh wave does not appear in (4.12) because, by definition, the angle of refraction is 90° and thus its sine is 1. An appreciation of the range of values of v_R, and therefore of θ_R, can be gained from Table 6.3 at the end of Chapter 6; for many metals the Rayleigh angle is about 30°, while for ceramics and semiconductors it is generally less. For all applications where Rayleigh wave contrast is important it is essential that the lens angle is large enough to include the Rayleigh angle.

Having specified the frequency of operation and the aperture diameter of the lens surface, it is now possible to consider the position and size of the transducer. This will be a disc and, ideally, it will operate as a so-called piston source. The amplitude in the far field of such a source can be calculated by the Fraunhofer approximation, and is the jinc function whose form (if not name) is familiar from the jinc^2 intensity of the diffraction pattern of a circular aperture, where $\text{jinc}(u) \equiv J_1(u)/u$. J_1 is a Bessel function of the first kind; it has a role in an axially symmetrical situation analogous to the sinc function for a slit (§3.1.1), and is defined as (Hecht 1987)

$$J_m(u) = \frac{i^{-m}}{2\pi} \int_0^{2\pi} e^{i\{mt + u\cos(t)\}}\, dt. \tag{4.13}$$

In this case $m = 1$ (t is simply a dummy variable of integration). If λ is the wavelength, a_T the radius of the transducer, L the distance from the transducer, and r the cylindrical radial coordinate at that distance, then $u = 2\pi L a_T/\lambda r$. Since $\text{jinc}(0) = \frac{1}{2}$, the amplitude distribution is $A + 2A_0\,\text{jinc}(2\pi L a_T/\lambda r)$, where A_0 is the amplitude on the axis at distance L. Whether the Fraunhofer approximation is valid is determined by the dimensionless distance s_F, which is defined as

$$s_F \equiv \lambda L/a_T^2. \tag{4.14}$$

The Fraunhofer approximation is useful when $s_F \gg 1$, i.e. well beyond the Fresnel distance $F \equiv a_T^2/\lambda$. Closer to the transducer the field must be calculated numerically (Zemanek 1971). If $a_T/\lambda > 1$, the amplitude and phase fluctuate considerably in the region between the face of the transducer and the plane $s_F = 1$; in particular there are nulls along the axis in the region $0 \le s \le 0.5$. The final maximum in the amplitude on the axis occurs at $s_F = 1$, i.e. $F = a_T^2/\lambda$, which is known as the Fresnel length. In the plane perpendicular to the axis at $s_F = 1$ the field is reasonably well behaved in both amplitude and phase.

Because of the behaviour of the near field of the transducer, the lens is often designed so that it is at the Fresnel length of the transducer. If the ratio a_T/λ is large, then in that plane the intensity falls by 6 dB (relative to its value on the axis) at a radial distance of $0.36a_T$, so that the transducer must be made somewhat larger than the lens aperture in order to increase the uniformity of the illumination. Having chosen the value of a_T and knowing λ in the lens material, the Fresnel length can be calculated to give the spacing from the transducer to the lens surface. For a 2 GHz sapphire lens with $a = 60\ \mu\text{m}$, $F = 0.65$ mm, giving a lens thickness a little less than a millimetre. With lower frequencies, since the lens radius can increase with square of the wavelength, the Fresnel length can become very large.

The effect of varying the distance between the lens and the transducer is to vary the illumination of the lens, and this in turn affects the point-spread function of the lens (Chou *et al.* 1988). Figure 4.3 shows the point-spread function calculated for three values of the transducer-lens spacing for the same lens with a radius of curvature $r_0 = 47\lambda_0$, where λ_0 is the wavelength in the coupling fluid, and $N.A. = 0.3$. The transducer radius was $2r_0$ in each case, and the distances from the transducer to the lens surface corresponded to the values $s_F = 0.5$, 1, 10. A paraxial approximation would give a jinc point-spread function, but this is modified by the high angle of convergence (Sheppard and Wilson 1981). At the Fresnel distance, $s_F = 1$, the illumination has an approximately Gaussian profile: it is greatest at the centre of the lens and decreases towards the edge (Kino 1987). This leads to considerable apodization: the side-lobes are reduced by over 10 dB and there is an increase of about 20 per cent in the radius of the first minimum of the point-spread function.

When the radius of the transducer is a large number of wavelengths, then there is a final null in the field along the axis of the transducer which occurs close to $s = 0.5$. Therefore, when the lens surface is at that distance from the transducer, there is reduced illumination of the centre

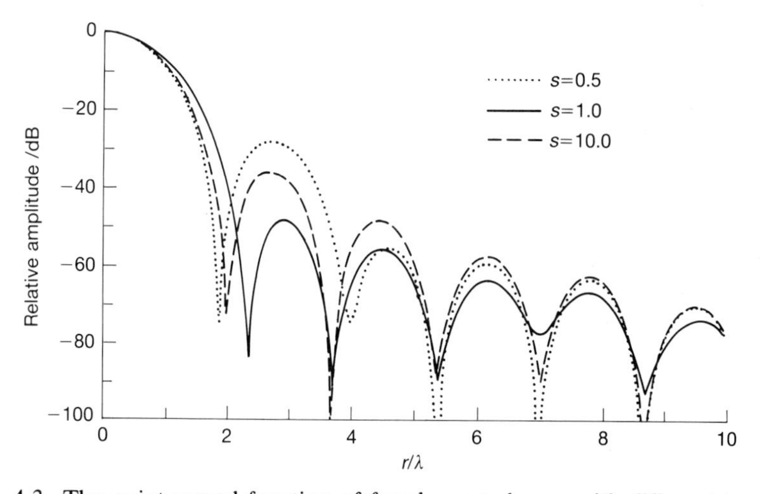

FIG. 4.3. The point-spread function of fused quartz lenses with different lengths L between the transducer and the lens, expressed as dimensionless distance $s \equiv \lambda L/a_T^2$ (4.14): · · · ·, $s = 0.5$; ——, $s = 1$; – – –, $s = 10$. Each lens had numerical aperture $N.A. = 0.3$, lens aperture radius $r_0 = 47\lambda_0$, where λ_0 is the wavelength in the fluid, and transducer radius $a_T = 2r_0$ (Chou *et al.* 1988).

of the lens, and relatively greater illumination of outer parts of the lens. This slightly improves the resolution, at the expense of increasing the side-lobe levels. But there is another effect that may be of more significance than either of these for acoustic microscopy. In applications where the most interesting contrast comes from the excitation of Rayleigh waves in the surface of the specimen, a lens whose surface is at $s = 0.5$ may have a greater proportion of the acoustic energy around the Rayleigh angle, thus enhancing the sensitivity to factors that affect the propagation of the Rayleigh waves (Atalar 1987, 1989). Special axicon lenses have been designed which can be tuned in to the Rayleigh angle of the specimen being examined and which, therefore, give maximum sensitivity to Rayleigh wave scattering (Atalar and Köymen 1987), and conical lenses have been developed for imaging specimens that support layered acoustic wave modes (Atalar and Köymen 1989).

There are thus competing criteria in lens design, and compromises must be made. For high resolution, it is desirable to have a small radius of curvature to allow high frequencies to be used, but not so small that the specimen echo cannot be separated from the unwanted lens echoes. The smallest radius for comfortable use is 40 μm, allowing operation up to 2 GHz in hot water. At a given frequency it is desirable to have the longest possible working distance, both for ease of operation and, as will be seen later, for maximum range of defocus, but this is limited by the need to keep the attenuation acceptable and to have an adequate numerical aperture both for good resolution and for Rayleigh angle excitation. For surface imaging a lens semi-angle (measured from the centre of curvature) of 50°–60° is usual. Finally, the lens illumination must represent a compromise between resolution, apodization, and response at the Rayleigh angle. For general-purpose high-resolution work the lens is usually designed so that the lens surface is at the Fresnel focus of the transducer.

The development of lenses for quantitative elastic measurements with spatial resolution is an area of great activity in acoustic microscopy. The classic lens for quantitative work is the line-focus-beam lens, which has a cylindrical surface and so generates cylindrical wavefronts that come to a line focus (Kushibiki *et al.* 1981*b*). This enables elastic measurements to be made in different directions on anisotropic specimens such as wafers of electronic and optoelectronic materials. The design of a line-focus-beam lens and the principles of its use will be described in §8.2.1. But because the line-focus-beam lens generates a focus along a line, it has poor spatial resolution in that direction. So other developments to enable angular resolution to be obtained without degrading the spatial resolution too much have used shear wave transducers (Chou *et al.* 1987; Chou and Khuri-Yakub 1989), or directional transducer patterns (Davids *et al.*

1988; Kushibiki *et al.* 1989) or directional lens apertures (Ishikawa *et al.* 1990). These lenses will be mentioned again in the context of stress measurement in §8.2.4, but their importance is much wider than that, and is applicable to all specimens with any anisotropy (e.g. Fig. 11.13). Another area of active development in lenses is that of lenses for specimens with surface layers. Axicon lenses can be made that generate conical wavefronts which have a single angle of incidence on the specimen (Atalar and Köymen 1989). By careful choice of frequency, these can have great sensitivity to the kinds of properties considered in §10.2-3, such as changes in the composition or thickness of a coating, or the extent of its adhesion to the substrate. It may be anticipated that many new and important possibilities will open up with the growing availability of specialized quantitative lenses.

4.3 Wanted and unwanted signals

The material for an acoustic lens should have a low attenuation, and a high velocity to minimize aberrations. Sapphire is an excellent material in both these respects. But the high velocity has a less desirable consequence. An acoustic impedance can be defined, which is equal to the product of the velocity and the density. The impedance of sapphire for longitudinal waves travelling parallel to the c-axis is thus 44.3 Mrayl, compared with the impedance of water which at room temperature is about 1.5 Mrayl, rising to 1.525 Mrayl at 60°C. When sound is transmitted across an interface between two materials of different impedance, the stress amplitude transmission coefficient is (§6.4.1; Auld 1973; Brekhovskikh and Godin 1990)

$$T = \frac{2Z_2/\cos\theta_2}{Z_1/\cos\theta_1 + Z_2/\cos\theta_2}. \tag{4.15}$$

The relationship between θ_1 and θ_2 is given by Snell's law

$$\frac{\sin\theta_1}{\sin\theta_2} = \frac{v_1}{v_2}. \tag{4.16}$$

These relationships can be extended to allow for mode conversion and for anisotropy. For normal incidence (4.15) reduces to the simple form familiar by analogy with transmission line theory (Bleaney and Bleaney 1983). Thus an axial ray in the acoustic microscope would have only a little over 3 per cent of its energy transmitted across the sapphire-water interface in each direction and, even if it were totally reflected at the specimen, only 0.1 per cent of its energy would eventually return to the transducer. This 30 dB loss would be very serious. To increase the power transmission though the lens surface, it is necessary to use a matching layer. The matching layer should be a quarter wavelength thick and have an impedance $Z = \sqrt{(Z_1 Z_2)}$. Borosilicate glass is often used, because it can be sputtered and is mechanically, chemically, and thermally robust. Sputtered SiO_2 is an alternative (Kushibiki *et al.* 1980). For the most accurate work chalcogenide glass is preferable because, although it does not have the robustness of borosilicate glass, its impedance can be tailored by adjusting the composition to give an almost perfect match (Kushibiki *et al.* 1981*a*). For cryogenic microscopy special techniques must be used (Rugar 1981), because the mismatch to helium is even more severe than to water. For complete accuracy the thickness of the matching layer should be a controlled function of the radial coordinate. Various attempts have been made to achieve this, but usually the simple cosine function that is obtained by deposition along the axis is acceptable.

The matching layer is sometimes called an acoustic antireflection coating. As well as increasing the transmission across the lens surface, and thereby increasing the wanted signal, as a corollary it also reduces unwanted echoes within the lens. Some of the significant acoustic reflections that contribute to the signal at the transducer are illustrated in Fig. 4.4(a). The wanted echo is the one that comes from a wave that is transmitted across the lens–fluid interface, is reflected by the specimen, and returns through the lens to the transducer. The next largest echo in a well designed lens is the longitudinal wave reflection from the lens surface. In the absence of a matching layer, this echo would be far larger than the specimen echo. Much more energy would be reflected than transmitted at the interface, and the specimen echo would then undergo further attenuation in the fluid; the only mitigating factor would be the effect of the convex reflecting surface on the internally reflected wave. But with a good chalcogenide matching layer this echo can be made substantially smaller than the specimen echo. There is usually also an echo from the lens surface that travels in one direction as a longitudinal wave and in the other direction as a shear wave. Although to first order this wave should not exist, both because mode-conversion at the lens should only occur off the axis, so that the wave thus generated should not travel back to the lens, and because the transducer should only generate and be sensitive to longitudinal waves, nevertheless experience shows that this echo can often be significant. The illumination inevitably spills over to the area around the lens, which is usually made flat to give resistance to accidental chipping, and this can lead to an echo from this surface which arrives at the transducer slightly after the echo from the lens-surface echo. These two echoes may reverberate within the lens rod, giving subsequent echoes at multiples of the first echo time.

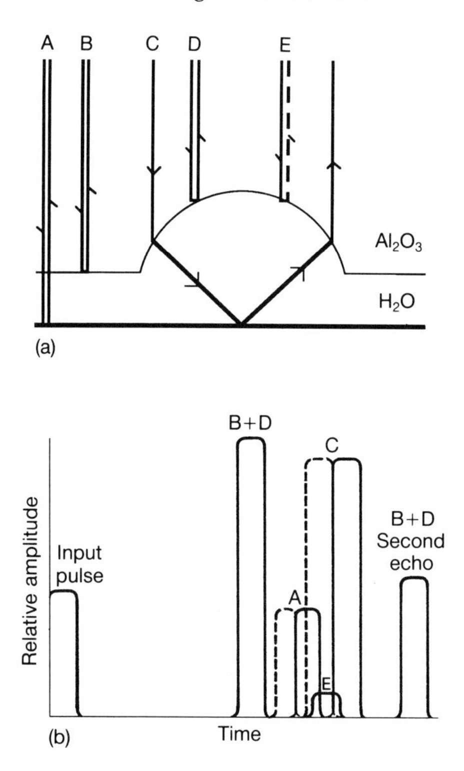

FIG. 4.4. Reflections from various interfaces in an acoustic microscope. (a) Paths travelled by the rays; the broken line shows a ray that returns as a shear wave after mode conversion from a longitudinal wave (or vice versa). (b) The arrival time and magnitude of each reflected ray in (a); the broken lines show the earlier arrival times of reflections from the specimen surface when the specimen is defocused towards the lens. (After Weaver 1986).

There may also be an echo from the waves that pass through this surface and are reflected from the specimen; this arrives a little before the echo from the waves that are focused on the specimen because the velocity is so much faster in the lens material than in the fluid. The arrival times of these echoes are illustrated schematically in Fig. 4.4(b) (in practice the profiles of very short pulses may be rather less rectangular). The echoes from the specimen surface will arrive earlier if the specimen is defocused towards the lens; this is indicated by the broken lines.

It is a basic principle of good lens design to ensure that the specimen echo arrives at the transducer before the second internal echo from the lens surface. If the refractive index is n, the focal length is q, and the

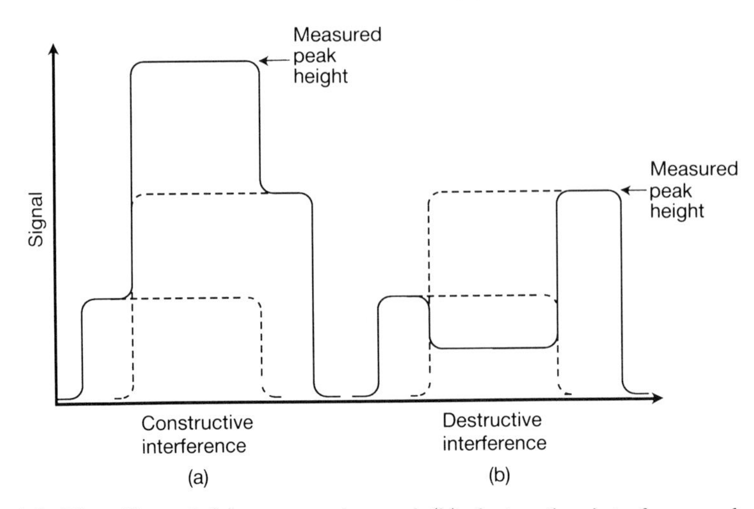

FIG. 4.5. The effect of (a) constructive and (b) destructive interference due to separate components in the specimen echo C in Fig. 4.4(b) on the measured peak value when the whole signal lies within the gate of the peak-detector. When the interference is constructive, the net signal is measured correctly, but when it is destructive, the magnitude of whichever component happens to be bigger is measured instead. (After Weaver 1986).

transducer-lens separation is D, then this requires that $D > (q - \tau v_0/2)/n$, where τ is the pulse length and v_0 is the velocity in the fluid. A lens of 40 μm radius of curvature with $n = 0.134$ has a focal length $q = 46\ \mu$m, and with a pulse length of 20 ns the required length is $D > 400\ \mu$m, which is indeed satisfied by the value of D selected by diffraction considerations; the value of D would put the specimen echo rather nicely halfway between the first and second lens echoes. Because of the f^2 dependence of attenuation, at lower frequencies the timing requirement is usually much easier to satisfy than the diffraction requirement, so that there may then have to be a sacrifice in either the working distance or the uniformity of illumination.

The converted longitudinal-shear echo, the one that in principle should not exist, arrives slightly before the mid-point in time between the first and second pure longitudinal lens surface echo. In many materials the shear velocity is slightly over half the longitudinal velocity. For propagation along the c-axis of sapphire the longitudinal velocity is 11 100 m s^{-1} and the shear velocity is 6031 m s^{-1}. The shear velocity is thus 54 per cent of the longitudinal velocity, and so the mode-converted lens-surface echo arrives after a delay of 1.46 × the first lens echo delay. This echo can prove particularly troublesome in quantitative $V(z)$ analysis. The problem is illustrated in Fig. 4.5. Measurements of the signal as the specimen is moved towards the lens, so-called $V(z)$ measurements, will be described in Chapter 8. If the specimen echo and an internal echo partially overlap, then there will be interference between them that varies as the separation between the lens and the specimen changes. When the interference is constructive, then the signal measured by a peak detector circuit is the sum of the amplitudes of the two components. But when the components are in antiphase, then the circuit measures whichever of the two happens to be larger, if the whole signal falls within the gate of the peak detector. For some qualitative work this may not matter, but for any quantitative analysis, especially when the signal is recorded as the lens is scanned along its axis, the consequences would be devastating. After the lens geometry and matching layer have been optimized to minimize such effects, attention must be turned to the electronic circuitry.

5

Electronic circuits for quantitative microscopy

Rubin came out at the top of the main staircase of the old building; the two flights curved outwards and converged again at the bottom; he walked across a marble-paved landing, past two old, unlit wrought-iron lanterns, into the laboratory corridor, and pushed open a door with a sign saying 'ACOUSTICS'. . . . The Acoustics Laboratory was a wide, high-ceilinged room with several windows. It was untidy and crammed with electronic instruments on wooden shelves, shiny aluminium stands, assembly benches, new plywood cabinets made in a Moscow factory and comfortable desks requisitioned in Germany. . . .

Little by little he had ceased to do original research himself and begun instead to supervise the work of others. . . . Another reason he had not been able to get on with his own research in recent years was that he constantly had to attend meetings and was overwhelmed by paperwork. . . . How he would have liked to sit over blueprints, handle a soldering iron again and watch the flickering pulse on the green screen of the oscilloscope. . . . If only he could throw off the shackles of high office and sit down himself at the drawing board, and find time to think. (*The First Circle,* Alexander Solzhenitsyn 1968)

5.1 Time and frequency domains

Any pulse can be described both in the time domain and in the frequency domain. In the time domain a signal may be oscillatory. The time domain behaviour is what is seen on an oscilloscope screen, because an oscilloscope is essentially an instrument for displaying a signal as a function of time. But a varying signal may also be described in terms of the components of each frequency present. This is a frequency domain description and is what is displayed by a spectrum analyser, just as an optical spectrum indicates the amount of each frequency (or wavelength) in a source of light. These two descriptions are related by a Fourier transform (Bracewell 1978), which may be written

$$\mathrm{F}(f) = \int_{-\infty}^{\infty} \mathrm{f}(t)\mathrm{e}^{\mathrm{i}2\pi ft}\,\mathrm{d}t. \tag{5.1}$$

It is often useful to abbreviate this as

$$\mathrm{F}(f) = \mathbb{F}\{\mathrm{f}(t)\} \tag{5.2}$$

where $\mathbb{F}$ denotes the Fourier transform. It follows from the nature of the Fourier transform that if a pulse is of finite duration, then it cannot have a single frequency, but must rather consist of a finite spread or spectrum of frequency components. If, for example, the envelope of a pulse is of rectangular form (as suggested schematically in Fig. 4.4), then in the frequency domain it will have the profile of the Fourier transform of a rectangular function, namely a sinc function (the sinc function is defined as $\operatorname{sinc}(x) = \sin(x)/x$; it is familiar in its squared form as the diffracted intensity pattern of a single slit). For a rectangular pulse of length t_0, the width of the central maximum in the frequency spectrum is $2/t_0$; this gives a measure of the spread of frequencies present. Indeed, as a rule of thumb, $\Delta t \approx 1/\Delta f$; the length of a pulse cannot be less than the reciprocal of spread of frequencies present. An ideally monochromatic signal would have only a single frequency present, and therefore would have to be of infinite duration.

The convolution of two functions of the same variable, for example $\mathrm{G}(f)$ and $\mathrm{H}(f)$, can be written

$$\mathrm{F}(f) = \int_{-\infty}^{\infty} G(f')\mathrm{H}(f - f')\,\mathrm{d}f'. \tag{5.3}$$

The convolution operation is a way of describing the product of two overlapping functions, integrated over the whole of their overlap, for a given value of their relative displacement (Bracewell 1978; Hecht 1987). The symbol $\otimes$ is often used to denote the operation of convolution. The convolution theorem states that the Fourier transform of the product of two functions is equal to the convolution of their separate Fourier transforms

$$\mathbb{F}\{\mathrm{G}(f) \times \mathrm{H}(f)\} = \mathbb{F}\{\mathrm{G}(f)\} \otimes \mathbb{F}\{\mathrm{H}(f)\}. \tag{5.4}$$

If a pulse is produced by taking the output of a monochromatic oscillator of amplitude A_0 and applying to it a rectangular gate of width t_0, the resulting spectrum will be the convolution of the spectrum of the oscillator (a delta function at f_0) and the spectrum of the gate, giving

$$\mathrm{F}(f) = A_0 t_0 \frac{\sin\{\pi(f - f_0)t_0\}}{\pi(f - f_0)t_0}. \tag{5.5}$$

If the pulses are repetitive, with a repetition frequency f_1, then the gated signal is convolved in the time domain with a comb function (a comb function is a series of delta functions at constant spacing). This means that the Fourier transform of a single gated pulse (5.5) is multiplied by the Fourier transform of the pulse repetition, which is in turn a comb function starting at the origin and repeating at intervals of f_1 (there is a close analogy with the diffraction pattern of a large diffraction grating). Thus the frequency spectrum of a series of gated pulses is the Fourier transform of the envelope of an individual pulse, centred at the frequency of the oscillator, composed not of a continuous distribution but of a series of discrete spikes (i.e. delta functions) within that envelope.

An example of such a spectrum is given in Fig. 5.1. The vertical axis is the logarithm of the square of the power per unit frequency interval, with a scale of 10 dB per division. Figure 5.1(a) has a horizontal scale of 10 MHz per division. The form of the spectrum is approximately a sinc^2 function; the bandwidth of the central lobe is about 25 MHz, corresponding to a pulse length of 80 ns. Measurement of the spectrum at a frequency resolution of 50 kHz per division in Fig. 5.1(b) shows how it is composed of a series of individual delta functions. These are spaced at 75 kHz, the pulse repetition frequency, but they are not multiples of 75 kHz because they are centred at the frequency of the r.f. oscillator. Smaller spectral lines are present just above the noise level, displaced in frequency by about 25 kHz relative to the main components; these are harmonics of spurious transient signals given out by the r.f. switches and they are at multiples of the pulse repetition frequency, extending right up to and beyond the frequency range of the r.f. signal itself.

In the acoustic microscope the required signal can be selected not only in the time domain but also in the frequency domain. The ability to select the specimen echo and separate it in time from the unwanted lens echo was the basis of the design considerations of the focal length of the lens, and hence the resolution available (§3.2 and §4.3). But it is a fact of experience that smaller lens reverberations are always present that cannot be separated in time from the echo from the specimen. These lead to the

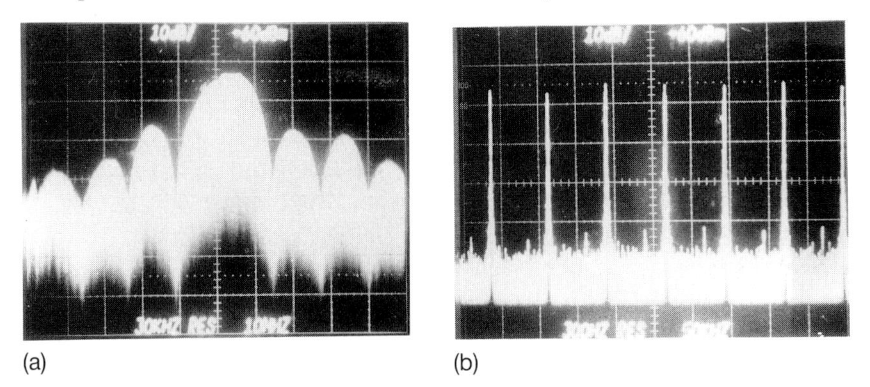

(a) (b)

Fig. 5.1. Spectra of the r.f. pulses in an acoustic microscope: the vertical axis is power per unit frequency interval, on a logarithmic scale of 10 dB/division. The horizontal axis is frequency: (a) 10 MHz/division; (b) 50 kHz/division (Weaver 1986).

kind of problem that was illustrated schematically in Fig. 4.5. Moreover, even if such reverberations could be eliminated, there would still be a problem arising from the fact that different parts of the acoustic wave from the specimen may be reflected with different time shifts. This could occur, for example, if the specimen has two closely spaced layers, each of which reflects some energy, or, as is often the case, if some of the energy is reflected back after coupling into surface waves. If the signal is measured by a peak detector circuit, a reasonable value will be obtained when the interference is constructive, but when it is destructive the value will be quite wrong, corresponding simply to the largest component present. If very long pulses could be used it might be possible to measure the middle where the overlap is adequate, but this is not practicable when there are constraints on the pulse length, as there are in all high-frequency microscopes.

A better approach is to use a narrow-band filter after the gate (this can be achieved in practice by gating the local oscillator and filtering the i.f. signal). It must be after the gate, because the effect of the filter will be to spread the information out in time, and the selection in the time domain should be performed with the maximum available bandwidth, and hence the best possible resolution in the time domain. But after the time gating has been performed, a narrow band filter can be applied to confine the frequency spread to, say, approximately half the width of the central lobe of the pulse spectrum. This has two effects. First, it improves the signal-to-noise ratio, because it removes noise in parts of the spectrum where there is little signal anyway. It is apparent from Fig. 5.1 that, provided the bandwidth of the filter lies between the first nulls in the signal spectrum, further narrowing of the bandwidth makes little difference if it is greater than the pulse repetition frequency, because it will filter out roughly equal amounts of signal and noise. Second, because of the Fourier relationship, it has the effect of smearing out everything in the time domain over a time comparable to twice the length of the pulse. This would mean that two reflected components would effectively interfere with each other over approximately the whole length of the pulse, and the awkward effects at each end would be largely eliminated. Also, there will be a measure of rejection of the switching transients. For normal operation of an acoustic microscope this narrow band filter after the time gating and before the peak detection (or sample-and-hold) is a satisfactory way of measuring the strength of the reflected echo (Atalar and Hoppe 1986).

5.2 Quasi-monochromatic systems

In much of the interference theory of the acoustic microscope, and especially the theory of how the contrast varies with defocus, it is assumed that the microscope is perfectly monochromatic. Of course, the monochromatic theory could be summed over a frequency spectrum actually used in a given system, but that would be messy, and would be almost impossible to invert when interpreting measured results. For much qualitative imaging the narrow-band detected signal may be adequate, but for accurate measurements something better is needed.

For a signal whose spectrum is like the one illustrated in Fig. 5.1, if the bandwidth of the detection system can be narrowed to a fraction of the spacing between adjacent spikes, it is possible to select one of them and to reject all other signals at spacings of multiples of the pulse repetition frequency. In this way it is possible to isolate and measure a single frequency component, and thus achieve a measurement that is monochromatic, subject only to the phase noise of the signal source. This is done using a heterodyne circuit, with the local oscillator phase-locked to the signal source and having a frequency that differs from the signal source by some small amount (Liang *et al.* 1985*a,* 1986). It is desirable to reject the components, also spaced at multiples of the pulse repetition frequency, due to switching transients. In order to achieve this the pulse repetition clock should be phase-locked to the signal source. If the frequency of the signal source were set to an exact multiple of the pulse repetition frequency, then the components of the switching spikes would sit exactly on the signal components and could not be separated. So it might be thought that the best separation would be achieved by setting the signal frequency at some half-integral multiple of the pulse repetition frequency. In fact it should be set close to a third-integral multiple.

The frequency component to be measured is best selected by using a heterodyne circuit with a local oscillator of frequency differing by a small amount Δf. This will have the effect of subtracting the local oscillator frequency from each of the frequency components in the signal. Frequencies that would become negative by this process are reflected about the origin so that they become positive. In order that switching components reflected about the origin fall on top of unreflected components, to make it as easy as possible to avoid both, the local oscillator frequency should be a multiple of the pulse repetition frequency. In order that a single signal component can then be separated both from other components reflected about the origin, and from switching components, the signal source frequency should be a third-integral multiple of the pulse repetition frequency.

Schematic diagrams for radio-frequency (r.f.) electronics are shown in Fig. 5.2. The circuit in Fig. 5.2(a) is a simple heterodyne circuit. The pulse length is defined by the switch S1. The speed of this switch determines the minimum pulse length, and hence the minimum lens focal length, and hence the highest frequency of the microscope; thus the limit

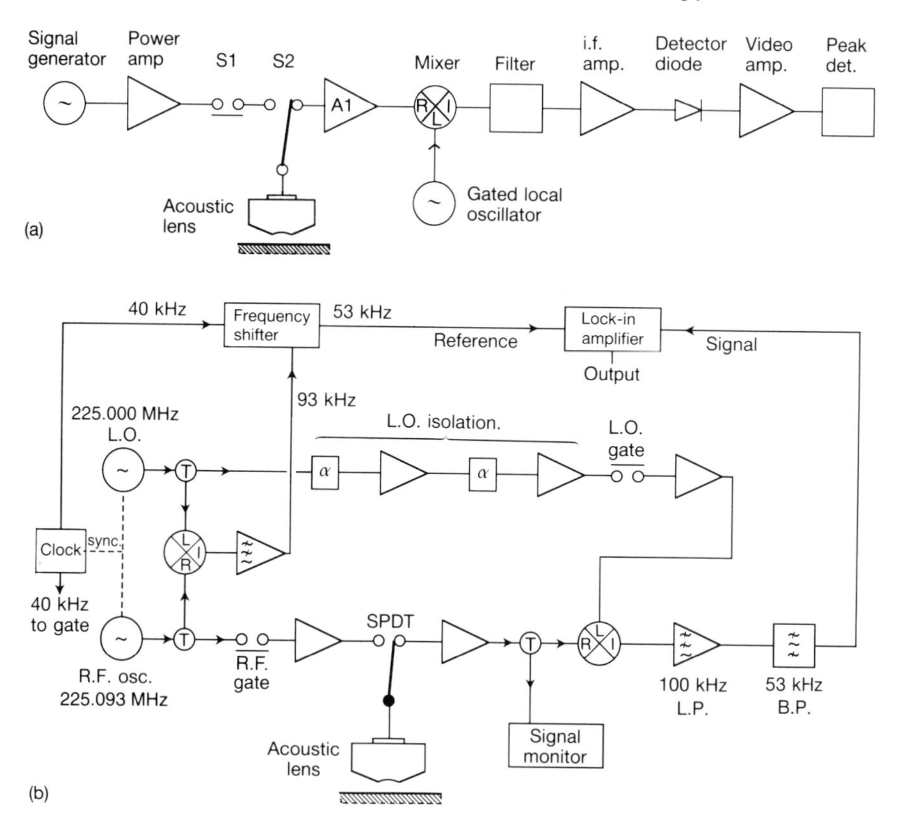

FIG. 5.2. Schematic r.f. systems. (a) Simple heterodyne circuit, S1 determines the pulse length, S2 switches the lens from transmit to receive, and A1 amplifies the reflected signal; (b) quasi-monochromatic circuit; the two oscillators and the pulse repetition frequency are phase-locked, and the final signal is lock-in detected (after Weaver 1991).

to the resolution is ultimately determined by the highest useful speed of this switch. S2 is the single-pole-double-throw (s.p.d.t.) switch, and A1 the low-noise preamplifier. The mixer is a device that takes two inputs, the radio-frequency signal and the local oscillator signal, and gives outputs at the sum and difference frequencies of these two (Henderson 1990). The local oscillator frequency is chosen so that the difference frequency is in the middle of the pass band of the narrow band filter, which is placed in the circuit at the intermediate frequency output of the mixer. Gating of the preamplifier output is achieved by gating the local oscillator of the heterodyne circuit. A heterodyne circuit of this type is used in most imaging microscopes (Atalar and Hoppe 1986).

For quantitative applications the quasi-monochromatic circuit of

Fig. 5.2(b) is better (Weaver 1991). The basic principles of this circuit are similar to those of Fig. 5.2(a), but there are some important differences. The r.f. oscillator and the local oscillator are two frequency synthesizers that are phase-locked to one another to give a difference frequency of precisely defined phase. This difference frequency is very much lower than in the simple heterodyne circuit, in order to make it comparable with the pulse repetition frequency. The pulse repetition frequency is derived from the local oscillator by division of the local oscillator frequency in order to define the relationship between these frequencies. In order for the system to behave as a truly monochromatic one, it is important that the receiver gate should not introduce any distortion of the frequency spectrum of the reflected pulse. The reflected pulse will be multiplied in the time domain by the receiver gate, and that corresponds in the frequency domain to a convolution with the (repetitive) pulse gate transform. In general that would result in a mixing of the different frequency components, which would be most undesirable. This can be avoided by ensuring that in the time domain the gate makes no difference to the desired pulse, i.e. that it is big enough to let the whole pulse through without clipping either end. It is then equivalent (so far as that pulse is concerned) to an infinite gate, whose transform is a single delta function of zero frequency. Provided this condition is fulfilled, there is no restriction on the removal of other unwanted pulses by the gate. It is also important not to introduce other frequency components into the signal by unwanted non-linear effects in the receiving amplifier or the mixer. The receiving amplifier is a fast-recovery limiting amplifier, to minimize any distortions following switching transients from the s.p.d.t. switch. The mixer is a type III double balanced mixer, which needs more local oscillator power than a type I mixer but gives less distortion (specifically, it has very small third-order intermodulation products of the form $f_1 + f_2 - f_3$). Finally, the amplifier on the output of this mixer, which need only amplify in the near-audio range, has nevertheless an input that is carefully matched to 50 Ω all the way up to the sum of frequencies of the signal source and the local oscillator.

After the output of the mixer has been amplified and low pass filtered, it may be fed to a lock-in amplifier. This is a phase-sensitive detector that performs at near-audio frequencies rather as a mixer does at radio frequencies, but with more sophisticated controls. In particular, an integration constant can be set on the output, so that, by the kind of Fourier relationship discussed above, only signals of frequency arbitrarily close to the reference frequency are detected. This means that the signal-to-noise ratio that can be ultimately obtained is essentially limited only by the phase noise of the synthesizer. For the reference signal, the difference frequency (obtained by simple mixing of the two synthesizer outputs) could be used, which would effectively give detection of the original r.f. frequency. But, because there may be spurious signals of that difference frequency present for various reasons, it is preferable to shift the difference frequency by summing it with the pulse repetition frequency (a simple active analogue circuit can be used for this purpose, and by summing sine and cosine components phase tracking can be maintained more easily than with filters even when transient changes in frequency occur). With a dual-phase lock-in amplifier, the in-phase and quadrature components or, equivalently, the modulus and phase can be measured simultaneously. A little manipulation of cosine sum and difference formulae reveals that the phase indicated by the lock-in amplifier is indeed the same as the phase of the corresponding r.f. component. If only the modulus of the detected component is required, then a readily available alternative to the circuit illustrated in Fig. 5.2(b) is to use a spectrum analyser with a measurement output in place of the heterodyne detection system in Fig. 5.2(a), with its own tracking generator as the r.f. oscillator and the swept frequency range set to zero (Kushibiki and Chubachi 1985).

Typical operating frequencies might be as follows. For a lens designed for quantitative work at 225 MHz, a pulse repetition frequency of 40 kHz might be suitable. The local oscillator frequency would be set to 225.000 MHz, and the signal frequency could be 225.093 MHz. The difference frequency of 93 kHz could be summed with the pulse repetition frequency to give a reference frequency for the lock-in amplifier of 53 kHz. This would mean that the quasi-monochromatic frequency that was being measured would be 225.053 MHz. It might appear a little untidy to be measuring the response at that frequency, rather than for example some exact number of megahertz, but in this way it is possible to make a measurement at a single frequency, and that frequency alone.

The quasi-monochromatic system has great advantages in terms of signal-to-noise ratio, dynamic range, and linearity over the quasi-c.w. system. It also allows the effect of the troublesome lens reverberations that occur within the time gate to be at last eliminated. Since this too now appears as a monochromatic contribution, it can be removed by subtracting a uniform signal of compensating amplitude and phase at the reference frequency from the input to the lock-in amplifier. But most important of all, it allows measurements to be made at effectively a single frequency, so that they can be analysed accurately in terms of the monochromatic theory of $V(z)$ (the video signal as an explicit function of defocus; see Chapter 7).

If two separate echoes are available from the lens, then one of them can be used to provide the reference signal for the lock-in amplifier

(Liang *et al.* 1986). In this case, the two pulses are separated by selective time gating, and passed to the reference and signal channels of a lock-in amplifier via separate narrow-band crystal filters. The output of the lock-in amplifier then gives the amplitude of the second signal, and also its phase relative to the first. An example of the use of this two-pulse monochromatic system is when the lens is designed so that some of the acoustic energy from the transducer falls on a flat annular surface around the lens, as illustrated in Fig. 5.3. One ingenious system for differential phase contrast used a transducer that was simultaneously excited at its fundamental frequency and at its third harmonic (Smith and Wickramasinghe 1982). Assuming that the size of the point-spread function at each frequency was proportional to the wavelength, the signal at the lower frequency gave a reference signal averaged over nine times the area of the higher frequency spot, so that the difference between the two gives a sensitive image of local variations in the specimen. An alternative differential imaging system, more analogous to Nomarski differential phase contrast in light microscopy, uses two slightly inclined transducers to produce two focal spots separated by a little over a wavelength (Nikoonahad 1987; Nikoonahad and Sivers 1989; Routh *et al.* 1989).

It is possible to make lenses with transducers that generate both longitudinal and shear waves in the sapphire, for example, by growing ZnO with the substrate tilted (Khuri-Yakub and Chou 1986; Chou *et al.* 1987; Jen *et al.* 1988; Chou and Khuri-Yakub 1989). In order to exploit

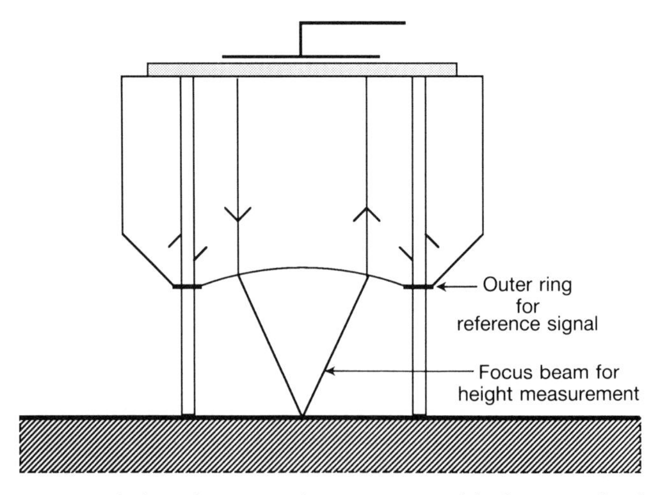

FIG. 5.3. An acoustic lens for two-pulse response, with the second echo coming from a flat annular region around the lens (after Liang *et al.* 1986).

the full capabilities of this kind of mixed-mode lens, a double-pulse arrangement is necessary, in which two pulses at the imaging frequency are sent to the lens (Meeks *et al.* 1989). The first pulse is given a dynamic phase modulation of $\pm 180°$ at audio frequency (8 kHz in the definitive implementation). The phase of the second pulse is controlled by a phase-locked feedback loop as follows. The time interval between the two transmitted pulses is set to be equal to the difference in the two-way transit time of longitudinal and shear waves through the lens. The echo from the specimen due to the shear wave from the first pulse will therefore coincide with the echo due to the longitudinal wave from the second pulse, and the detection gating can be set to select these two signals arriving together. The shear wave echo will be phase-modulated, while the longitudinal wave echo has a steady phase. The two will therefore interfere to produce a signal at the modulation frequency. This signal is measured by a lock-in amplifier whose output controls the phase of the second pulse, thus completing the phase-locked feedback loop. The phase signal is the output of the system, and it can be used for Rayleigh wave velocity measurements using the theory and analysis to be described in Chapters 7 and 8. The first echo from the specimen, due to the longitudinal wave excited by the phase-modulated pulse, can be used to give a measurement of height. The phase noise of this measurement system can be as low as 0.1°, which would correspond to a velocity measurement accuracy of 50 p.p.m. or 0.005 per cent.

5.3 Very short pulse techniques

While for some purposes it may be necessary to have accurate frequency definition, for others good time discrimination is useful. These are opposite requirements. Because of the Fourier relationship between frequency and time, the more precisely the time of a signal is known, the greater bandwidth of frequencies is necessary (there is a close analogy here with Heisenberg's uncertainty principle). Approximately, the time resolution τ is the reciprocal of the bandwidth B_W, so that their product $B_W \tau \approx 1$.

It is not necessary that all the frequency components be reflected from the specimen at the same time, and pulse compression techniques can be used (Yue *et al.* 1982; Nikoonahad *et al.* 1985). A delta-function excitation pulse can be passed through a filter that introduces a delay that is a linear function of frequency, so that, for example, the low-frequency components of the pulse emerge from the filter first. Such a filter can be fabricated using surface acoustic wave technology; it is sometimes called a chirp filter because of the way that a bird's chirp also rises in frequency with time. The power in the output of the chirp filter is lower than the pulse power, because the energy has become spread out in time. But, if the chirp is subsequently passed through a matched filter, the original sharp pulse can, in principle, be reconstructed. The product of the bandwidth of the filter and the length of the chirp pulse is known as the processing gain of the system. It gives a measure of the ratio of the peak power in the reconstructed pulse (neglecting losses in the filter) to the power in the expanded pulse. Such a pulse compression system can therefore be useful in acoustic microscopy if the maximum power that can be applied to the transducer is limited, and in water-coupled instruments processing gains, taking into account losses, of 12 dB or better have been achieved at centre frequencies up to 750 MHz. Its use has been suggested for subsurface imaging applications where the poor transmission across the water–solid interface in both directions makes the signal from a subsurface feature relatively weak (Nikoonahad *et al.* 1985), but in some demonstrations of the use of the technique the power to the transducer would not have been a limitation, thus making comparison a little artificial. The main limitation of pulse compression techniques arises from the presence of side-lobes, so that improvement in signal-to-noise ratio is achieved at the expense of dynamic range. A Fourier transform method ideally requires integration between infinite limits, but the filters have a finite bandwidth, causing truncation in the time-to-frequency conversion, and the action of the s.p.d.t. switch can cause truncation in the frequency-to-time conversion. These truncations give rise to side-lobes in the reconstructed pulse, analogous to side-lobes due to a finite aperture in optics. It is difficult to reduce the side-lobes below about −35 dB relative to the main pulse, so pulse compression cannot be used if a small signal of interest is more than this amount below an adjacent larger signal. If signal-to-noise ratio of more than that is available then pulse compression is not necessary, and if a dynamic range of more than that is required it will not be of any use anyway. But pulse compression has proved useful in liquid helium acoustic microscopy to overcome the severe limitation on the peak power that can be propagated through the focal region (Foster and Rugar 1983).

Pulse compression was originally developed for radar, and seemed applicable to acoustic microscopy because of the many similarities between the two, allowing for the fact that acoustic waves in water travel a factor of 2×10^5 more slowly than electromagnetic waves in air, so that the distances are correspondingly smaller. Another technique that has been used in radio altimeters for some time involves the use of a swept-frequency continuous-wave signal, and this too has been adapted for acoustic microscopy in the way mentioned in §2.3 (Faridian and Somekh 1986). If the frequency of the transmitted signal is varied linearly with time, then there will be a frequency shift in any reflected signal that is proportional to its travel time. Thus frequency demodulation techniques can be used to measure both height and strength of reflection over the surface of a specimen. Usually a sawtooth function of time is chosen for the frequency and, if s.p.d.t. switching is used for the lens, then that can be synchronized to the sawtooth. There are similar processing gain advantages, with similar limitations, to those of a pulse compression system. Of course, the Fourier transform does not have to be performed by analogue devices or circuits, and a slowly swept system can be used to make measurements over a spectrum of frequencies which can subsequently be manipulated digitally (Kulik *et al.* 1989, 1990). But it is also possible to use straightforward short pulses.

The design of a system for working with short pulses follows the same principles as other pulsed ultrasonic systems such as ultrasonic flaw detectors, but in this case it is necessary to achieve very much greater stability and higher bandwidth. A schematic circuit is shown in Fig. 5.4 A very short impulse is generated by a step recovery diode. The pulse has a width of half the period of the centre frequency of the lens; if it is shorter than that the energy in the pulse is reduced without any improvement in the signal bandwidth. Thus the lens acts as a sort of matched filter with poor time resolution but optimal signal-to-noise ratio (Thompson and Hsu 1988). The maximum voltage of the pulse, after amplification, is limited to the breakdown voltage of the transducer. This depends on the quality of the piezoelectric layer: as a rough guide a safe voltage for high-quality ZnO is about 5 V per micron

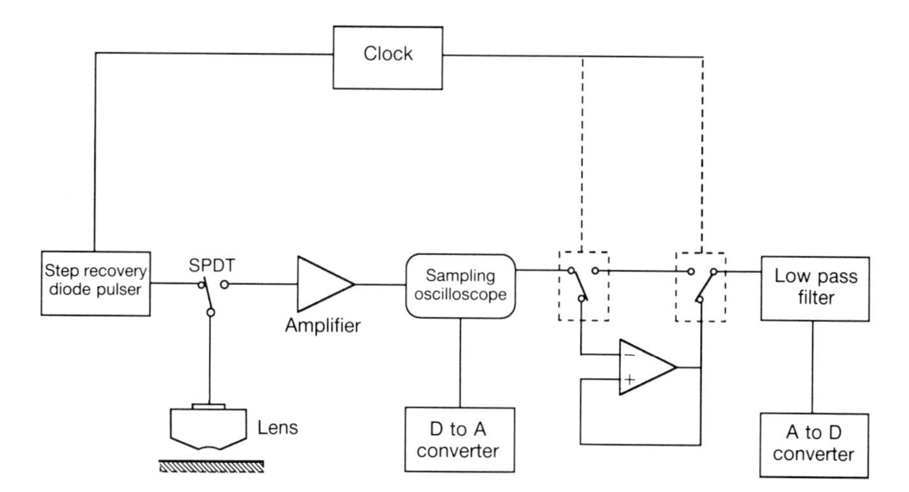

FIG. 5.4. Schematic r.f. system for very short pulses (after Weaver *et al.* 1989).

of thickness. The limitation in ZnO is imposed by breakdown through defects in the film, so the safe operating voltage can be increased by depositing a thin insulating layer of high breakdown field strength, such as SiO_2, between the piezoelectric and the electrode. The signal passes through a high bandwidth s.p.d.t. switch to the lens, and then back through the s.p.d.t. switch to the receiving amplifier. Throughout this part of the system the elimination of electrical reflections is of paramount importance. There are reverberations between the lens and the s.p.d.t. switch because of impedance mismatches, and it is difficult to eliminate these without compromising the bandwidth and lengthening the signal. Therefore, it is best simply to keep the cable between the lens and the s.p.d.t. switch as short as possible. On the other hand, the cables connecting the amplifiers to the s.p.d.t. switch can be made long, and then reflections in them are sufficiently delayed that they can be separated from the signals of interest (the speed in the cables is about $0.2\,\mathrm{m\,ns^{-1}}$, so that each metre of cable introduces a round trip delay of 10 ns). Multistage GaAs FET amplifiers can introduce large spurious pulses, presumably due to mismatches between stages, and better results have been obtained with bipolar amplifiers, which have poorer noise figures (by about 3 dB) but better reverse isolation. It is not easy to digitize a signal directly at the frequencies of interest, so a sampling oscilloscope is used. An analogue voltage generated by the controlling computer determines the time at which a measurement is made of the instantaneous value of the signal from the lens, with a time resolution much shorter than the period of the highest frequency present. That value is held for long enough to digitize it at a conventional speed. The measurement delay time is then incremented, and the next point on the waveform is digitized, and so on. In this way the whole waveform can be digitized with a stability of a small fraction of a nanosecond. By alternately measuring the signal with and without energizing the impulse generator, spurious signals due to switching transients can be subtracted out (Yamanaka 1983). There may also be small remaining acoustic or electrical reverberations associated with the impulse from the generator; these can be subtracted later from the recorded waveform using a reference waveform measured with no specimen present. When such a system is used to observe the reflection from a good reflector at focus, the signal-to-noise ratio is adequate for direct observation of the trace on the sampling oscilloscope, but in many applications greater accuracy and also greater dynamic range are required, and then both hardware and software signal averaging can be used.

Impulse excitation in acoustic microscopy is doubly expensive in terms of the signal-to-noise ratio. For a given peak voltage or power, the integrated energy exciting the transducer decreases as the pulse length, and the detected noise power is proportional to the bandwidth. Therefore the signal-to-noise ratio suffers as the square of the pulse length. In §8.3 and §9.2–3 techniques and applications of time-resolved quantitative measurements will be given in which impulse excitation is essential. But many uses of acoustic microscopy call primarily for an image, and for that purpose conventional gated continuous wave electronics are quite adequate, and indeed are to be preferred because of the good signal strength that they give for obtaining pictures at a reasonable speed with good contrast. Contrast must always be interpreted, and to do that it is necessary to understand a little elementary acoustics.

6
A little elementary acoustics

Pfuhl was one of those theoreticians who are so fond of their theory that they lose sight of the object of that theory—its application in practice. His passion for theory made him hate all practical considerations, and he would not hear of them. He even rejoiced in failure, for failures only proved to him the accuracy of his theory. (*War and Peace*, L. N. Tolstoy 1869)

6.1 Scalar theory

6.1.1 *Acoustic waves*

The propagation of linear acoustic waves in solids depends on two laws discovered by two of the most illustrious physicists of the seventeenth century, one from Cambridge and the other from Oxford. Consider a volume element of an isotropic solid subjected to shear, as shown in Fig. 6.1. If the displacement in the transverse direction is ξ, and the component of shear stress in that direction is σ_s, then Newton's third law may be written

$$\rho \frac{\partial^2 \xi}{\partial t^2} = \frac{\partial \sigma_s}{\partial x} \tag{6.1}$$

where ρ is the density. Hooke's law states that, provided the system remains linear, stress is proportional to strain, so that

$$\sigma_s = \mu \frac{\partial \xi}{\partial x} \tag{6.2}$$

where μ is the shear modulus. Combining these gives

$$\rho \frac{\partial^2 \xi}{\partial t^2} = \mu \frac{\partial^2 \xi}{\partial x^2}. \tag{6.3}$$

This is the equation of a wave, as can be seen by substituting

$$\xi = \xi_0 e^{i(\omega t \pm kx)} \tag{6.4}$$

which gives

$$\rho \omega^2 = \mu k^2. \tag{6.5}$$

Phase velocity is given by $v = \omega / k$; thus, for shear (or transverse) waves

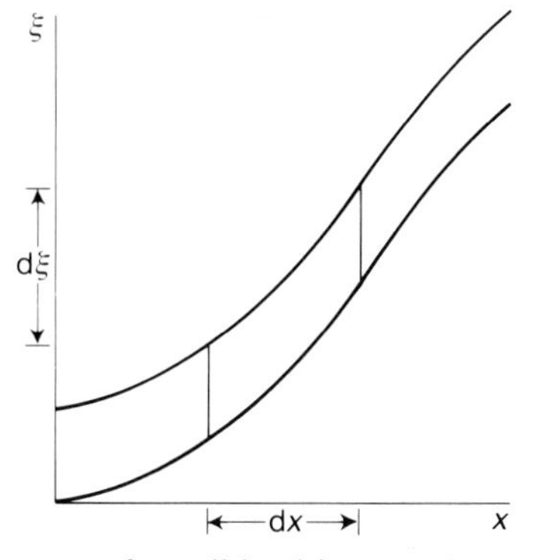

FIG. 6.1. A volume element of a solid subject to shear. The difference between the stresses at the two ends of the volume element is given by Hooke's law as $\mu(\partial^2\xi/\partial x^2)\,\mathrm{d}x$, and by Newton's third law this must be equal to $\rho(\partial^2\xi/\partial t^2)\,\mathrm{d}x$.

the velocity is

$$\boxed{v_s = \sqrt{\frac{c_{44}}{\rho}}}, \tag{6.6}$$

where $c_{44} \equiv \mu$. The longitudinal velocity may be similarly derived, except that in this case a different elastic modulus is appropriate. The relationships between different moduli are given in Table 6.1 in §6.2. The longitudinal velocity is

$$\boxed{v_l = \sqrt{\frac{c_{11}}{\rho}}}. \tag{6.7}$$

A fluid cannot support shear waves over any appreciable distance, so only longitudinal waves can propagate in a fluid. In this case the appropriate modulus is the adiabatic bulk modulus B. Extending to three dimensions with the Laplacian operator

$$\nabla^2 \equiv \frac{\partial^2}{\partial x^2} + \frac{\partial^2}{\partial y^2} + \frac{\partial^2}{\partial z^2},$$

and using as the variable not displacement but pressure p, the wave equation is

$$\rho \frac{\partial^2 p}{\partial t^2} = B\nabla^2 p. \tag{6.8}$$

In this case the velocity is

$$\boxed{v_0 = \sqrt{\frac{B}{\rho}}}. \tag{6.9}$$

Putting $\partial^2/\partial t^2 \to -\omega^2$, and using $\omega/k = v_0$, (6.8) may be written as the Helmholtz equation,

$$(\nabla^2 + k^2)p = 0. \tag{6.10}$$

The Helmholtz equation is widely used as the starting point for much theoretical acoustics in fluids (Morse and Ingard 1987).

6.1.2 *Attenuation in fluids*

If a medium is not perfectly elastic or perfectly fluid, but has viscous properties, then the equation for the acoustic displacement u contains an extra term (Lindsay 1982)

$$\rho \frac{\partial^2 u}{\partial t^2} = c\nabla^2 u + d\nabla^2 \frac{\partial u}{\partial t} \tag{6.11}$$

where c describes the elastic bulk modulus as before, and d relates to the viscous relaxation processes associated with the rate of compression. This equation is satisfied by a damped harmonic wave in which the particle displacement, for propagation in the x-direction, is

$$u = u_0 \mathrm{e}^{-\alpha x} \mathrm{e}^{\mathrm{i}(\omega t - kx)}. \tag{6.12}$$

Substitution into (6.11) gives

$$-\rho\omega^2 = c(\alpha + \mathrm{i}k)^2 + \mathrm{i}\omega d(\alpha + \mathrm{i}k)^2. \tag{6.13}$$

Separating real and imaginary parts,

$$\begin{aligned} -\rho\omega^2 &= c(\alpha^2 - k^2) + 2\omega d\alpha k, \\ 0 &= 2c\alpha k + d\omega(\alpha^2 - k^2). \end{aligned} \tag{6.14}$$

Hence the attenuation is

$$\alpha = \frac{\dfrac{\omega}{k}\dfrac{\rho}{c}\,\omega^2\dfrac{d}{c}}{2(1 + \omega^2 d^2/c^2)}. \tag{6.15}$$

Writing the relaxation time, i.e. the time taken after an instantaneous change in stress for the strain to relax by $(1 - 1/\mathrm{e})$ of the difference in equilibrium values, as $\tau \equiv d/c$, and the undamped velocity as $v_0 \equiv \omega/k = \sqrt{(c/\rho)}$, yields

$$\alpha = \frac{\omega^2 \tau}{2v_0(1+\omega^2\tau^2)}. \tag{6.16}$$

In acoustic microscopy, viscous attenuation is most significant in the coupling fluid. In water (and all the other fluids in Table 3.1 except CS_2 and superfluid ^4He) at the frequencies used in acoustic microscopy, the relaxation time is much less than the period of the wave. Thus $\omega\tau \ll 1$ and, to a good approximation,

$$\alpha \approx \frac{\omega^2 \tau}{2v_0}. \tag{6.17}$$

This is the origin of the f^2 attenuation that is found in these fluids. There is also a velocity shift, the modified phase velocity being (Lindsay 1982)

$$v_p = \frac{\sqrt{2}v_0}{\omega\tau}\{(1+\omega^2\tau^2)[(1+\omega^2\tau^2)^{1/2}-1]\}^{1/2}. \tag{6.18}$$

For $\omega\tau \ll 1$, this may be approximated by a second-order term,

$$v_p = v_0(1+\omega^2\tau^2)^{1/2}. \tag{6.19}$$

If the dominant relaxation is described by the shear viscosity η (the Stokes assumption), then

$$d = 4\eta/3 \tag{6.20}$$

and

$$\tau = 4\eta/(3\rho v_0^2). \tag{6.21}$$

Hence the attenuation can be related directly to the shear viscosity,

$$\alpha = 2\omega^2\eta/(3\rho v_0^3). \tag{6.22}$$

In water the shear viscosity at 20°C is $\eta = 1.0019 \times 10^{-3}$ N s m^{-3}, giving a relaxation time $\tau = 0.6086 \times 10^{-12}$ s. This would predict an attenuation $\alpha_0 = 8.081 \times 10^{-15}$ Np m^{-1} Hz^{-2} = 0.070 dB μm^{-1} GHz^{-2}, compared with the measured value of $\alpha_0 = 0.217$ dB μm^{-1} GHz^{-2}. The discrepancy is due primarily to bulk viscous effects, attributed to the time taken for local reconfiguration of molecules following the pressure changes due to the acoustic wave (Hall 1948; Pinkerton 1949). The value of τ corresponding to the observed attenuation would change the velocity at 1 GHz by +0.018 per cent. The definitive measurements of the velocity of sound in pure water were made at 5 MHz, at which frequency the change would be less than 0.005 p.p.m., in experiments with an accuracy of 2 p.p.m. The change in velocity could be important in quantitative acoustic microscopy at higher frequencies. But the most accurate measurements are often made at 225 MHz (Kushibiki and Chubachi 1985), where the error due to the finite relaxation time would be less than 0.001 per cent.

The attenuation may be expressed by making the wavenumber complex (this would be $k - i\alpha$ in eqn (6.12)), and the velocity ($=\omega/k$) may also be written as a complex quantity. This in turn corresponds to a complex modulus, so that the relationship $v = \surd(B/\rho)$ is preserved; indeed the acoustic wave equation may be written as a complex-valued equation, without the need for the extra term in (6.11). Complex-valued elastic moduli are frequency-dependent, and the frequency-dependent attenuation and the velocity dispersion are linked by a causal Kramers–Kronig relationship (Lee *et al.* 1990).

Attenuation in solids due to viscosity may be treated by a similar analysis. There may well be other damping mechanisms, such as heat conduction (i.e. imperfectly adiabatic conditions) which also gives an f^2 law, and other phenomena associated with solid state defects that may have more complicated frequency and temperature dependence. In polycrystalline solids, especially metals and alloys and also ceramics, elastic grain scattering may cause much greater attenuation than any inelastic damping (Papadakis 1968; Stanke and Kino 1984).

6.2 Tensor derivation of acoustic waves in solids

In tensor notation the three Cartesian directions x, y, and z are designated by suffixed variables i, j, k, l, etc. (Landau and Lifshitz 1970; Auld 1973). Thus the force acting per unit area on a surface may be described as a traction vector with components τ_j; $j = x, y, z$. The stress in an infinitesimal cube volume element may then be described by the tractions on three of the faces, giving nine elements of stress σ_{ij} $(i, j = x, y, z)$, where the first suffix denotes the normal to the plane on which a given traction operates, and the second suffix denotes the direction of a traction component.

A kinetic argument shows that $\sigma_{ij} = \sigma_{ji}$ always. Any imbalance between these two would lead to an angular acceleration of a volume element. If this volume element were shrunk, then the torque would reduce in proportion to the linear dimension cubed, but the moment of inertia would reduce in proportion to the fifth power of the linear dimension, so that the angular acceleration would increase as the reciprocal of the square of the size of the volume element, becoming infinite in the limit. Thus *reductio ad absurdum*, $\sigma_{ij} = \sigma_{ji}$. Hence there are only six independent components of the stress tensor.

Strain is expressed in terms of the derivatives of the displacements, u_k, of points in a solid, in such a way that a rigid rotation gives zero strain. The components of the strain tensor are

$$\varepsilon_{kl} = \frac{1}{2}\left(\frac{\partial u_k}{\partial r_l} + \frac{\partial u_l}{\partial r_k}\right) \tag{6.23}$$

where r_l and r_k are components of the position vector.

In tensor notation Hooke's law becomes

$$\sigma_{ij} = c_{ijkl}\varepsilon_{kl} \tag{6.24}$$

where c_{ijkl} describes a contribution to the stress component σ_{ij} due to the strain component ε_{kl}. To obtain the value of the component σ_{ij} summation must be carried out over repeated suffixes on the right-hand side that do not also appear on the left-hand side, and this summation is always implied by the tensor notation in a tensor equation such as (6.24) (This is known as the Einstein convention; in this particular case it means summation over all k, $l = x, y, z$, i.e. nine terms altogether). A corollary of this is that the equation cannot simply be inverted. If compliance is defined by

$$\varepsilon_{kl} = s_{klij}\sigma_{ij}, \tag{6.25}$$

then $s_{klij} \neq 1/c_{ijkl}$.

The stiffness tensor c_{ijkl} is a tensor of fourth rank, with 81 elements. To enable it to be written as a matrix, a reduced notation for the independent elements of stress and strain is used,

$$\begin{pmatrix} \sigma_{xx} & \sigma_{xy} & \sigma_{xz} \\ \sigma_{yx} & \sigma_{yy} & \sigma_{yz} \\ \sigma_{zx} & \sigma_{zy} & \sigma_{zz} \end{pmatrix} = \begin{pmatrix} \sigma_1 & \sigma_6 & \sigma_5 \\ \sigma_6 & \sigma_2 & \sigma_4 \\ \sigma_5 & \sigma_4 & \sigma_3 \end{pmatrix}. \tag{6.26}$$

The abbreviated suffixes are obtained by counting along the diagonal and then either way around two sides of the matrix thus: $\sigma_{xx} \to \sigma_{yy} \to \sigma_{zz} \to \sigma_{yz} \to \sigma_{xz} \to \sigma_{xy}$. Since there are only six independent components, the stress may be written in the abbreviated notation as a single-column vector. The strain is similarly written in abbreviated notation, but by convention with some factors of $\frac{1}{2}$ introduced that simplify the statement of Hooke's law;

$$\begin{pmatrix} \varepsilon_{xx} & \varepsilon_{xy} & \varepsilon_{xz} \\ \varepsilon_{yx} & \varepsilon_{yy} & \varepsilon_{yz} \\ \varepsilon_{zx} & \varepsilon_{zy} & \varepsilon_{zz} \end{pmatrix} = \begin{pmatrix} \varepsilon_1 & \frac{1}{2}\varepsilon_6 & \frac{1}{2}\varepsilon_5 \\ \frac{1}{2}\varepsilon_6 & \varepsilon_2 & \frac{1}{2}\varepsilon_4 \\ \frac{1}{2}\varepsilon_5 & \frac{1}{2}\varepsilon_4 & \varepsilon_3 \end{pmatrix}. \tag{6.27}$$

Hence Hooke's law (6.24) may be written in abbreviated notation as

$$\sigma_I = c_{IJ}\varepsilon_J. \tag{6.28}$$

The factors $\frac{1}{2}$ in (6.27) take care of the fact that two terms such as $\varepsilon_{xy}+\varepsilon_{yx}$ are only counted once in the abbreviated notation. In this notation the compliance matrix s_{IJ} is the inverse of the stiffness matrix c_{IJ}.

The equivalence of $\sigma_{ij}=\sigma_{ji}$ and $\varepsilon_{ij}=\varepsilon_{ji}$ has enabled the number of elements in the stiffness and compliance tensors to be reduced to 36. It turns out, because of symmetry, that also $c_{IJ}=c_{JI}$ (Auld 1973), so that for an arbitrary anisotropic medium there are only 21 independent constants. For many crystalline materials this number is further reduced by the crystal symmetry; for example, for cubic symmetry there are only three independent constants, generally taken as c_{11} $(=c_{22}=c_{33})$, c_{44} $(=c_{55}=c_{66})$, and c_{12} $(=c_{IJ},\ I\neq J;\ I$ and $J\leq 3)$, all other elements being zero. Thus the stiffness tensor for a material with cubic symmetry may be written in matrix form using abbreviated notation,

$$c_{IJ}=\begin{pmatrix} c_{11} & c_{12} & c_{12} & 0 & 0 & 0 \\ c_{12} & c_{11} & c_{12} & 0 & 0 & 0 \\ c_{12} & c_{12} & c_{11} & 0 & 0 & 0 \\ 0 & 0 & 0 & c_{44} & 0 & 0 \\ 0 & 0 & 0 & 0 & c_{44} & 0 \\ 0 & 0 & 0 & 0 & 0 & c_{44} \end{pmatrix}. \tag{6.29}$$

For isotropic materials there are only two independent constants, which may be taken as c_{11} and c_{44} (the relationship between the various isotropic elastic constants is given in Table 6.1 at the end of this section). The isotropic stiffness tensor may be obtained by substituting $c_{12}=c_{11}-2c_{44}$ in the cubic stiffness matrix.

For an isotropic medium Hooke's law (6.24), taking account of the zero matrix elements, becomes

$$\begin{aligned} \sigma_{ij} &= c_{11}\varepsilon_{ij}+\sum c_{12}\varepsilon_{kl}, \qquad && \text{for } i=j\neq k=l; \\ &= 2c_{44}\varepsilon_{ij}, && \text{for } i\neq j. \end{aligned} \tag{6.30}$$

The summation sign has been included for emphasis. In the first line of (6.30) there do not appear to be any repeated subscripts, but that is because they have been subsumed in the abbreviated notation. If the stiffness tensor elements are written out in full, as in (6.24), this is immediately apparent. The selection rules in (6.30) correspond directly to the components of stress and strain that are related by each of the three stiffness constants.

In an isotropic medium $c_{11}-c_{12}=2c_{44}$, and eqn (6.30) may be written

$$\sigma_{ij} = (c_{11} - c_{12})\varepsilon_{ij} + c_{12}(\varepsilon_{11} + \varepsilon_{22} + \varepsilon_{33})\delta_{ij}$$
$$= 2c_{44}\varepsilon_{ij} + (c_{11} - 2c_{44})\Theta\delta_{ij}. \tag{6.31}$$

The strain terms may be related to displacement via equation (6.23). The dilation $\Theta \equiv \varepsilon_{11} + \varepsilon_{22} + \varepsilon_{33} \equiv \mathbf{\nabla} \cdot \mathbf{u}$, which is the divergence (or colloquially the 'outspoutingness') of the displacement. The vector operator del is

$$\mathbf{\nabla} \equiv \left(\frac{\partial}{\partial x}, \frac{\partial}{\partial y}, \frac{\partial}{\partial z}\right);$$

the displacement vector is $\mathbf{u} \equiv (u_x, u_y, u_z)$; and δ_{ij} is the Kronecker delta function, $\delta_{ij} = 1 \Leftrightarrow i = j$, $\delta_{ij} = 0 \Leftrightarrow i \neq j$. Thus in (6.31), when $i \neq j$ only the first term applies; and, when $i = j$, the c_{44} coefficients multiplying ε_{ij} cancel, leaving only $c_{11}\varepsilon_{ij}$, while the two remaining uniaxial strain elements in Θ give the $c_{12}\varepsilon_{kl}$ terms in the first line of (6.30).

Newton's law may be written

$$\rho\frac{\partial^2 u_i}{\partial t^2} = \frac{\partial \sigma_{ij}}{\partial r_j} + F_i. \tag{6.32}$$

Summation over repeated subscripts is again implied, in this case over the subscript j. F_i are body forces (such as gravity), which will be neglected.

Combining Newton's law with Hooke's law, and swapping the order of the terms in (6.31),

$$\rho\frac{\partial^2 \mathbf{u}}{\partial t^2} = (c_{11} - c_{44})\mathbf{\nabla}(\mathbf{\nabla} \cdot \mathbf{u}) + c_{44}\nabla^2\mathbf{u}. \tag{6.33}$$

Using the standard vector identity

$$\mathbf{\nabla} \times (\mathbf{\nabla} \times \mathbf{u}) \equiv \mathbf{\nabla}(\mathbf{\nabla} \cdot \mathbf{u}) - \nabla^2\mathbf{u}, \tag{6.34}$$

(6.33) may be rewritten as

$$\rho\frac{\partial^2 \mathbf{u}}{\partial t^2} = c_{11}\mathbf{\nabla}(\mathbf{\nabla} \cdot \mathbf{u}) - c_{44}\mathbf{\nabla} \times (\mathbf{\nabla} \times \mathbf{u}). \tag{6.35}$$

Using Helmholtz's theorem, the displacement vector $\mathbf{u}$ may be written in terms of a scalar and a vector potential,

$$\mathbf{u} = \mathbf{\nabla}\phi + \mathbf{\nabla} \times \boldsymbol{\psi}. \tag{6.36}$$

From the further standard vector identities

$$\mathbf{\nabla} \times (\mathbf{\nabla}\phi) \equiv 0 \tag{6.37}$$

and

$$\nabla \cdot (\nabla \times \boldsymbol{\psi}) \equiv 0, \tag{6.38}$$

substituting (6.36) into (6.35) gives

$$\rho \frac{\partial^2 \nabla \phi}{\partial t^2} + \rho \frac{\partial^2 \nabla \times \boldsymbol{\psi}}{\partial t^2} = c_{11} \nabla (\nabla^2 \phi) - c_{44} \nabla \times (\nabla^2 \boldsymbol{\psi}). \tag{6.39}$$

The last term was obtained using the identity (6.34) again, this time with $\boldsymbol{\psi}$ as the vector and then recognizing that the term $\nabla \times \{\nabla(\nabla \cdot \boldsymbol{\psi})\}$ vanishes by identity (6.37). At last, (6.39) separates into two uncoupled equations

$$\nabla\left(\rho \frac{\partial^2 \phi}{\partial t^2} - c_{11} \nabla^2 \phi\right) = 0, \quad \boxed{\Rightarrow \rho \frac{\partial^2 \phi}{\partial t^2} = c_{11} \nabla^2 \phi}; \tag{6.40}$$

and

$$\nabla \times \left(\rho \frac{\partial^2 \boldsymbol{\psi}}{\partial t^2} - c_{44} \nabla^2 \boldsymbol{\psi}\right) = 0, \quad \boxed{\Rightarrow \rho \frac{\partial^2 \boldsymbol{\psi}}{\partial t^2} = c_{44} \nabla^2 \boldsymbol{\psi}}. \tag{6.41}$$

The first equation is scalar, and has a wave solution with velocity $v_l = \surd(c_{11}/\rho)$. This is the longitudinal wave of eqn (6.7). It is sometimes called an irrotational wave, because $\nabla \times \mathbf{u} = 0$ and there is no rotation of the medium. The second equation is vector, and has two degenerate orthogonal solutions with velocity $v_s = \surd(c_{44}/\rho)$. These are the transverse or shear waves of eqn (6.6); the degenerate solutions correspond to perpendicular polarization. They are sometimes called divergence-free waves, because $\nabla \cdot \mathbf{u} = 0$ and there is no dilation of the medium. Waves in fluids may be considered as a special case with $c_{44} = 0$, so that the transverse solutions vanish, and $c_{11} = B$, the adiabatic bulk modulus.

In an isotropic solid, $c_{11} - 2c_{44} = c_{12}$ and cannot be negative. Therefore, c_{44} cannot be greater than $c_{11}/2$, so the shear wave velocity cannot exceed $1/\surd 2$ times the longitudinal wave velocity. In some materials, notably fused silica and fine-grained ceramics, the ratio of shear to longitudinal velocity can approach $1/\surd 2$. The limit $c_{12} = 0$ (Poisson ratio $\sigma = 0$) would correspond to a cylinder that undergoes no change in cross section when it is stretched elastically. The other limit $c_{44} = 0$ ($\sigma = 0.5$) corresponds to a cylinder that undergoes no change in volume when it is stretched elastically. In rubbery polymers where the Poisson ratio approaches 0.5, and also in biological soft tissue, the transverse velocity may indeed be a small fraction of the longitudinal velocity. In common metals the shear velocity is usually about half the longitudinal velocity, corresponding to a Poisson ratio $\sigma = 1/3$. The ability of solids to support

Table 6.1
Relationships between isotropic elastic constants

Lamé constants

$$\lambda \equiv c_{12}; \qquad \mu \equiv c_{44}.$$

for an isotropic material, $c_{11} - c_{12} = 2c_{44}$, hence $\lambda = c_{11} - 2c_{44}$; μ is also called the shear modulus, sometimes designated G.

Poisson ratio

$$\sigma \equiv -\frac{s_{12}}{s_{11}} \equiv \frac{c_{12}}{2(c_{11} - c_{44})} \equiv \frac{1 - 2c_{44}/c_{11}}{2(1 - c_{44}/c_{11})} \equiv \frac{\lambda}{2(\lambda + \mu)} \equiv \frac{E}{2\mu} - 1$$

Young modulus

$$E \equiv \frac{1}{s_{11}} \equiv \frac{c_{44}(3c_{11} - 4c_{44})}{c_{11} - c_{44}} \equiv \frac{\mu(3\lambda + 2\mu)}{\lambda + \mu} \equiv 2\mu(1 + \sigma)$$

Bulk modulus

$$B \equiv c_{11} - \tfrac{4}{3}c_{44} \equiv \lambda + \tfrac{2}{3}\mu \equiv \frac{E\mu}{3(3\mu - E)} \equiv \frac{E}{3(1 - 2\sigma)}$$

Longitudinal wave modulus

$$c_{11} \equiv \lambda + 2\mu \equiv \frac{\mu(4\mu - E)}{3\mu - E} \equiv \frac{(1 - \sigma)E}{(1 + \sigma)(1 - 2\sigma)} \equiv \frac{2(1 - \sigma)}{1 - 2\sigma}\mu \equiv B + \tfrac{4}{3}\mu$$

Thin plate wave modulus (plane strain or two-dimensional Young modulus)

$$\frac{E}{1 - \sigma^2} \equiv c_{11} - \frac{c_{12}^2}{c_{11}}$$

both longitudinal and transverse waves leads to the phenomenon of Rayleigh waves which are of such importance in acoustic microscopy.

6.3 Rayleigh waves

The idea of surface acoustic waves was derived theoretically by Rayleigh (1885; Auld 1985). They consist of a superposition of longitudinal and shear waves travelling along the surface with a common phase velocity, v_R, that is slower than the velocity of either kind of wave in the bulk (Auld 1973; Kino 1987; Brekhovskikh and Godin 1990). The wavenumber along the surface, $\beta \equiv \omega/v_R$, is therefore greater than the longitudinal and shear wavenumbers $k_l \equiv \omega/v_l$ and $k_s \equiv \omega/v_s$. Since the component of wavevector along the surface is also β for the longitudinal and shear waves (this is equivalent to Snell's law), they must both have imaginary components of wavevector into the solid, corresponding to exponential decay. The components into the solid may be written α_l and α_s, so that by Pythagoras' theorem,

$$\beta^2 + \alpha_l^2 = k_l^2, \tag{6.42}$$

$$\beta^2 + \alpha_s^2 = k_s^2. \tag{6.43}$$

Figure 6.2(a), which was calculated using the results to be derived here, shows the amplitudes of the longitudinal and shear components of a Rayleigh wave in fused silica, and their exponential decay below the surface.

The problem is treated as a two-dimensional one, with x-coordinate parallel to the surface and z-coordinate normal to the surface, with the z-axis negative into the solid. Since $\partial/\partial y$ and u_y both vanish, the only non-zero component of the vector potential is ψ_y. Suppose that there is a solution whose longitudinal and shear components each decay exponentially away from the surface, and that these are described by the scalar and vector potentials respectively. Then the potentials may be written

$$\phi = \phi_0 \exp(\mathrm{i}\beta x - \mathrm{i}\alpha_l z), \tag{6.44}$$

$$\psi_y = \psi_0 \exp(\mathrm{i}\beta x - \mathrm{i}\alpha_s z). \tag{6.45}$$

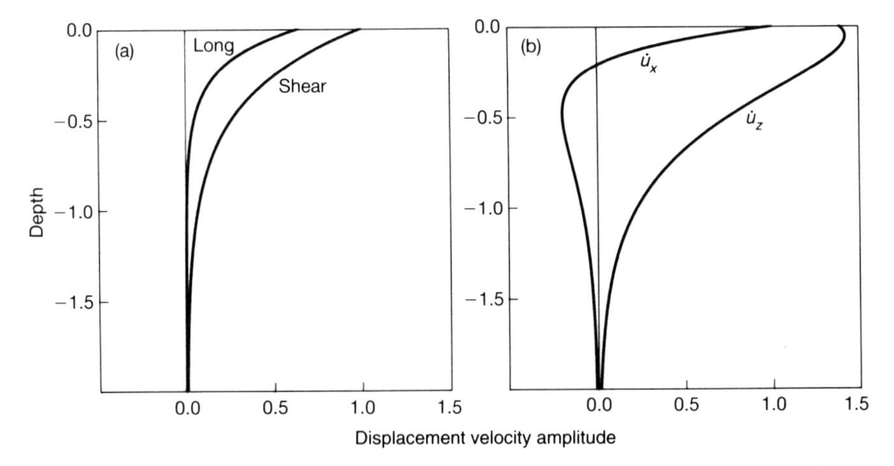

FIG. 6.2. Rayleigh wave displacement velocity components as a function of depth from the surface, measured in Rayleigh wavelengths: (a) longitudinal and shear components (eqns (6.44), (6.45), (6.51), and (6.52)); (b) components parallel and perpendicular to the surface (eqns (6.59) and (6.60)). The curves have been normalized to give the shear component at the surface a value of unity. The Poisson ratio $\sigma = 0.17$, corresponding to fused silica, was used to calculate the curve.

From (6.36) the displacement is

$$\begin{aligned}\mathbf{u} = (u_x, u_z) &= (\mathrm{i}\beta, -\mathrm{i}\alpha_l)\phi + (\mathrm{i}\alpha_s, \mathrm{i}\beta)\psi_y \\ &= (\mathrm{i}\beta\phi + \mathrm{i}\alpha_s\psi_y, -\mathrm{i}\alpha_l\phi + \mathrm{i}\beta\psi_y),\end{aligned} \tag{6.46}$$

and the dilation is

$$\Theta = \nabla \cdot \mathbf{u} = -(\alpha_l^2 + \beta^2)\phi. \tag{6.47}$$

The boundary conditions require that at the free surface each component of traction should vanish. The traction can be found from Hooke's law (6.30), with the strain components obtained from (6.46) using (6.23), and the dilation from (6.47). The normal component of the traction is

$$\begin{aligned}\sigma_{zz} &= 2c_{44}\frac{\partial u_z}{\partial z} + (c_{11} + 2c_{44})\Theta \\ &= 2c_{44}(-\alpha_l^2\phi + \mathrm{i}\beta\alpha_s\psi_y) + (c_{11} - 2c_{44})(-\alpha_l^2 - \beta^2)\phi = 0.\end{aligned} \tag{6.48}$$

Since $c_{11}/c_{44} = k_s^2/k_l^2 = (\beta^2 + \alpha_s^2)/(\beta^2 + \alpha_l^2)$, this may be rewritten as

$$\sigma_{zz} = c_{44}\{(-\alpha_s^2 + \beta^2)\phi + 2\beta\alpha_s\psi_y\} = 0, \qquad z = 0. \tag{6.49}$$

The tangential component of the traction is

$$\begin{aligned}\sigma_{xz} = 2c_{44}\varepsilon_{xy} &= c_{44}\left(\frac{\partial u_x}{\partial z} + \frac{\partial u_z}{\partial x}\right) \\ &= c_{44}\{2\beta\alpha_l\phi - (-\alpha_s^2 + \beta^2)\psi_y\} = 0, \qquad z = 0.\end{aligned} \tag{6.50}$$

In (6.49) and (6.50) the x-dependence of ϕ and ψ_y is the same (this is one of the boundary conditions, equivalent to Snell's law), and the z-dependence is simply unity at the surface. The explicit dependence on the elastic constants has been eliminated, so two simultaneous equations remain. From (6.49),

$$\phi_0 = \frac{-2\beta\alpha_s\psi_0}{-\alpha_s^2 + \beta^2} \tag{6.51}$$

and, from (6.50),

$$\psi_0 = \frac{2\beta\alpha_l\phi_0}{-\alpha_s^2 + \beta^2}. \tag{6.52}$$

Hence,

$$4\beta^2\alpha_l\alpha_s + (\beta^2 - \alpha_s^2)^2 = 0. \tag{6.53}$$

This is one form of the equation for the Rayleigh velocity. If the two terms are squared and multiplied out, and the relations, $\beta^2 + \alpha_l^2 = (\omega/v_l)^2$, $\beta^2 + \alpha_s^2 = (\omega/v_s)^2$, $\beta = \omega/v_R$ (eqns (6.42), (6.43)) are used, then

(6.53) can be expressed as a sextic equation,

$$\boxed{\left(\frac{v_R}{v_s}\right)^6 - 8\left(\frac{v_R}{v_s}\right)^4 + 8\left\{3 - 2\left(\frac{v_s}{v_l}\right)^2\right\}\left(\frac{v_R}{v_s}\right)^2 - 16\left\{1 - \left(\frac{v_s}{v_l}\right)^2\right\} = 0.} \tag{6.54}$$

More simply, putting $X = (v_R/v_s)^2$, $Y = (v_s/v_l)^2$,

$$X^3 - 8X^2 + 8(3 - 2Y)X - 16(1 - Y) = 0. \tag{6.55}$$

In general a sextic equation has six roots, but an allowable solution for the Rayleigh velocity, v_R, must be both real and positive, and only one such solution exists (Auld 1973; Brekhovskikh and Godin 1990).

The value of the Rayleigh velocity depends primarily on the shear wave velocity of the material, and rather less strongly on the ratio of the shear velocity to the longitudinal velocity. The ratio of the Rayleigh velocity to the shear velocity may be considered as a function of the Poisson ratio σ. From Table 6.1, the Poisson ratio may be expressed as

$$\sigma = \frac{1 - 2(v_s/v_l)^2}{2\{1 - (v_s/v_l)^2\}}. \tag{6.56}$$

In terms of the Poisson ratio, eqn (6.55) takes the form

$$\frac{X^3}{8(1 - X)} + X = \frac{1}{1 - \sigma}. \tag{6.57}$$

For the permissible range of σ from 0 to 0.5, over which v_s/v_l ranges from $1/\sqrt{2}$ to 0, the Rayleigh velocity varies from about 87 to 95 per cent of the shear velocity, although at the latter limit both the shear and the Rayleigh velocities vanish. An approximate solution to (6.57) is (Scruby *et al.* 1987)

$$v_R \approx v_s(1.14418 - 0.25771\sigma + 0.12661\sigma^2)^{-1}. \tag{6.58}$$

The field distribution of a Rayleigh wave decays below the surface (z is negative into the surface in all the exponents here). The simplest way to visualize this is to plot the exponential decay with depth of the longitudinal and shear displacement amplitudes from eqns (6.44) and (6.45). By differentiating the scalar and vector potentials spatially (this is most easily achieved by multiplying by k_l and k_s, neglecting the phase term i), and using eqns (6.51) and (6.52) to relate ϕ_0 to ψ_0, the longitudinal and shear wave amplitudes for fused silica have been plotted in Fig. 6.2(a). The curves have been normalized by plotting the displacement divided by $\omega k_l \psi_0$ and $\omega k_s \psi_0$; the appropriate k in each case comes from the spatial differentiation, and the ω corresponds to differentiation with respect to time. For a given power density the two displacement amplitudes vary with frequency but the displacement velocity amplitudes do not, and so displacement velocity is the more fundamental quantity. The shear displacement dominates both in amplitude at the surface and in depth of penetration, and so it is sensible to use the shear amplitude at the surface as a parameter for the amplitude of the Rayleigh wave as a whole. The reason why the shear decay is so much smaller than the longitudinal decay is that $\beta^2 - (\omega/v_\mathrm{l})^2 \gg \beta^2 - (\omega/v_\mathrm{s})^2$, so that $|\alpha_\mathrm{l}^2| \gg |\alpha_\mathrm{s}^2|$; it is the α terms that determine the rates of decay of both the longitudinal and shear components in (6.44) and (6.45) and also their relative amplitudes in (6.51) and (6.52).

The displacement amplitude field is obtained by combining the fields of the longitudinal and shear waves, i.e. by substituting (6.44) and (6.45) into (6.46), and again using (6.51) and (6.52) to express ϕ_0 in terms of ψ_0. The component parallel to the surface is

$$u_x = -\mathrm{i}\alpha_\mathrm{s}\left\{\frac{2\beta^2}{\beta^2 - \alpha_\mathrm{s}^2}\exp(-\mathrm{i}\alpha_\mathrm{l}z) - \exp(-\mathrm{i}\alpha_\mathrm{s}z)\right\}\psi_0 \mathrm{e}^{\mathrm{i}\beta x}, \qquad (6.59)$$

and the component normal to the surface is

$$u_z = \mathrm{i}\beta\left\{\frac{-2\alpha_\mathrm{l}\alpha_\mathrm{s}}{\beta^2 - \alpha_\mathrm{s}^2}\exp(-\mathrm{i}\alpha_\mathrm{l}z) - \exp(-\mathrm{i}\alpha_\mathrm{s}z)\right\}\psi_0 \mathrm{e}^{\mathrm{i}\beta x}. \qquad (6.60)$$

These components are both non-zero at the surface. Since $|\alpha_\mathrm{s}| < |\alpha_\mathrm{l}|$, at large depths the second term in brackets dominates in each case. Also, since $\beta > -\mathrm{i}\alpha_\mathrm{s}$, u_x reverses sign within a small distance below the surface (approximately $z < -0.2\lambda_\mathrm{R}$, depending on the value of σ). The displacement amplitudes for fused silica are plotted in Fig. 6.2(b), normalized in the same way as the longitudinal and shear wave amplitudes in Fig. 6.2(a). The decay of the stress components has a simpler double exponential dependence. From (6.49)–(6.52),

$$\sigma_{zz} = c_{44}\phi_0(-\alpha_\mathrm{s}^2 + \beta^2)\{\exp(-\mathrm{i}\alpha_\mathrm{s}z) - \exp(-\mathrm{i}\alpha_\mathrm{l}z)\} \qquad (6.61)$$

and

$$\sigma_{xz} = 2c_{44}\beta\alpha_\mathrm{l}\phi_0\{\exp(-\mathrm{i}\alpha_\mathrm{s}z) - \exp(-\mathrm{i}\alpha_\mathrm{l}z)\}. \qquad (6.62)$$

For both of these stress components there is a maximum at a depth given by

$$z_{\sigma_{\max}} = \frac{-\mathrm{i}\ln(\alpha_\mathrm{l}/\alpha_\mathrm{s})}{\alpha_\mathrm{l} - \alpha_\mathrm{s}}. \qquad (6.63)$$

A characteristic depth can be defined for the stresses whose depth profiles are described by (6.61) and (6.62),

$$z_c = \frac{\int_{-\infty}^{0} z\{\exp(-i\alpha_s z) - \exp(-i\alpha_l z)\}\, dz}{\int_{-\infty}^{0} \{\exp(-i\alpha_s z) - \exp(-i\alpha_l z)\}\, dz}$$

$$= i\frac{1/\alpha_s^2 - 1/\alpha_l^2}{1/\alpha_l - 1/\alpha_s} = i\frac{\alpha_l + \alpha_s}{\alpha_l \alpha_s}. \tag{6.64}$$

Rayleigh wave parameters for the permissible range of values of the Poisson ratio are summarized in Table 6.2. The dependence is very weak. In particular, the characteristic depth is about $0.6\lambda_R$ for values of σ corresponding to a wide range of materials. The characteristic depth gives an indication of the depth into the material that the Rayleigh wave penetrates. Thus, as a simple guide, the stresses and displacements associated with the Rayleigh wave may be considered as being mainly confined to within a Rayleigh wavelength of the surface. So, when the contrast in the acoustic microscope depends on factors that affect the

Table 6.2
Rayleigh wave parameters

σ	v_s/v_l	v_R/v_s	$-i\alpha_l/\beta$	$-i\alpha_s/\beta$	$z_{\sigma_{max}}/\lambda_R$	z_c/λ_R
0	0.7071[a]	0.8740[b]	0.7862	0.4859	0.2551	0.5300
0.05	0.6882	0.8837	0.7938	0.4681	0.2581	0.5405
0.1	0.6667	0.8931	0.8034	0.4498	0.2611	0.5519
0.15	0.6417	0.9022	0.8154	0.4313	0.2639	0.5642
0.17	0.6305	0.9058	0.8209	0.4238	0.2650	0.5695
0.2	0.6124	0.9110	0.8299	0.4124	0.2666	0.5777
0.25	0.5774[c]	0.9194[d]	0.8475	0.3933	0.2690	0.5924
0.3	0.5345	0.9274	0.8685	0.3740	0.2712	0.6088
1/3	0.5	0.9325	0.8846	0.3611	0.2724	0.6207
0.35	0.4804	0.9350	0.8934	0.3546	0.2729	0.6269
0.4	0.4082	0.9422	0.9231	0.3351	0.2743	0.6474
0.45	0.3015	0.9490	0.9582	0.3154	0.2751	0.6707
0.5	0	0.9553	1	0.2956	0.2754	0.6976

In terms of the variables in (6.55) columns 2–5 are $v_s/v_l = \sqrt{Y}$, $v_R/v_s = \sqrt{X}$, $-i\alpha_l/\beta = \sqrt{(1-XY)}$, $-i\alpha_s/\beta = \sqrt{(1-X)}$. Where less than four decimal places are given, the value is exact. Other exact values are: [a] $1/\sqrt{2}$; [b] $\sqrt{(3-\sqrt{5})}$; [c] $1/\sqrt{3}$; [d] $\sqrt{(2-2/\sqrt{3})}$. The value $\sigma = 0.17$ corresponds to fused silica in Figs. 6.2 and 6.3(b); $\sigma = 1/3$ is included because it gives $v_s/v_l = 1/2$; $\sigma = 0.5$ represents a limit which may be approached but not reached by solids, materials that approach this limit (such as rubber or tissue) generally have high shear wave damping.

propagation of Rayleigh waves, the information may be considered to be coming from within a depth λ_R from the surface of the specimen.

6.4 Reflection

6.4.1 *Impedance*

Reflection of acoustic waves incident on a planar interface between two isotropic media is most easily considered in terms of impedances. Acoustic characteristic impedance is defined as minus the ratio of traction to particle displacement velocity,

$$Z \equiv -\frac{\boldsymbol{\sigma}_{\mathbf{T}}}{\partial \mathbf{u}/\partial r}. \tag{6.65}$$

The traction $\boldsymbol{\sigma}_{\mathbf{T}}$ is the vector force per unit area acting on a surface. The unit of impedance is the rayl, where $1\,\text{rayl} = 1\,\text{kg}\,\text{m}^{-2}\,\text{s}^{-1}$; the usual multiple is the megarayl, abbreviated Mrayl. The impedance is equal to the product of the density ρ of a medium and the velocity v of a given wave propagating in it; this is how impedance is usually calculated,

$$\boxed{Z = \rho v}. \tag{6.66}$$

When a wave is incident upon an interface, there is a requirement of continuity of traction and displacement across the interface. Traction is the force per unit area acting on an interface: it is a vector with three components. If the stress amplitude (or pressure amplitude if the medium is a fluid) of a wave incident normally on an interface is unity, and the stress amplitudes of the reflected and transmitted waves are R and T, then continuity of traction requires that

$$1 + R = T, \tag{6.67}$$

and continuity of displacement velocity requires that

$$(1 - R)/Z_1 = T/Z_2 \tag{6.68}$$

where Z_1, Z_2 are the impedances of the two media. The change in sign of R arises from the reversal of the propagation direction of the reflected wave. Combining (6.67) and (6.68),

$$(1 + R)Z_1 = (1 - R)Z_2, \tag{6.69}$$

giving a stress amplitude reflection coefficient

$$\boxed{R = \frac{Z_2 - Z_1}{Z_2 + Z_1}}. \tag{6.70}$$

Similarly, the stress amplitude transmission coefficient is

$$\boxed{T = \frac{2Z_2}{Z_2 + Z_1}}. \tag{6.71}$$

Power flux is the scalar product of displacement velocity and traction. In an isotropic medium with no viscosity these are parallel, and the mean power flow is

$$\langle P \rangle = \left\langle \frac{\partial \mathbf{u}}{\partial t} \cdot \boldsymbol{\sigma}_{\mathrm{T}} \right\rangle \equiv Z \left\langle \left(\frac{\partial \mathbf{u}}{\partial t} \right)^2 \right\rangle \equiv \frac{1}{Z} \langle \boldsymbol{\sigma}_{\mathrm{T}}^2 \rangle. \tag{6.72}$$

From (6.70)–(6.72), for unit incident intensity the reflected and transmitted powers are

$$\langle P_{\mathrm{R}} \rangle = \left(\frac{Z_2 - Z_1}{Z_2 + Z_1} \right)^2 = \frac{Z_2^2 - 2Z_2 Z_1 + Z_1^2}{(Z_2 + Z_1)^2}, \tag{6.73}$$

$$\langle P_{\mathrm{T}} \rangle = \frac{1}{Z_2} \left(\frac{2Z_2}{Z_2 + Z_1} \right)^2 = \frac{4Z_2 Z_1}{(Z_2 + Z_1)^2}, \tag{6.74}$$

and power is conserved. These results are exactly analogous to transmission line and electromagnetic theory (Bleaney and Bleaney 1983).

When an acoustic wave is incident on a planar boundary at an angle other than normal, each refracted ray obeys Snell's law

$$\boxed{\frac{\sin \theta_1}{\sin \theta_2} = \frac{v_1}{v_2}}. \tag{6.75}$$

Snell's law may alternatively be expressed as the requirement that the tangential (i.e. parallel to the boundary) component of the wavevector be conserved across a boundary, so that $k_1 \sin \theta_1 = k_2 \sin \theta_2$. Acoustic normal impedance at a boundary is

$$Z' \equiv - \frac{\sigma_{\mathrm{T}}}{\partial u_z / \partial t} = \frac{\rho v}{\cos \theta}. \tag{6.76}$$

Continuity of normal impedance follows immediately from continuity of traction and displacement. When one or both media is a solid, any incident wave will generate a longitudinal and one or two transverse waves in each solid, so that a fuller analysis is required (Auld 1973). A case of particular importance in acoustic microscopy is that of a wave in a fluid incident on a solid (Brekhovskikh 1980; Brekhovskikh and Godin 1990). The case of a uniform isotropic solid will be considered here, layered materials will be dealt with in Chapter 10, and anisotropic materials in Chapter 11.

6.4.2 *Reflection at a fluid–solid interface*

The coordinate system is the same as for the discussion of the Rayleigh wave in §6.3. The interface lies in the plane $z = 0$, with the fluid lying on the positive z side. The x-axis is parallel to the surface, and the problem is considered as a two-dimensional one, so that the field does not vary with y. The incident wave in the fluid in general will generate a reflected scalar wave in the fluid, and in the solid a longitudinal wave and a shear wave polarized in the x–z plane. These four waves are denoted respectively by the subscripts i, r, l, s. The z-component of their wavevectors are each denoted by α, with the appropriate subscript. The x-component of all the waves is the same, by Snell's law, and is denoted by β, and the densities of the fluid and solid are ρ_0 and ρ_1, respectively. In the fluid there is an incident wave propagating towards the interface and a reflected wave propagating away from the interface, and in the solid there are a longitudinal wave and a shear wave propagating away from the interface. The shear wave is polarized in the x–z plane. The fields in the two half spaces are thus, with the x dependence implicit,

$$\begin{aligned} z \geq 0, \quad & \phi_0 = \phi_i \exp(i\alpha_0 z) + \phi_r \exp(-i\alpha_0 z), \qquad \text{putting } \alpha_0 = \alpha_i = -\alpha_r; \\ z \leq 0, \quad & \phi_1 = \phi_l \exp(i\alpha_l z), \\ & \psi_1 = \psi_s \exp(i\alpha_s z). \end{aligned} \tag{6.77}$$

Of the six boundary conditions (continuity across the boundary of three components of displacement and three components of traction), those concerned with displacement and stress in the y-direction are not relevant, nor is displacement in the x-direction since the fluid can slide freely. Hence, the boundary conditions are continuity of displacement and traction normal to the surface and zero traction parallel to the surface.

In the fluid, at $z = 0$,

$$\begin{aligned} \mathbf{u}_0 &= (u_{0x}, u_{0z}) = i(\beta, \alpha_0)\phi_i - i(\beta, -\alpha_0)\phi_r, \\ \Theta_0 &= -(\beta^2 + \alpha_0^2)(\phi_i + \phi_r) = -k_0^2(\phi_i + \phi_r), \\ \sigma_{0zz} &= -Bk_0^2(\phi_i + \phi_r), \qquad \text{putting } B \text{ as the modulus}, \\ \sigma_{0xz} &= 0. \end{aligned} \tag{6.78}$$

In the solid, at $z = 0$,

$$\mathbf{u}_1 = (u_{1x}, u_{1z}) = \mathrm{i}(\beta, \alpha_1)\phi_1 + \mathrm{i}(\alpha_s, -\beta)\psi_s,$$

$$\Theta_1 = -(\beta^2 + \alpha_s^2)\phi_1 = -k_1^2\phi_1,$$

$$\sigma_{1zz} = -2c_{44}(\alpha_1^2\phi_1 - \alpha_s\beta\psi_s) - (c_{11} - c_{44})k_1^2\phi_1 = c_{44}\{(\beta^2 - \alpha_s^2)\phi_1 - 2\alpha_s\beta\psi_s)\},$$

$$\sigma_{1xz} = c_{44}\{-2\alpha_1\beta\phi_1 - (\beta^2 - \alpha_s^2)\psi_s\}. \tag{6.79}$$

At this point the substitutions are made

$$\rho_0\omega^2/k_0^2 = B, \qquad \rho_1\omega^2/k_s^2 = c_{44},$$

and

$$p \equiv (\beta^2 - \tfrac{1}{2}k_s^2)/\beta = (\beta^2 - \alpha_s^2)/(2\beta). \tag{6.80}$$

Then applying the boundary conditions, cancelling ω^2, and anticipating the third equation in the first,

$$\alpha_0(\phi_i - \phi_r) - (\alpha_1\phi_1 - \beta\psi_s) = \alpha_0(\phi_i - \phi_r) - (k_s^2/2\beta)\psi_s = 0, \tag{6.81}$$

$$\rho_0(\phi_i + \phi_r) + \frac{\rho_1}{k_s^2}\{(\beta^2 - \alpha_s^2)\phi_1 + 2\alpha_s\beta\psi_s\}$$
$$= \rho_0(\phi_i + \phi_r) + \frac{2\rho_1\beta}{k_s^2}(p\phi_1 + \alpha_s\psi_s) = 0, \tag{6.82}$$

$$\alpha_1\phi_1 + \frac{(\beta^2 - \alpha_s^2)\psi_s}{2\beta} = \alpha_1\phi_1 + p\psi_s = 0. \tag{6.83}$$

Writing the reflection coefficient as $R = \phi_r/\phi_i$, and the two stress amplitude transmission coefficients as $T_l = (\rho_1/\rho_0)\phi_1/\phi_i$ and $T_s = (\rho_1/\rho_0)\psi_s/\phi_i$, the three simultaneous equations can be solved to give

$$R = \frac{4\alpha_0\beta^2(\alpha_1\alpha_s + p^2)\rho_1/\rho_0 - \alpha_1k_s^4}{4\alpha_0\beta^2(\alpha_1\alpha_s + p^2)\rho_1/\rho_0 + \alpha_1k_s^4}, \tag{6.84}$$

$$T_l = \frac{-4\alpha_0p\beta k_l^2}{4\alpha_0\beta^2(\alpha_1\alpha_s + p^2) + \alpha_1k_s^4\rho_0/\rho_1}, \tag{6.85}$$

$$T_s = \frac{\alpha_1}{p}T_l. \tag{6.86}$$

The reflection and transmission coefficients can be expressed in terms of the angles θ, θ_l, θ_s that the incident and transmitted rays make with the normal to the surface. By Snell's law

$$\beta = k \sin \theta = k_1 \sin \theta_1 = k_s \sin \theta_s, \tag{6.87}$$

and hence

$$\alpha_0 = k \cos \theta, \quad \alpha_1 = k_1 \cos \theta_1, \quad \alpha_s = k_s \cos \theta_s, \quad p = -k_s \frac{\cos 2\theta_s}{2 \sin \theta_s}. \tag{6.88}$$

Defining the normal impedances as in (6.76),

$$Z = \frac{\rho_0 v_0}{\cos \theta}, \qquad Z_1 = \frac{\rho_1 v_1}{\cos \theta_1}, \qquad Z_s = \frac{\rho_1 v_s}{\cos \theta_s}, \tag{6.89}$$

the reflection and stress transmission coefficients become

$$R(\theta) = \frac{Z_1 \cos^2 2\theta_s + Z_s \sin^2 2\theta_s - Z}{Z_1 \cos^2 2\theta_s + Z_s \sin^2 2\theta_s + Z}, \tag{6.90}$$

$$T_1(\theta) = \frac{2Z_1 \cos 2\theta_s}{Z_1 \cos^2 2\theta_s + Z_s \sin^2 2\theta_s + Z}, \tag{6.91}$$

$$T_s(\theta) = \frac{-2Z_s \sin 2\theta_s}{Z_1 \cos^2 2\theta_s + Z_s \sin^2 2\theta_s + Z}. \tag{6.92}$$

By making the substitution

$$Z_{\text{tot}} \equiv Z_1 \cos^2 2\theta_s + Z_s \sin^2 2\theta_s, \tag{6.93}$$

the reflectance function can be written in the form

$$\boxed{R(\theta) = \frac{Z_{\text{tot}} - Z}{Z_{\text{tot}} + Z}} \tag{6.94}$$

which is already familiar from (6.70).

These reflection and transmission coefficients relate the pressure amplitude in the reflected wave, and the amplitude of the appropriate stress component in each transmitted wave, to the pressure amplitude in the incident wave. The pressure amplitude in the incident wave is a natural parameter to work with, because it is a scalar quantity, whereas the displacement amplitude is a vector. The displacement amplitude reflection coefficient has the opposite sign to (6.90) or (6.94); the displacement amplitude transmission coefficients can be obtained from (6.91) and (6.92) by dividing by the appropriate longitudinal or shear impedance in the solid and multiplying by the impedance in the fluid. The impedances actually relate force per unit area to displacement velocity, but displacement velocity is related to displacement by a factor ω which is the same for each of the incident, reflected, and transmitted waves, and so it all comes to the same thing in the end. In some mathematical texts the reflection and transmission coefficients are expressed in terms of the potentials ϕ and ψ of (6.36), which must be spatially differentiated (this is equivalent to multiplying by the appropriate wavenumber $k = \omega/v$) to yield the displacement amplitudes.

For incident energy flux E, the energy flux propagating away from the surface is

reflected: $$E_r/E = |R(\theta)|^2; \qquad (6.95)$$

transmitted longitudinal: $$E_l/E = \frac{\rho_0 \tan\theta}{\rho_1 \tan\theta_l}|T_l(\theta)|^2; \qquad (6.96)$$

transmitted shear: $$E_s/E = \frac{\rho_0 \tan\theta}{\rho_1 \tan\theta_s}|T_s(\theta)|^2. \qquad (6.97)$$

When the incident angle θ is greater than the longitudinal critical angle, then the angle of refraction θ_l in (6.96) becomes imaginary, so that no energy is propagated away from the surface into the solid by a longitudinal wave. Similarly, when the incident angle is greater than the shear critical angle, θ_s in (6.97) is imaginary, and then no energy is carried away from the surface by a shear wave either. But this does not means that nothing is going on in the solid in the vicinity; indeed it can seethe with activity. If all the energy is reflected back into the fluid, then the modulus of the reflectance function must be unity, but the near-surface excitations in the solid can nevertheless manifest themselves in the phase of the reflectance function.

Two sets of curves of stress amplitude functions $R(\theta)$, $T_l(\theta)$, and $T_s(\theta)$ calculated from (6.90)–(6.92) are shown in Fig. 6.3. The reflection and transmission coefficients are complex-valued functions, and in these figures the modulus is shown by the solid curve and the phase is shown by the broken curve. The curves in Fig. 6.3(a) are for waves incident from water on polymethylmethacrylate (PMMA, alias Perspex in England and Plexiglas or Lucite in America). At normal incidence, only a longitudinal wave is excited in the solid, so the reflection coefficient is simply $R = (Z_l - Z_0)/(Z_l + Z_0)$, the longitudinal transmission coefficient is $2Z_l/(Z_l + Z_0)$, and the shear transmission coefficient is zero. As the angle of incidence increases both longitudinal and shear waves are excited in the solid, until the critical angle for excitation of longitudinal waves is reached. At this angle $\sin\theta_l = 1$. Several interesting phenomena occur at this angle. First, the reflection coefficient becomes unity; that means that all of the incident energy is reflected. Second, the longitudinal transmission coefficient has a local maximum value. This is not in contradiction

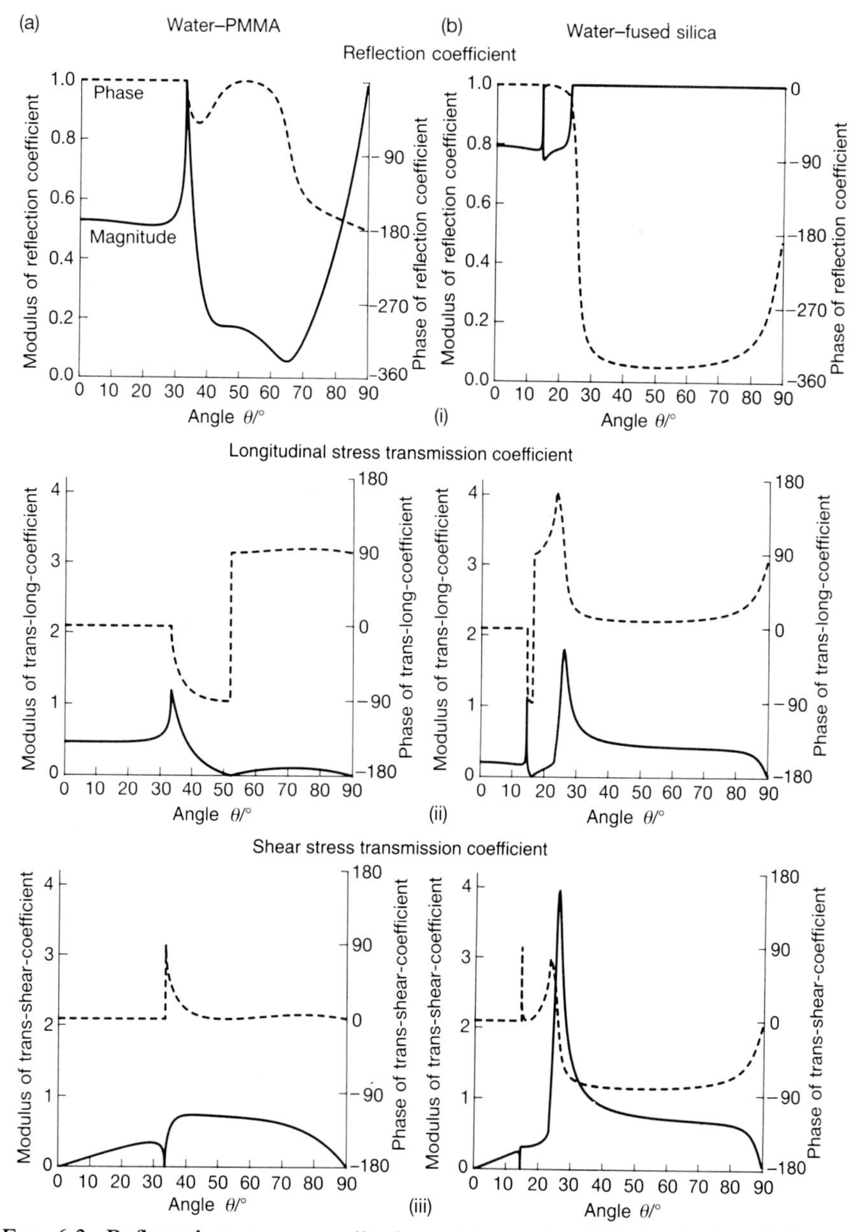

FIG. 6.3. Reflected pressure amplitude and transmitted longitudinal and shear stress amplitudes at a fluid–solid interface: (a) water–PMMA; (b) water–fused silica: (i) reflected wave in the fluid; (ii) transmitted longitudinal wave; (iii) transmitted shear wave. ——— Magnitude (left ordinate); ------ phase (right ordinate); eqns (6.90)–(6.92).

with all the energy being reflected, because the longitudinal wave is propagating parallel to the surface, and no energy is being carried away from the surface. The transmission coefficients correspond to the stress amplitudes in the material immediately below the surface, and not to the energy flux propagating away from the surface which would be given by eqns (6.96) and (6.97). Third, the shear transmission coefficient is zero at the longitudinal critical angle, beyond which the shear wave reappears with a 90° phase shift. Further changes occur smoothly up to glancing incidence, at which the reflection coefficient approaches -1 (i.e. unity with a 180° phase shift), and the two transmission coefficients become zero. The phase change of 180° in the longitudinal transmission coefficient at about 71° occurs where the modulus is zero and is equivalent to plotting the modulus as crossing the axis. The shear velocity in PMMA is slower than the velocity in water, and so no shear wave critical angle phenomena occur.

The curves for waves in water incident on fused silica, plotted in Fig. 6.3(b), show something extra. The behaviour at the longitudinal critical angle (~14°) is qualitatively similar to PMMA. The reflection coefficient is unity, the longitudinal transmission coefficient shows a maximum, and the shear transmission coefficient becomes zero. Beyond this angle no longitudinal waves are excited in the solid that carry energy away from the surface, but shear waves are still excited up to the shear critical angle. Beyond the shear critical angle the reflectance is unity, but this does not mean that nothing is happening in the solid, as can be seen by looking at the behaviour of the phase of the reflectance. At normal incidence the phase is zero. There is a slight wiggle at the longitudinal angle, but the greatest phase change occurs beyond the shear critical angle. Beyond that angle there can be no excitation of longitudinal or shear waves that propagate energy away from the surface, but there can be excitation of waves that decay exponentially into the bulk. These are just the kind of waves that formed the Rayleigh wave discussed above, and they lead to a phase change of nearly 2π that can be seen in the reflectance function just beyond the shear critical angle. When a Rayleigh wave is excited in a solid surface that is in contact with a fluid, it is a leaky Rayleigh wave, because it can re-radiate, or leak, energy back into the fluid in the form of waves. The angle at which this occurs is called the Rayleigh angle, and in the limit of very light loading it would be the angle at which the tangential component of the wavevector in the fluid matches the propagation vector of the Rayleigh wave. Because the Rayleigh wave is leaking energy into the fluid, its wavevector is complex, with the imaginary part describing the exponential attenuation due to the leaking. This has the effect of broadening the range of angle over which waves in the fluid can couple to leaky Rayleigh waves in the solid. But the angle at which the phase of the reflectance function is $-\pi$ is independent of the properties of the coupling fluid. This can be seen when the reflectance function is expressed in the form of eqn (6.94). The phase change of $-\pi$ occurs when $Z_{tot} = 0$, and by inspection Z_{tot} is a function only of the properties of the solid.

6.5 Materials constants

To apply all this splendid theory, you need to know the appropriate parameters for the materials that you are interested in. The density, the longitudinal, shear, and Rayleigh velocities, and the Poisson ratio for a number of isotropic materials are given in Table 6.3. Polycrystalline materials can be considered to be isotropic if their grain size is much smaller than a wavelength; anisotropic elastic constants will be given in Table 11.3 in the chapter on anisotropy, but there is plenty to say about isotropic materials meanwhile! The product of the density and the appropriate velocity from Table 6.3 gives the impedance for a given wave mode. Both the Rayleigh velocity and the Poisson ratio can be found from the longitudinal and shear velocities using (6.54) and (6.56), but they are sufficiently troublesome to calculate quickly, and so central to acoustic microscopy (v_R especially, and σ for Table 6.2), that it is useful to have them at your fingertips. Table 6.3 is by no means exhaustive (there are plenty of other materials) nor definitive (more accurate or reliable values may be available, and in many cases there may be variations, sometimes large, between different materials of nominally similar composition and processing route). But the table does enable useful calculations to be done in order to see what the appearance of a feature of interest in a specimen is likely to be, and how best to look for it. In order to be able to do that, it is necessary to understand how the contrast arises.

Table 6.3
Isotropic acoustic parameters

	Density	Velocity			Poisson ratio
	ρ (kg m^{-3})	Longitudinal v_l (m s^{-1})	Shear v_s (m s^{-1})	Rayleigh v_R (m s^{-1})	σ
Metals					
Aluminium (rolled)	2 700	6 420	3 040	2 844	0.355
Aluminium	2 698	6 374	3 111	2 906	0.345
Bearing Babbit	8 700	2 300			
Beryllium	1 846	12 890	8 880	7 844	0.048
Bismuth	9 803	2 200	1 100	1 026	0.333
Brass*	8 500	4 372	2 100	1 964	0.350
Cadmium	8 647	2 780	1 500	1 390	0.295
Chromium	7 194	6 608	4 005	3 655	0.210
Constantan	8 900	5 177	2 625	2 445	0.327
Copper	8 933	4 759	2 325	2 171	0.343
Duraluminium	2 795	6 398	3 122	2 916	0.344
Gold (hard drawn)	19 281	3 240	1 200	1 134	0.421
Hafnium	13 276	3 840			
Inconel	8 390	5 700	3 000	2 786	0.308
Indium	7 290	2 560			
Invar†	8 000	4 657	2 658	2 447	0.259
Iron (soft)	7 690	5 957	3 224	2 986	0.29
Iron (cast)	7 220	4 994	2 809	2 591	0.269
Lead	11 343	2 160	700	663	0.441
Magnesium	1 738	5 823	3 163	2 930	0.291
Manganese	7 473	4 600			
Molybdenum	10 222	6 475	3 505	3 247	0.293
Monel	8 820	5 350	2 720	2 533	0.326
Nickel (unmag. soft)	8 907	5 608	2 929	2 722	0.312
Nickel (unmag. hard)	8 907	5 814	3 078	2 857	0.306
Niobium	8 578	5 068	2 092	1 970	0.397
Platinum	21 450	3 260	1 730	1 605	0.300
Silver	10 500	3 704	1 698	1 592	0.367
Steel (mild)	7 900	5 960	3 235	2 996	0.291
Steel (tool) hardened		5 874	3 179	2 945	0.293
Steel (stainless)	7 800	5 980	3 297	3 048	0.282
Tantalum	16 670	4 159	2 036	1 901	0.342
Thorium	11 725	2 400	1 560	1 403	0.134
Tin	7 285	3 380	1 594	1 491	0.357
Titanium	4 508	6 130	3 182	2 958	0.321
Tungsten (annealed)	19 254	5 221	2 887	2 668	0.280
Tungsten (drawn)		5 410	2 640	2 464	0.344
Uranium	19 050	3 370	1 940	1 784	0.252
Vanadium	6 090	6 023	2 774	2 600	0.365
Zinc (rolled)	7 135	4 187	2 421	2 225	0.249
Zircaloy	9 360	4 720	2 360	2 201	0.333
Zirconium	6 507	4 650	2 250	2 102	0.347
Oxides and ceramics					
Alumina	3 970	10 822	6 163	5 676	0.26
Barium titanate	6 020	4 000			
Boron carbide	2 400	11 000			
Silicon carbide	3 210	12 099	7 485	6 806	0.19

Table 6.3 (*Continued*)

	Density	Velocity			Poisson ratio
	ρ (kg m^{-3})	Longitudinal v_l (m s^{-1})	Shear v_s (m s^{-1})	Rayleigh v_R (m s^{-1})	σ
Silicon nitride	3 185	10 607	6 204	5 694	0.24
Titanium carbide	5 150	8 270	5 160	4 684	0.181
Tungsten carbide‡	15 000	6 655	3 984	3 643	0.22
Uranium dioxide	10 960	5 180			
Zinc oxide (*c*-axis)	5 606	6 400	2 950	2 765	0.365
Sapphire (*c*-axis)	3 980	11 150	6 036		
Glasses					
Corning 0215 sheet	2 490	5 660			
Crown	2 240	5 660	3 420	3 127	0.22
Heavy flint	3 600	5 260	2 960	2 731	0.27
Pyrex	2 230	5 640	3 280	3 013	0.244
Quartz (fused)	2 200	5 970	3 765	3 410	0.170
Soda lime	2 500	6 000			
Silica (fused)	2 150	5 968	3 764	3 409	0.170
Plastics and rubbers					
Acrylic, Plexiglas	1 190	2 750			
Bakelite	1 400	1 590			
Butyl rubber	1 110	1 700			
Mylar	1 180	2 540			
Neoprene	1 310	1 560			
Nylon 66	1 140	2 620	1 100	1 035	0.39
Perspex (PMMA)	1 185	2 700	1 330	1 242	0.34
Polycarbonate	1 190	2 220			
Polyester casting resin	1 070	2 290			
Polyethylene (low density)	920	1 950	540	513	0.46
Polypropylene	904	2 740			
Polystyrene (Styron 666)	1 050	2 400	1 150	1 075	0.35
Polyvinyl chloride	1 350	2 300			
Polyvinyl chloride acetate		2 250			
Polyvinyl formal		2 680			
Polyvinylidene chloride	1 700	2 400			
Teflon (PTFE)	2 140	1 390	700	652	0.33
Other materials					
Bone (human tibia)	1 900	4 000	1 970	1 839	0.34
Carbon, vitreous	1 470	4 260	2 680	2 429	0.172
Epon 828, mpda	1 210	2 829	1 230	1 156	0.45
Sandstone		2 920	1 840	1 677	0.171
Silver epoxy, e-solder	2 710	1 900	980	912	0.319

In this table, the number of digits given does not necessarily indicate the accuracy of the values. Data from Kaye and Laby (1986), Selfridge (1985), Hung and Goldstein (1983), and other sources.

* 70%Cu, 30%Zn.

† 63.8%Fe, 36%Ni, 0.2%C.

‡ 6%Co.

7
Contrast theory

> It is proposed to investigate the behaviour of waves upon the plane free surface of an infinite homogeneous isotropic elastic solid, their character being such that the disturbance is confined to a superficial region, of thickness comparable with the wave-length. The case is thus analogous to that of deep-water waves, only that the potential energy here depends upon elastic resilience instead of upon gravity. (Lord Rayleigh 1885)

> KO-KO.......................Lord High Executioner of Titipu
> POOH-BAH.......................Lord High Everything Else
> (*The Mikado,* W. S. Gilbert and Sir Arthur Sullivan, also 1885)

A very small amount of experience with an acoustic microscope quickly reveals that the contrast varies very sensitively with the distance between the lens and the surface of the specimen. The first and strongest effect is that the signal is greatest when the specimen surface is at the focus of the lens. The second effect is more subtle; as the specimen is moved from the focal position towards the lens the contrast in an image varies and, in some cases, there can even be reversal of the relative contrast from different parts of an image. Moreover, if the lens is kept over a single point on the specimen and moved towards the specimen from the focal position, then rather than simply decreasing monotonically it can undergo a series of oscillations (Weglein and Wilson 1978). This behaviour is best visualized as a $V(z)$ curve. Here V refers to the video or envelope-detected signal that is used to modulate the brightness of the picture, and z refers to the amount by which the specimen surface is displaced from the focal plane of the lens. By convention the focal position is designated $z = 0$, and displacement of the specimen away from the lens is taken as positive. The process of decreasing the separation of the lens and specimen relative to the focal distance is often referred to as defocusing: it turns out that most of the interesting phenomena occur at negative defocus. A $V(z)$ curve measured on a glass specimen is given in Fig. 7.1. It is the purpose of this chapter to give two theoretical models for the $V(z)$ effect, and to explain their uses.

It is a great help in following the mathematics to have a clear picture of the behaviour of the waves that are being described. Figure 7.2 is a picture of the waves that have been excited in a large-scale model (Negishi and Ri 1987). The lens and the specimen were made of glass, and the coupling fluid was water. The instantaneous amplitude of the

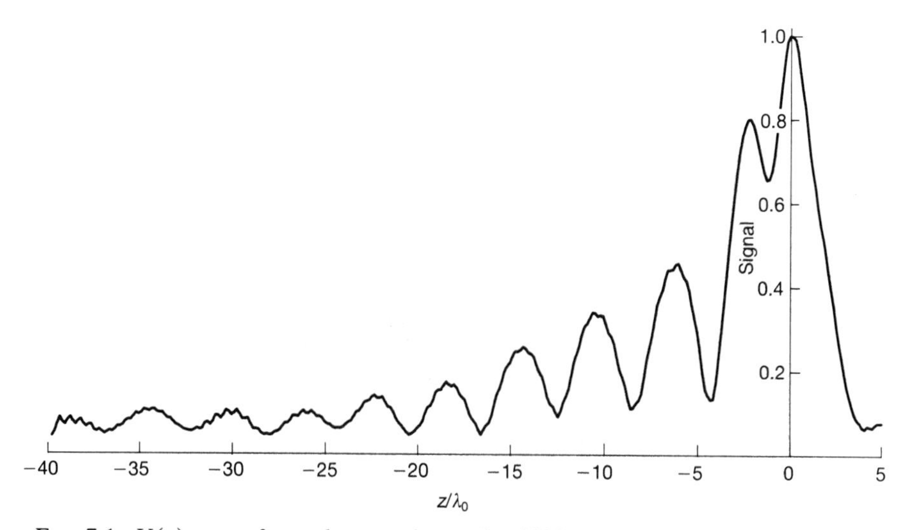

FIG. 7.1. $V(z)$ curve for a glass specimen; $f = 300$ MHz, $T = 70°C$, $\lambda_0 = 5.2\ \mu m$.

waves was photographed by stroboscopic photoelasticity with a Schlieren optical system. At the instant corresponding to this picture the tail end of the acoustic pulse has reached the surface of the specimen, and all of the pulse ahead of that has either propagated into the specimen or been reflected from its surface. In the specimen, a shear wave is propagating radially from the focus on the surface of the specimen; the greatest amplitude of the shear wave occurs around 40°, corresponding to a maximum value of E_s/E in (6.97). The shear wave vanishes in the normal direction, and the longitudinal wave is not visualized under the Schlieren conditions used. Close to the surface another wave is present in the solid; this is the Rayleigh wave. There is a null between the shear wave and the Rayleigh wave, and careful counting of wavefronts reveals that about 6.5 Rayleigh wavelengths correspond to six shear wavelengths, indicating that the Rayleigh velocity is about 92 per cent of the shear velocity. As the Rayleigh waves propagate away from the focus, they generate plane waves in the water whose wavefronts propagate along a direction corresponding to the Rayleigh angle θ_R. The amplitude of the Rayleigh waves decays by 1/e, due to the coupling into the water, over a distance of about five Rayleigh wavelengths. Within the arc subtended on each side by the Rayleigh angles from the focus, a cylindrical wave is present, corresponding to the geometrical reflection. At the join between this and the plane waves due to Rayleigh excitation there is a mismatch of a half a wavelength, corresponding to the phase change of π associated

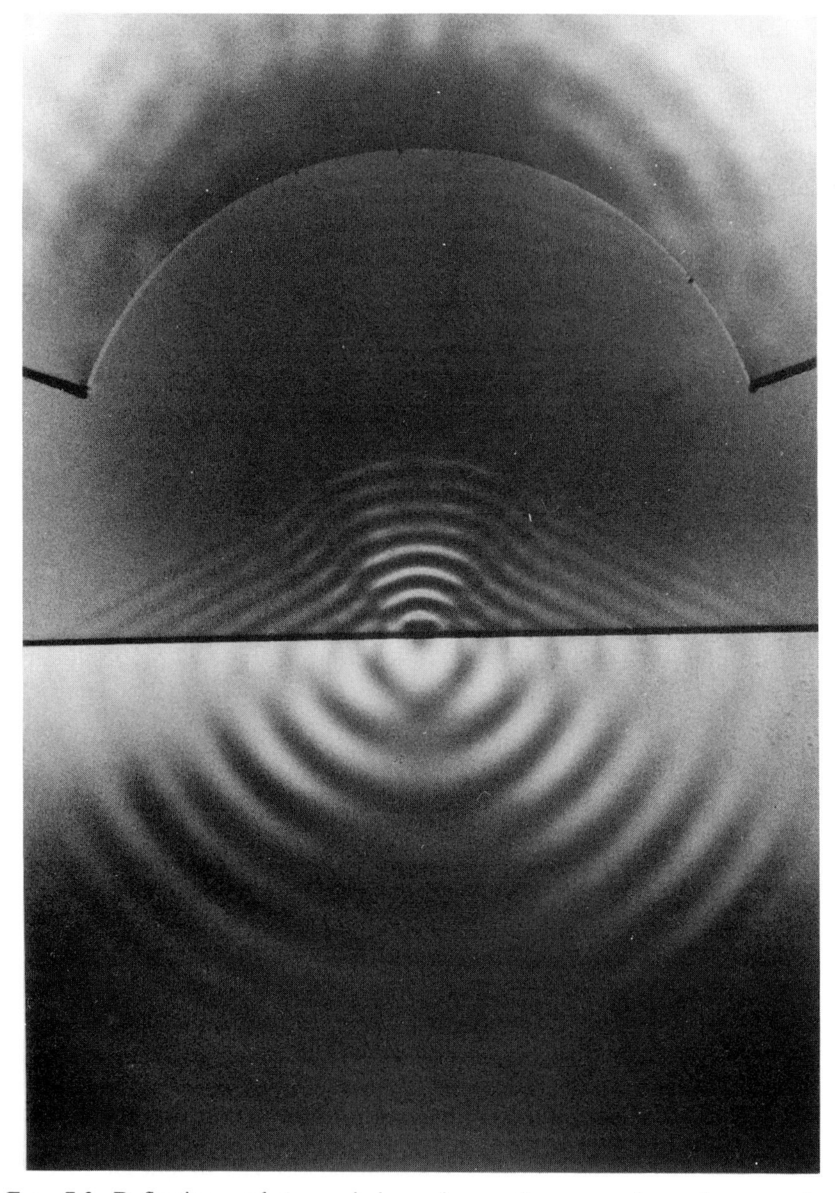

FIG. 7.2. Reflection and transmission of acoustic waves focused on a glass surface, visualized by stroboscopic photoelasticity with a Schlieren optical system. The coupling fluid is water, and the lens and specimen are made of optical glass (BK-7). The frequency of the acoustic pulse is 1.2 MHz, and the radius and focal length of the lens are 20 and 27 mm, respectively. The longitudinal and shear acoustic velocities in the glass are $v_l = 6000\ \mathrm{m\ s^{-1}}$ and $v_s = 3670\ \mathrm{m\ s^{-1}}$, giving Rayleigh velocity $v_R \approx 0.912 v_s \approx 3350\ \mathrm{m\ s^{-1}}$ and Rayleigh angle $\theta_R \approx 26°$ (Negishi and Ri 1987).

with Rayleigh reflection, together with a slight enhancement of intensity at this end of the Rayleigh-excited waves (Tew *et al.* 1988). Rayleigh waves and related surface waves are referred to in many important texts as surface acoustic waves, abreviated SAW; in the presence of fluid loading they are then leaky surface acoustic waves, abreviated LSAW. Out of loyalty to Cambridge, all of these will be designated Rayleigh waves here, the exact type being determined by the context. The confusion of waves inside the lens indicates some of the pathologies that may occur in practice. An understanding of the $V(z)$ effect is crucial to an understanding of the contrast in the acoustic microscope; in high-stiffness specimens the Rayleigh wave activity captured so beautifully in Fig. 7.2 plays a central role.

Consider the situation illustrated schematically in Fig. 7.3. This represents a microscope with a specimen in which Rayleigh waves can be excited, and which is defocused towards the lens. Most rays from the lens are reflected specularly from the specimen, and then pass through the lens with an inappropriate angle to contribute significantly to the excitation of the transducer (ray aa′). Two rays are, however, significant. The first is the ray (bb′) that propagates along the axis of the lens (taken

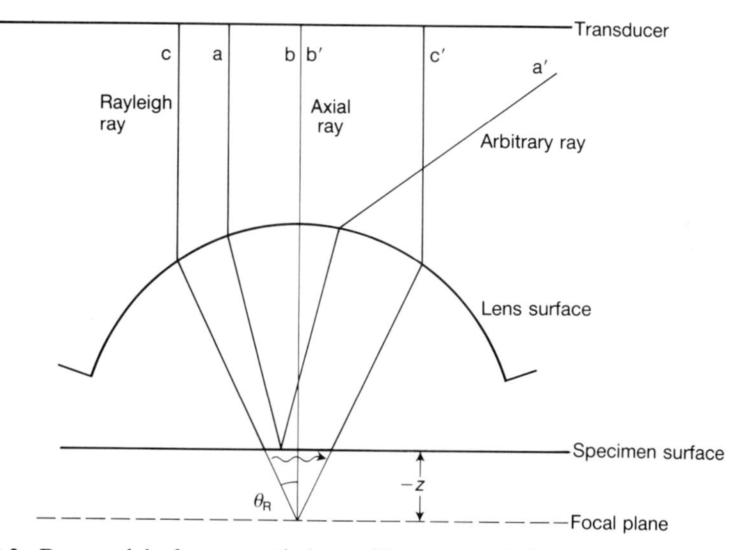

FIG. 7.3. Ray model of an acoustic lens with negative defocus: aa′ is an arbitrary ray, which is reflected at such an angle that is misses the transducer (or else hits the transducer obliquely and therefore contributes little to the signal because of phase cancellation across the wavefront); bb′ is the axial ray, which goes straight down and returns along the same path; cc′ is the symmetrical Rayleigh propagated wave, which returns to the transducer normally and so also contributes to the signal. The wavy arrow indicates the Rayleigh wave.

to be normal to the specimen surface), is reflected, and then propagates back along the same path. The second is a ray (cc′) that is incident on the specimen at the Rayleigh angle, $\theta_R \equiv \sin^{-1}(v_0/v_R)$, and excites a Rayleigh wave in the surface of the specimen. The Rayleigh wave in turn excites waves in the fluid at the Rayleigh angle. The particular ray of importance is the one that propagates back to the lens along a path symmetrical to the initial ray responsible for exciting the Rayleigh wave. Both the axial ray and the Rayleigh ray contribute to the signal at the transducer. Although they are incident at different places on the transducer, the piezoelectric voltages that they excite are summed with respect both to amplitude and to phase, so that their complex-valued sum is detected, and therefore interference effects between them are observed (Weglein 1979*a*; Parmon and Bertoni 1979).

As z changes, the phases of these two rays change at different rates, so that they will alternate between constructive and destructive interference. The phase ϕ_G of the geometrically reflected normal ray is

$$\phi_G = -2kz, \tag{7.1}$$

where k is the wavenumber in the fluid. The phase ϕ_R of the Rayleigh ray is a little more complicated. It advances by virtue of the shortening of the path in the fluid, but against this must be set the path as a Rayleigh wave in the specimen surface. The overall phase is

$$\phi_R = -2(k \sec \theta_R - k_R \tan \theta_R)z - \pi \tag{7.2}$$

$$= -2kz\left(\frac{1-\sin^2 \theta_R}{\cos \theta_R}\right) - \pi, \tag{7.3}$$

since by Snell's law $k_R = k \sin \theta_R$ (6.75). The phase of π even at focus is associated with the phase change of π in the reflection coefficient at the Rayleigh angle in Fig. 6.3(b)i. The phase of the Rayleigh ray can be further simplified using trigonometrical identities. The final expression is

$$\phi_R = -2kz \cos \theta_R - \pi, \qquad z < 0. \tag{7.4}$$

If the output of the transducer is detected by a phase-insensitive circuit (as is usually the case), then it is the difference between the phases of the two rays that is important. This is

$$\phi_G - \phi_R = -2kz(1 - \cos \theta_R) + \pi. \tag{7.5}$$

As the specimen is moved towards the lens, the two rays will alternate between being in phase and being out of phase. The period Δz of the resulting oscillations in $V(z)$ is the movement in the z-direction needed for a change of 2π in the relative phase,

$$\Delta z = \frac{2\pi}{2k(1 - \cos \theta_R)}. \tag{7.6}$$

If this is expressed in terms of the wavelength in water, rather than the wavenumber, it becomes

$$\boxed{\Delta z = \frac{\lambda_0}{2(1 - \cos \theta_R)}} \tag{7.7}$$

with θ_R given by Snell's law,

$$\boxed{\sin \theta_R = \frac{v_0}{v_R}} \tag{7.8}$$

where v_0 is the wave velocity in the fluid, and v_R is the velocity of the Rayleigh wave in the surface of the specimen. The expression for the period of the oscillations in $V(z)$ is of fundamental importance, and must be understood by every acoustic microscopist.

A derivation analogous to that of (7.7) gives the change in the total attenuation suffered by the Rayleigh ray,

$$\boxed{\Delta \alpha = 2z(\alpha_0 \sec \theta_R - \alpha_R \tan \theta_R)} \; . \tag{7.9}$$

Thus, at negative defocus, the amplitude of the Rayleigh ray will vary as $\exp\{-2z(\alpha_0 \sec \theta_R - \alpha_R \tan \theta_R)\}$. If the Rayleigh contribution to $V(z)$ is smaller than the geometrical contribution, the amplitude of the oscillations in $V(z)$ will follow the same exponential decay.

The simple picture of two interfering rays can be a great help in visualizing what is happening in an acoustic microscope as it is defocused (Quate 1980), and for many intuitive purposes this is quite sufficient. For analytical purposes more substantial theory is needed, and this can be derived both from diffraction optics and from a more rigorous development of ray theory. The simple ray model is vindicated both by the more detailed theoretical models presented in this chapter, and also by quantitative measurements based on them, which will be described in Chapter 8.

7.1 Wave theory of $V(z)$

7.1.1 *Reflection*

A simplified picture of the transducer, lens surface, and reflecting object is shown in Fig. 7.4. The waves radiated by the transducer are refracted

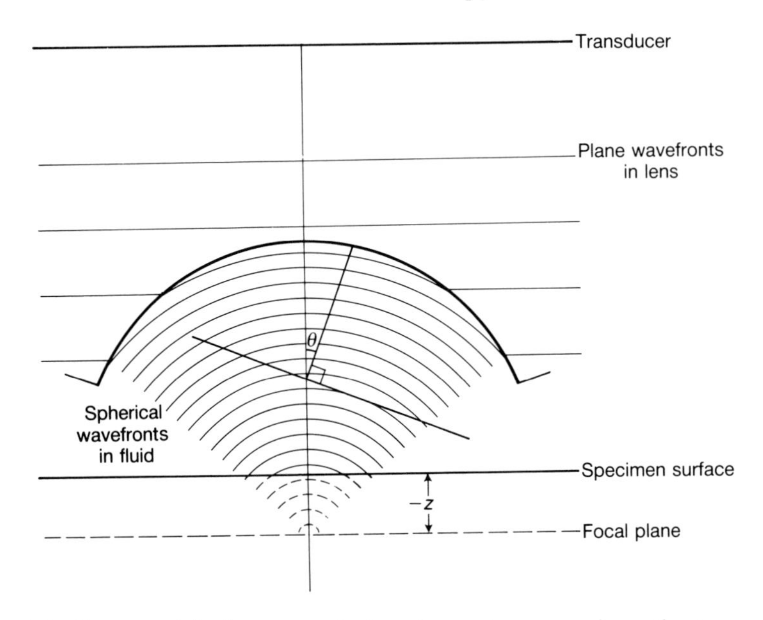

FIG. 7.4. Wave model of an acoustic lens focused on a surface; the tangent to one of the wavefronts illustrates one of the family of plane waves of which the wavefront is considered to be composed.

by the lens so as to form a spherical wavefront centred on the focal point of the lens. Each point on this wavefront can be described by its angular coordinates from the focus; let these be θ for the zenithal angle (i.e. the angle to the lens axis, again taken to be normal to the specimen surface) and ϕ for the azimuthal angle. Thus the spherical wave emerging from the lens can be described by a function $L_1(\theta, \phi)$. The wave is reflected from the specimen surface with a reflectance function $R(\theta, \phi)$, which also depends on these angles. Finally, the wave returns through the lens to the transducer, where it is detected with a sensitivity $L_2(\theta, \phi)$. The total received signal at focus can therefore be calculated by integrating over θ and ϕ,

$$V = \int_0^{\pi/2} \int_{-\pi}^{\pi} L_1(\theta, \phi) R(\theta, \phi) L_2(\theta, \phi) \sin\theta \, \mathrm{d}\phi \, \mathrm{d}\theta, \qquad z = 0. \tag{7.10}$$

Since $L_1(\theta, \phi)$ and $L_2(\theta, \phi)$ depend only on the geometry and material of the lens, it is convenient to combine them in a single pupil function, which may be defined as

$$P(\theta, \phi) = L_1(\theta, \phi) L_2(\theta, \phi)/\cos\theta. \tag{7.11}$$

The $\cos\theta$ term in the definition allows for the obliqueness of the excitation of the spherical waves by plane waves incident on the refracting surface. It is analogous to the reduction in the amount of sunlight per unit area falling on the earth at high angles, and means that a lens with uniform illumination and perfect transmission up to a given cut-off angle would have a pupil function of unity up to that angle and zero above it (this is sometimes described as a top hat pupil function because of its appearance when plotted in polar coordinates with the third axis denoting the value of the function). With this definition of the pupil function, the response is

$$V = \int_{0}^{\pi/2} \int_{-\pi}^{\pi} P(\theta, \phi) R(\theta, \phi) \sin\theta \cos\theta \, \mathrm{d}\phi \, \mathrm{d}\theta. \tag{7.12}$$

Most imaging lenses are designed to have axial symmetry, so that P is independent of ϕ. Anisotropic specimens will be considered in Chapter 11, but if the specimen is isotropic then the ϕ integration may be absorbed in P, so that the expression for V simplifies to

$$V = \int_{0}^{\pi/2} P(\theta) R(\theta) \sin\theta \cos\theta \, \mathrm{d}\theta. \tag{7.13}$$

This expression describes the response of the system to a specimen at focus. If there are no aberrations, then at focus the phase of all the plane wave contribution at different angles is the same. But the functions in the integrand can be complex-valued. It was seen in Chapter 6 that $R(\theta)$ has both amplitude and phase, and this can be represented by making it complex, with the angle in the complex plane representing the phase. Similarly, any lens aberrations can be described by making $P(\theta)$ complex though, as was discussed in Chapter 2, such phase aberrations can be very small even for lenses of large opening angle.

If the specimen is moved away from the focal position, then this will cause a phase shift that depends on θ. If the wavenumber in the coupling fluid is $k \equiv 2\pi/\lambda_0$, then the z component of the wavevector is $k_z = k \cos\theta$. Defocusing the specimen by an amount z causes a phase delay of $2zk_z$, or $2zk \cos\theta$ (the factor of two arises because both the incident wave and the reflected wave suffer a change in path length). Expressing this phase delay as the complex exponential of a phase angle, the response of the microscope with a defocus z is

$$\boxed{V(z) = \int_{0}^{\pi/2} P(\theta) R(\theta) \mathrm{e}^{-\mathrm{i}2zk\cos\theta} \sin\theta \cos\theta \, \mathrm{d}\theta}. \tag{7.14}$$

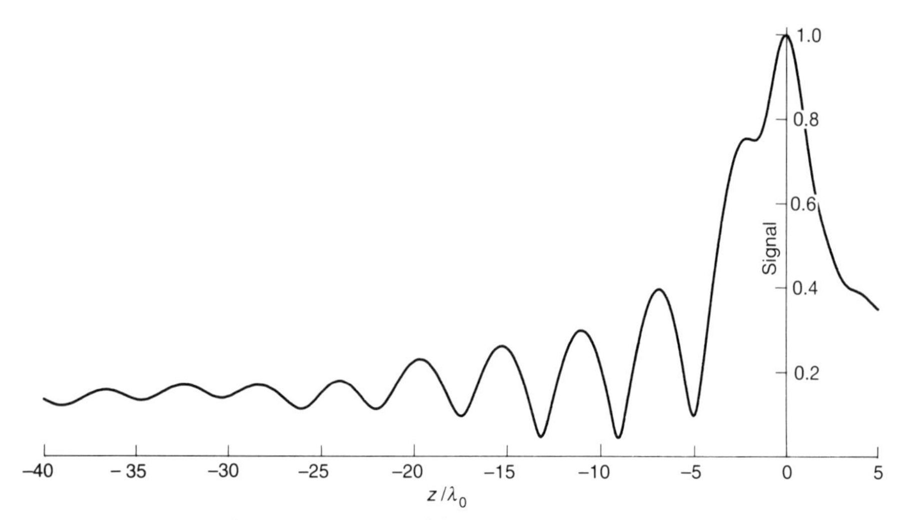

FIG. 7.5. A theoretical $V(z)$ curve for a glass specimen.

This is the basic expression for the response of the microscope to a uniform isotropic specimen at a defocus z. It should be permanently imprinted on the mind! A $V(z)$ curve for glass calculated using this expression is shown in Fig. 7.5. The primary difficulty in achieving exact agreement with measured curves lies in knowing the exact pupil function of the lens in the experiment.

The integral expression for $V(z)$ has a long history. The early formulae involved a paraxial approximation (Atalar 1978, 1979; Wickramasinghe 1978, 1979; Quate *et al.* 1979). Their work was extended to high lens angles by Sheppard and Wilson (1981), who derived the result given here. It was later derived, in greater detail, by Liang *et al.* (1985*b*), who analyzed the approximations made and also applied their result to a spherical transducer (in which case the cos θ term due to obliquity is not needed), and evaluated for certain cases with great accuracy by Chou and Kino (1987).

The equation for $V(z)$ can be expressed as a Fourier transform by a suitable change of variables (Hildebrand *et al.* 1983; Ilett *et al.* 1984; Liang *et al.* 1985*b*; Fright *et al.* 1989). The new variables are defined as

$$u \equiv kz$$

and

$$t \equiv \frac{1}{\pi} \cos \theta. \tag{7.15}$$

Then (7.14) may be written

$$V(u) = \int_0^{1/\pi} P(t)R(t)\mathrm{e}^{-\mathrm{i}2\pi ut} t \, \mathrm{d}t. \tag{7.16}$$

With the further substitution

$$Q(t) \equiv P(t)R(t)t, \tag{7.17}$$

$V(u)$ may be written

$$V(u) = \int_0^{1/\pi} Q(t)\mathrm{e}^{-\mathrm{i}2\pi ut}\,\mathrm{d}t. \tag{7.18}$$

The significance of the relationship should now become clear. Equation (7.18) describes a Fourier transform (Bracewell 1978), with $V(u)$ and $Q(t)$ as the transform pair. Of course, the limits of the integration in (7.18) should be from $-\infty$ to $+\infty$, but since $Q(t)$ vanishes outside the given limits this makes no difference.

There are two particular discontinuities in $R(t)$ that are of great importance for materials in which Rayleigh waves are excited. The first occurs at $t_0 = 1/\pi$, because beyond that value $Q(t)$ changes discontinuously to zero. The second is at $t_R = \cos\theta_R/\pi$, because around the Rayleigh angle θ_R there is a phase change of 2π in $R(\theta)$, cf. Fig. 6.3(b)i, and hence in $Q(t)$. The Fourier relationship gives oscillations in $V(u)$ of periodicity

$$\Delta u = \frac{1}{t_0 - t_R}. \tag{7.19}$$

There is a direct analogy with the fringe pattern that is seen in a Young's double slit experiment, in which the diffraction pattern from two slits produces periodic fringes whose spacing varies inversely with the separation of the slits. The oscillations can also be interpreted in terms of the distortions of the reflected wavefronts in Fig. 7.2 at the Rayleigh angle (Atalar 1979).

In terms of the original variables the period of the oscillations in $V(z)$ is

$$\boxed{\Delta z = \frac{\lambda_0}{2(1 - \cos\theta_R)}} \tag{7.20}$$

which is identical to (7.7).

7.1.2 *Transmission*

An expression similar to (7.14) can be developed for a transmission acoustic microscope (Bukhny *et al.* 1990; Levin *et al.* 1990; Maev and Levin 1990). In the general case it can be quite complicated if the lenses are not aligned, but in the co-axial approximation the dependence of the signal, given a new variable to distinguish it from the reflection case, is

$$A(z) = \int_0^{\pi/2} P(\theta)T(\theta)e^{-i(z-d)k\cos\theta} \sin\theta \cos\theta \, d\theta. \qquad (7.21)$$

$P(\theta)$ is the lens function for the two lenses and $T(\theta)$ gives the magnitude and phase of waves emerging from a parallel-sided specimen of thickness d relative to the amplitude and phase of the wave incident at the same angle θ on the other side of the specimen. $T(\theta)$ can be a complicated function of θ, often with resonance phenomena, containing information about the elastic properties and thickness of the specimen (Wang *et al.* 1980; Tsai and Lee 1987); sometimes additional information can be gained by tilting the specimen (Wang and Tsai 1985). In some important cases, notably biological soft tissue and some polymers with Poisson ratio $\sigma \approx 0.5$, shear waves play little role, and the longitudinal velocity and density are close to those of water. The transmission function can then be approximated by a phase function containing simply the phase shift due to a single path through the specimen at the angle determined by Snell's law. With the refractive index $n = v_0/v_l$, the ratio of the velocity in the fluid to the longitudinal velocity in the specimen, the phase function is

$$T(\theta) = \exp\{ikd(n - \sin^2\theta)\}. \qquad (7.22)$$

If $P(\theta) \propto 1/\cos\theta$ up to some maximum angle θ_{max}, and is zero beyond that, (7.21) with (7.22) can be integrated using Fresnel integrals. To a good approximation, the longitudinal velocity in a specimen of known thickness can be deduced from the phase shift in $A(0)$ alone (Bennett 1982). But generally it is easier to measure the amplitude of a signal than its phase, and the most prominent effect of the specimen is to shift the maximum in $A(z)$. Unlike the paraxial result of (2.7), this more detailed treatment gives a shift in the maximum that depends explicitly on the aperture of the lens. The lens movement required to recover the maximum signal is

$$\delta z = d\{(1 - 1/n) + (1 - 1/n^2)(1 - \cos\theta_{max})/2n\}. \qquad (7.23)$$

By measuring δz the longitudinal velocity can be found in a specimen of known thickness. An example is given in Fig. 7.6. The broken curves are experimental $A(z)$ curves without (curve 1) and with (curve 2) a 12 μm thick collagen specimen. From the shift in the position of the maximum signal and longitudinal velocity $v_l = 2400$ m s^{-1} was deduced using (7.23). From that value, and the other parameters in the caption, calculated $A(z)$ curves were plotted for the situations with and without the specimen.

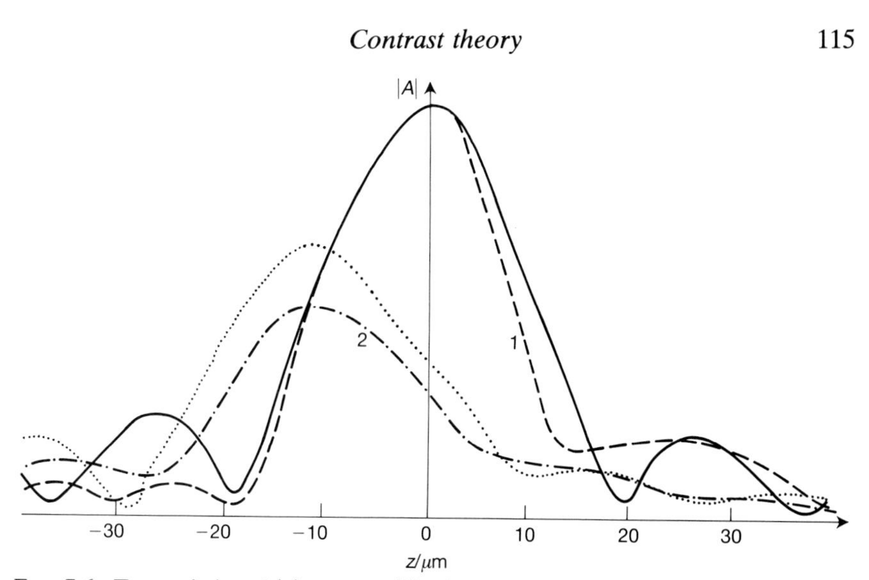

FIG. 7.6. Transmission $A(z)$ curves: (1) without a sample; (2) with a collagenous tissue section 12 μm thick. ----, ·—· Measured; ——, ····· calculated (7.13); 450 MHz.

In transmission microscopy of specimens with properties not too different from those of water, Rayleigh waves may safely be disregarded. But in reflection microscopy of specimens of higher stiffness, Rayleigh waves generally play a dominant role. This is recognized explicitly in the ray theory treatment.

7.2 Ray model of $V(z)$

The expression for the period of the oscillations in $V(z)$ can also be derived by ray theory. The ray model is particularly powerful in the analysis of $V(z)$ data.

7.2.1 *Separation of the Rayleigh pole*

In the simple form given at the start of this chapter, the ray picture correctly predicts the period of the oscillations in $V(z)$, but it does not enable any other aspect of $V(z)$ to be calculated. To do that it is necessary to look more carefully at the form of the reflectance function (Bertoni 1984; Brekhovskikh and Godin 1990). Writing the expression for the reflectance function in (6.90) explicitly in terms of the wavenumber in the fluid k and the longitudinal and shear wavenumbers in the solid k_l and k_s, and with the tangential component of the wavevector $\beta \equiv k_x$, which, by Snell's law, is the same in both media, and with the densities of the fluid and the solid ρ_0 and ρ_1, the reflectance function is

$$R(k_x) = \frac{[(2k_x^2 - k_s^2)^2 - 4k_x^2\{(k_x^2 - k_l^2)(k_x^2 - k_s^2)\}^{\frac{1}{2}}] - \mathrm{i}k_s^4\{(k_x^2 - k_l^2)(k^2 - k_x^2)\}^{\frac{1}{2}}\rho_1/\rho_0}{[(2k_x^2 - k_s^2)^2 - 4k_x^2\{(k_x^2 - k_l^2)(k_x^2 - k_s^2)\}^{\frac{1}{2}}] + \mathrm{i}k_s^4\{(k_x^2 - k_l^2)(k^2 - k_x^2)\}^{\frac{1}{2}}\rho_1/\rho_0}. \quad (7.24)$$

When $k_x < k_l$ the last term in the numerator and the denominator are real, and the reflectance function is real, so that there is no variation of phase with incident angle. When $k_x = k_l$ all terms except the first term in the numerator and the denominator vanish, and $R(k_x) = 1$. When $k_l < k_x < k$, the form of (7.24) makes it explicit that the last term in the numerator and the denominator are pure imaginary. The term in square brackets depends only on k_x and the properties of the solid. Provided that there is no dissipation in the solid, it vanishes when k_x is equal to the Rayleigh wavenumber $k_R = \omega/v_R$, as can be seen by substituting $X = (k_s/k_R)^2$, $Y = (k_l/k_s)^2$, and comparing with (6.55).

As k_x is, in general, a complex wavenumber, with the imaginary part representing an exponential decay in amplitude, $R(k_x)$ may be considered throughout the complex k_x plane, and eqn (7.24) gives poles (denominator going to zero) and zeros (numerator going to zero) at $k_x = \pm k_p$ and $k_x = \pm k_0$, respectively. In the absence of dissipation in the solid $k_0 = k_p^*$, the asterisk denoting complex conjugate, and it is convenient to write $k_p \equiv \beta + \mathrm{i}\alpha_R$, with α_R describing the attenuation of the Rayleigh wave due to radiation into the fluid. For most solids in contact with water (the exceptions being those with low density and shear modulus, for instance polymers), $|(\beta - k_R)/k_R| \ll \alpha_R/k_R \ll 1$, where k_R is the Rayleigh wavenumber in the absence of any fluid loading. In that case the poles and zeros are so close to k_R that the behaviour of $R(k_x)$ for k_x real and close to $\pm k_R$ can be approximately described by

$$R_R(k_x) \approx \frac{k_x - k_0}{k_x - k_p} \cdot \frac{k_x + k_0}{k_x + k_p} = \frac{k_x^2 - k_0^2}{k_x^2 - k_p^2}. \quad (7.25)$$

This expression has the properties associated with the behaviour of the reflectance function around the Rayleigh angle, namely an amplitude of unity and a phase change of nearly -2π as $|k_x|$ increases past k_R. If waves from only one side of normal were of interest, corresponding to one or other sign of k_x, then only one of the fractions in the approximation of (7.25) would be needed; including the differences of the squares ensures that both signs of k_x near the Rayleigh poles and zeros are catered for. By taking out the factor $R_R(k_x)$, everything else can be described by $R_0(k_x)$, so that

$$R(k_x) \equiv R_0(k_x)\frac{k_x^2 - k_0^2}{k_x^2 - k_p^2} = R_0(k_x)\left(1 + \frac{k_p^2 - k_0^2}{k_x^2 - k_p^2}\right). \tag{7.26}$$

This equation is, of course, simply a definition of $R_0(k_x)$. Its usefulness arises from the explicit separation of the geometrical and Rayleigh components of the reflectance. Moreover, the fraction $(k_p^2 - k_0^2)/(k_x^2 - k_p^2)$ in (7.26) contributes significantly only when $|k_x|$ is close to $|k_p|$. The Rayleigh wavenumber is always greater than the bulk shear wavevector k_s by 5–10 per cent or so (Table 6.2), and since $|R(k_x)| = 1$ for $k_x \geq k_s$, again cf. Fig. 6.3(b)i, the approximation may be made that

$$\boxed{R(k_x) \approx R_0(k_x) + \frac{k_p^2 - k_0^2}{k_x^2 - k_p^2}}. \tag{7.27}$$

Like (7.26), this equation separates the geometrical and Rayleigh components of the reflectance; it has the further advantage that the Rayleigh component is amenable to analytic manipulation. The second term on the right describes the Rayleigh pole singularity; the first term, like Pooh-Bah, describes everything else.

If the field incident on the surface of the specimen is described by $p_{\text{inc}}(x)$, and $P_{\text{inc}}(k_x)$ is its Fourier transform, then the reflected field at the surface can be found by taking the inverse Fourier transform of the product of $P_{\text{inc}}(k_x)$ and $R(k_x)$. Lower case denotes real-space fields, upper case spatial frequency (or reciprocal-space) fields. A time dependence $\exp(\mathrm{i}\omega t)$, corresponding to a frequency $\omega/2\pi$, is implicit throughout. If the reflected field is separated into the geometrical part $p_G(x)$ and the leaky Rayleigh wave part $p_R(x)$,

$$p_{\text{ref}}(x) \equiv p_G(x) + p_R(x), \tag{7.28}$$

then the two terms may be calculated using the separation of $R(k_x)$ in (7.27). The geometrical term is

$$p_G(x) = \frac{1}{2\pi}\int_{-\infty}^{\infty} R_0(k_x)P_{\text{inc}}(k_x)\exp(\mathrm{i}k_x x)\,\mathrm{d}k_x. \tag{7.29}$$

The Rayleigh component is

$$p_R(x) = \frac{1}{2\pi}\int_{-\infty}^{\infty} \frac{k_p^2 - k_0^2}{k_x^2 - k_p^2} P_{\text{inc}}(k_x)\exp(\mathrm{i}k_x x)\,\mathrm{d}k_x. \tag{7.30}$$

The k_x-field in the integrand can be expressed as the Fourier transform of the spatial field by a further integration. For the leaky wave part,

$$p_{\mathrm{R}}(x) = \frac{1}{2\pi} \int_{-\infty}^{\infty} p_{\mathrm{inc}}(x) \left\{ \int_{-\infty}^{\infty} \frac{k_{\mathrm{p}}^2 - k_0^2}{k_x^2 - k_{\mathrm{p}}^2} \exp[\mathrm{i}k_x(x - x')] \, \mathrm{d}k_x \right\} \mathrm{d}x'. \quad (7.31)$$

The integration over k_x may be performed by deforming the path of integration into the complex k_x plane. For $x - x' > 0$, the path is deformed into the upper half-plane, capturing the pole at k_{p}. When $x - x' < 0$, deforming into the lower half plane captures the pole at $-k_{\mathrm{p}}$. When the effect of the fluid can be regarded as a light perturbation on the propagation of Rayleigh waves on the surface of the solid, so that $\alpha_{\mathrm{R}} \ll k_{\mathrm{R}}$, then

$$\frac{\mathrm{i}(k_{\mathrm{p}}^2 - k_0^2)}{2k_{\mathrm{p}}} \approx -2\alpha_{\mathrm{R}}. \quad (7.32)$$

The re-radiated field is then

$$p_{\mathrm{R}}(x) = -2\alpha_{\mathrm{R}} \int_{-\infty}^{\infty} p_{\mathrm{inc}}(x') \exp[\mathrm{i}k_{\mathrm{p}} \, |x - x'|] \, \mathrm{d}x'. \quad (7.33)$$

This equation has an extremely important interpretation. In its differential form it means that, in two dimensions (with the x-axis lying in the plane of the interface between the solid and the fluid and the z-axis lying normal to the plane), if a pressure $p_{\mathrm{inc}}(x')$ with implicit frequency dependence $\exp(\mathrm{i}\omega t)$ acts along a strip in the y-direction at x' of width $\mathrm{d}x'$, then the Rayleigh wave that is excited will propagate and the response in the fluid immediately above the surface at x will be

$$\mathrm{d}p_{\mathrm{R}}(x) = p_{\mathrm{inc}}(x')\{-2\alpha_{\mathrm{R}} \exp[\mathrm{i}k_{\mathrm{p}} \, |x - x'|]\} \, \mathrm{d}x'. \quad (7.34)$$

The term in the curly brackets describes the response at x due to the excitation at x', transmitted by the Rayleigh wave mechanism; it can therefore be thought of as a kind of Green function. The results described by eqns (7.33) and (7.34) are central to the theory of the contrast from cracks and interfaces that will be presented in Chapter 12.

7.2.2 *Rayleigh ray contrast*

The reflected fields of eqns (7.29) and (7.30) enable the signal at the transducer to be calculated (Bertoni 1984). The calculation of the geometrically reflected contribution at the transducer falls into three domains. Let D be the distance between the transducer and the lens surface; q, the focal length of the lens (both of these being measured from the lens surface at the axis); a_T, the radius of the transducer; a_0, the radius of the lens aperture; n, the refractive index of the lens (the ratio of the velocity in the coupling fluid to the longitudinal velocity in the lens material); and λ_0, the wavelength in the coupling fluid. The following values of defocus can be defined:

$$z_{\min} = -q^2\left[\frac{(a_T/a_0)-1}{2(D/n-q)}\right] \approx -z_0\left[\frac{a_T}{a_0}-1\right]; \tag{7.35}$$

$$z_0 = \frac{q^2}{2(D/n-q)} \approx \left(\frac{q}{q_T}\right)^2 \frac{\lambda_0}{2F_T}; \tag{7.36}$$

$$z_{\max} = \frac{q^2(1+a_T/a_0)}{2[D/n-(1+a_T/a_0)f]} \approx z_0\left[\frac{a_T}{a_0}-1\right]. \tag{7.37}$$

The limits $z_{\min}$ and $z_{\max}$ are the values of defocus at which the reflected rays just fill the area of the transducer; thus for $z_{\min} \leq z \leq z_{\max}$ all the reflected rays that enter the lens fall on the transducer, while for values of defocus outside that range some of the rays miss it altogether. The value z_0 is the defocus at which the geometrically reflected rays are focused on the transducer; at this point, as indeed at $z = 0$, ray optics breaks down because it does not allow for diffraction, although it does correctly predict the position of a minimum in $V(z)$ at z_0. The approximate expressions are valid when $D/n \gg q$, as is usually the case in a high-resolution acoustic microscope. In the approximation for z_0, the quantity F_T is the ratio of the separation between the transducer and the back focal plane of the lens $(D - q/n)$ to the Fresnel distance for the transducer (na_T^2/λ_0),

$$F_T = (D/n - q)\lambda_0/a_T^2. \tag{7.38}$$

In most high-resolution lenses, $F_T \approx 1$ so that z_0, and hence $-z_{\min}$ and $z_{\max}$, are of the order of λ_0. With the further notation that the wavenumber and attenuation in the coupling fluid are k and α_0, respectively, and that the response of the transducer to a uniform field of unit amplitude would be V_0, the response to the geometrically reflected field is

$$V_G(z) = V_0AR_0(0)\left(\frac{-2z_0F_T}{\pi z}\right)\sin X_R \exp\{2(ik-\alpha_0)z + iX_R\} \tag{7.39}$$

where

$$A = -\exp[i2k(Dn+q) - 2\alpha_0 q],$$

and

$$X_R = \frac{\pi}{2F_T} \times \begin{cases} \dfrac{z}{z - z_0}, & z \le z_{min} \text{ or } z \ge z_{max}, \\ \left(\dfrac{a_0}{a_T}\right)^2 \dfrac{z(z - z_0)}{z_0^2}, & z_{min} < z \le 0, \\ \left(\dfrac{a_0}{a_T}\right)^2 \dfrac{z(z - z_0)}{z_0^2(1 + 2z/q)^2}, & 0 < z < z_{max}. \end{cases} \tag{7.40}$$

In these expressions an absolute amplitude is implicit in A, including transmission coefficients each way for the lens surface. Also, for simplicity, the normal reflection coefficient $R_0(0)$ has been used in (7.39); this is appropriate in accounting for $V(z)$ oscillations because, as the lens is defocused in the negative z direction, only rays increasingly close to the normal contribute to the geometrical term.

The Rayleigh reflection at the surface of the specimen is obtained from eqn (7.31). This leads to a family of rays at the Rayleigh angle propagating towards the lens. Because in any plane through the axis they are parallel, they are brought to a focus at the back focal plane of the lens, and, if there is axial symmetry, they will be focused in a ring, of radius $a_R \approx q \sin \theta_R$; the self-focusing of the Rayleigh wave ring is the reason why the resolution is often scarcely degraded even at considerable defocus (§3.6). Although Rayleigh waves can propagate backwards in some circumstances when there is a periodic interface so that an acoustic equivalent of Bragg scattering can occur (Breazeale and Torbett 1976; De Billy *et al.* 1983; Nagy and Adler 1989), only forward propagation is significant here, so there is a contribution from the symmetrically excited ray only when the specimen is defocused towards the lens. Thus, for the purpose of understanding the oscillations in $V(z)$ only the range $z < z_{R1}$ is of interest, where z_{R1} is a small negative defocus given by

$$z_{R1} = -\lambda_0/(4 \cos \theta_R \sin^2 \theta_R). \tag{7.41}$$

For z more negative than this the situation can be described by considering the incident ray at the Rayleigh angle as exciting a leaky Rayleigh wave in the specimen, which in turn excites rays at the Rayleigh angle in the liquid. The contribution to the signal at the transducer is then

$$V_R(z) = V_0 A K \left\{ \frac{2\alpha_R \lambda_0 (\sqrt{\pi q^3} \sin \theta_R)^{1/2}}{(\lambda_0 D/n)^{3/4}} \right\} e^{i\pi} \exp\{i2k_0 z \cos \theta_R\} \\ \times \exp\{2z(\alpha_T \tan \theta_R - \alpha_0 \sec \theta_R)\}; \tag{7.42}$$

where

$$K = e^{-i\pi/4} \frac{2}{\eta_T^2} \int_0^{\eta_T} \{\exp[i(\eta - \eta_R)^2] + e^{-i\pi/2} \exp[i(\eta + \eta_R)^2]\} \sqrt{\eta}\, d\eta, \tag{7.43}$$

with

$$\eta_T = a_T \surd(k_0 n/2D) \approx \surd(\pi/F_T),$$

and

$$\eta_R = a_R \surd(k_0 n/2D) \approx (a_R/a_T)\surd(\pi/F_T)). \tag{7.44}$$

The value of K is close to unity, and its magnitude and phase vary only weakly with a_R. For $z < z_1$ the dependence of V_R on z is rather simple. There is a phase variation given by the second exponential term in (7.42), the first describing the constant phase change of π associated with the phase change in the reflectance function at the Rayleigh angle. The final exponential term in (7.42) describes the change in attenuation due to the decrease in water path and increase in Rayleigh wave path as the defocus increases. The total Rayleigh wave attenuation per unit distance, α_T, may contain contributions from elastic scattering, inelastic damping, and leaking into the coupling fluid. The attenuation due to leaking into the coupling fluid, α_R, may be calculated approximately from (Dransfeld and Salzmann 1970)

$$\alpha_R \lambda_0 \approx \frac{\rho_0 v_0^2}{\rho_1 v_R^2} = \frac{\rho_0}{\rho_1} \sin^2 \theta_R \tag{7.45}$$

where ρ_0 and ρ_1 are the densities of the fluid and the solid and θ_R is the Rayleigh angle as usual. The value of (7.45) has the form of a ratio of impedances multiplied by a ratio of velocities, and it provides a means of relating Rayleigh wave attenuation (due to radiation into the fluid) to the density of the specimen. With the approximation of (7.45), the term in the large curly brackets in (7.42) becomes

$$\left\{\frac{2\alpha_R \lambda_0 (\surd \pi q^3 \sin \theta_R)^{1/2}}{(\lambda_0 D/n)^{3/4}}\right\} \approx \frac{2\pi^{1/4}}{F_T^{3/4}} \left(\frac{q}{a_T}\right)^{3/2} \frac{\rho_0}{\rho_1} \sin^{5/2} \theta_R. \tag{7.46}$$

Both the geometrical contribution and the Rayleigh contribution can thus be expressed in terms of material constants and the geometry of the lens, and therefore be directly compared. The complex summation of (7.39) and (7.42) enables $V(z)$ to be computed and leads immediately to oscillations with period

$$\Delta z = \frac{2\pi}{2k_0 - 2k_0 \cos \theta_R} = \frac{\lambda_0}{2(1 - \cos \theta_R)} \tag{7.47}$$

as before. In (7.42), the decay of $V_R(z)$ has the same form as predicted by (7.9), and once again if $V_R(z) < V_G(z)$ the oscillations will have an identical decay.

7.2.3 *Lateral longitudinal waves*

In a small number of materials of intermediate stiffness (mainly polymers of relatively high moduli such as PMMA) lateral longitudinal waves can take the place of Rayleigh waves. These waves, also known as surface skimming compressional waves or SSCW, propagate parallel to the fluid–solid interface when the angle of refraction is 90° (Tamir 1972), so that the longitudinal critical angle replaces the Rayleigh angle in calculating the period of the oscillations. In this case the dominant effect comes from the branch point singularities at $\pm k_l$ due to the terms in $\kappa_l = \sqrt{(k_x^2 - k_l^2)}$ in eqn (7.24). The lateral longitudinal contribution in the reflectance function can be separated from the rest in a manner analogous to (7.26), by expanding the reflectance function to first order in κ_l/k_l (Chan and Bertoni 1991),

$$R(k_x) \approx A(k_x) + B(k_x)\kappa_l/k_l. \tag{7.48}$$

In the vicinity of k_l to first approximation the values of $A(k_x)$ and $B(k_x)$ may be taken as their values at $k_x = k_l$, which are

$$\begin{aligned} A(k_l) &= 1, \\ B(k_l) &= -2\left(\frac{c_{11}}{c_{12}}\right)^2 \frac{\rho_0}{\rho_1} \tan\theta_l. \end{aligned} \tag{7.49}$$

The elastic constants c_{11} and c_{12} for an isotropic solid are set out in Table 6.1, ρ_0 and ρ_1 are again the densities of the fluid and the solid, respectively, and the longitudinal critical angle is $\theta_l = \sin^{-1}(v_0/v_l)$, where v_0 is the velocity in the fluid and v_l is the longitudinal velocity in the solid. The results for Rayleigh waves can be adapted for lateral longitudinal waves by substituting $k_p \rightarrow k_l$, $\theta_p \rightarrow \theta_l$, and multiplying by the factor

$$M_l \equiv \frac{B(k_l)}{2\alpha_R} \frac{1}{\sqrt{(2\pi k_l)}} \frac{e^{i\pi/4}}{|x - x'|^{3/2}}. \tag{7.50}$$

Since k_l is a real number, whereas k_p is complex, the exponential decay of the Rayleigh wave goes out, to be replaced by the algebraic decay of 3/2 power in amplitude. Moreover, whereas the Rayleigh wave decay was determined directly by the fluid loading, the algebraic decay of the lateral longitudinal wave is independent of loading and is entirely a geometrical effect. Any remaining attenuation, α_l, is due to damping or scattering in the solid. With these substitutions, the response in the fluid just above the surface at x caused by excitation at x' and transmitted by lateral longitudinal waves in the solid is, from (7.34),

$$dp_l(x) = p_{inc}(x')\left\{B(k_l)\frac{1}{\surd(2\pi k_l)}\frac{e^{i\pi/4}}{|x-x'|^{3/2}}\exp[(ik_l-\alpha_l)\,|x-x'|]\right\}dx'. \tag{7.51}$$

Similarly, the contribution to the response in the microscope due to lateral longitudinal waves is, by analogy with (7.42),

$$V_l(z) = V_0 AKB(k_l)\frac{1}{k}\left\{\frac{(\surd\pi\lambda_0 f^3)^{1/2}}{(2\,|z|\tan\theta_l)^{3/2}(\lambda_0 D/n)^{3/4}}\right\}e^{i5\pi/4}\exp[i2kz\cos\theta_l]$$
$$\times\exp[2(\alpha_l\sin\theta_l-\alpha_0)z\sec\theta_l]. \tag{7.52}$$

The variable K is defined as in (7.43), but with $a_l = q\sin\theta_l$ in place of a_R in (7.44). The whole derivation of the contribution of lateral longitudinal waves to the reflectance function depends on the approximations of (7.48) and (7.49). These are valid when k_x is close to k_l, so that $|B(k_l)\kappa_l(k_x)/k_l| \ll 1$. In a microscope at negative defocus, this range of validity corresponds to (Chan and Bertoni 1991)

$$z \ll -\frac{2}{\pi}\left(\frac{c_{11}}{c_{12}}\right)^4\left(\frac{\rho_0}{\rho_1}\right)^2\frac{\lambda_0}{\cos\theta_l}. \tag{7.53}$$

To give three examples, this requirement would correspond to: for PMMA, $z \ll -3\lambda_0$; for aluminium, $z \ll -\lambda_0$; but for fused silica, $z \ll -77\lambda_0$. It is found that the lateral rays make an important contribution to $V(z)$ for polymers in which $v_l \geq v_0/\sin\theta_0$ (where θ_0 is the lens angle) and in which Rayleigh waves are not excited. In faster materials lateral longitudinal waes with their algebraic decay are still significant at larger values of $-z$ after the Rayleigh waves with their punishing exponential decay have expired. Lateral waves give rise to oscillations in $V(z)$ described by the same relationship as (7.7) and (7.47) but with the longitudinal critical angle. The periodicity is therefore

$$\Delta z_l = \frac{\lambda_0}{2(1-\cos\theta_l)}. \tag{7.54}$$

An experimental $V(z)$ curve measured on PMMA is given in Fig. 7.7. The oscillations are entirely due to lateral longitudinal waves. At 225 MHz, eqn (7.54) would predict a periodicity of 20 μm; this can be compared with the spacing of the oscillations in the measured data.

7.3 Tweedledum or Tweedledee?

Much of acoustic microscopy is concerned with—nay, obsessed by—leaky Rayleigh waves. This is because of their tremendous importance in

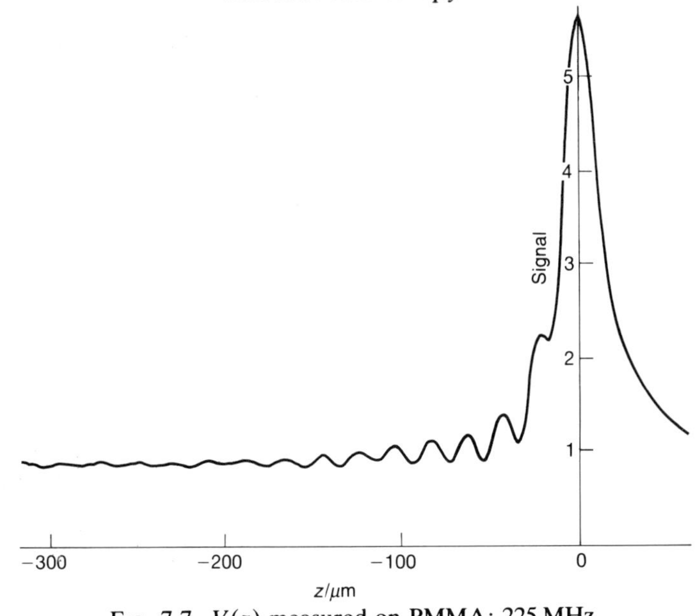

FIG. 7.7. $V(z)$ measured on PMMA; 225 MHz.

high modulus materials, including most metals and alloys, and almost all ceramics and semiconductors. But there are a few solids, such as polymers with intermediate modulus like PMMA, in which lateral longitudinal waves can play a role similar to that of Rayleigh waves. There are certain differences that arise because the lateral waves are a surface manifestation of a bulk wave, rather than a wave that is bound to the surface. Thus whereas in the presence of the fluid the Rayleigh wave velocity increases slightly, because it makes the surface feel stiffer than a free surface, the lateral longitudinal wave velocity is unchanged from the bulk velocity. Again, whereas the Rayleigh wave is subject to an exponential decay due to energy radiating in to the fluid, lateral waves are subject to algebraic decay. In two dimensions (which curiously enough is what is relevant for both spherical and cylindrical lenses), the algebraic decay of the amplitude has the form $r^{3/2}$. An exponential decay will always eventually lose against an algebraic decay, and so in a material in which both Rayleigh waves and lateral longitudinal waves are excited the $V(z)$ curve will show a periodicity associated with longitudinal waves at sufficiently large defocus. This periodicity will correspond to (7.7), but with the longitudinal critical angle θ_l in place of the Rayleigh angle θ_R. In materials such as PMMA in which Rayleigh waves are not excited at all in an acoustic microscope (because the Rayleigh angle would be too large or complex), good oscillations may be found in $V(z)$, exhibiting the longitudinal periodicity throughout the range of negative defocus. In lower modulus polymers (i.e. those in which the longitudinal velocity is lower than about 2500 ms^{-1}), and in most biological tissues except bones and teeth, neither lateral waves nor Rayleigh waves matter, and so many biologists and polymer scientists can be gloriously free from having to master these! But in most materials of high stiffness the interference between the Rayleigh rays and the specular rays is all important.

The Fourier theory and the ray model offer quite different insights into imaging theory. Both enable $V(z)$ to be calculated *ab initio*. Figure 7.8 shows $V(z)$ curves calculated from field theory and from ray theory, compared with a measured curve. In the Fourier model it is not necessary to know about Rayleigh waves or any other surface wave modes, because these are all contained implicitly in the reflectance function. Thus in any

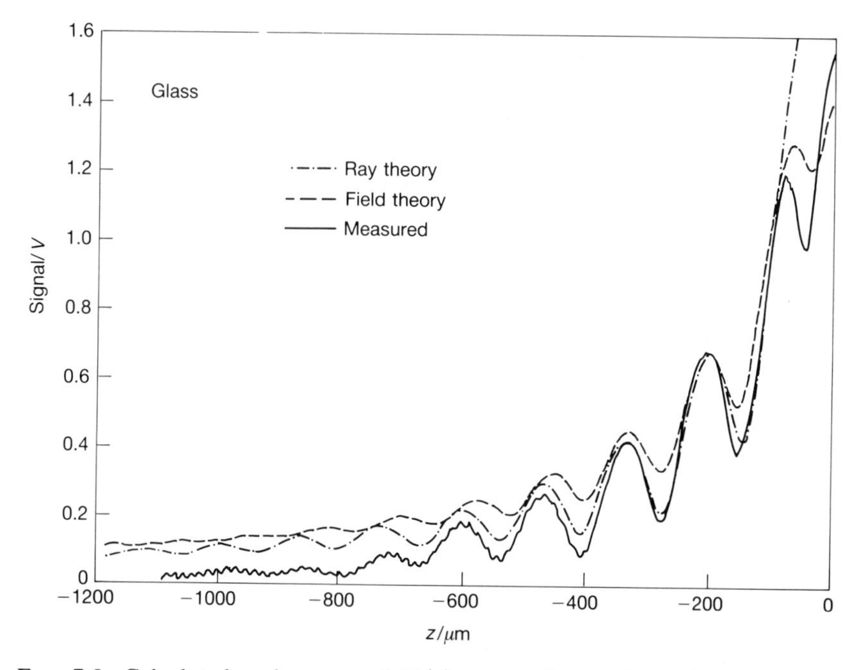

FIG. 7.8. Calculated and measured $V(z)$ curves for a glass microscope slide: ---- field theory; -·-·- ray theory; —— measured. The lens was of fused quartz, with focal length $q = 4$ mm, aperture radius $a_0 = 2.5$ mm (giving $N.A. = 0.625$), and length from transducer to lens surface $D = 60$ mm; the frequency was $f = 49$ MHz. The field theory calculations were made by computing the field in the back focal plane of the lens due to the transducer, truncating the field at the lens aperture and multiplying by the two-way transmission coefficient at the lens–water interface, and then performing the integration in eqn (7.14): the ray theory calculations were made using the equations in §7.2.2 (Chiznik 1991).

situation in which the reflectance function of a surface can be calculated (even for a complicated layered structure), and for which the pupil function of the lens is known, the complete $V(z)$ curve can immediately be computed without any approximations. In practice accurate measurement of the pupil function can be very difficult, but since it is constant for a given lens at a given frequency it is at least possible to make comparisons between $V(z)$ curves for different specimen parameters. It is invaluable to set up programs to enable $R(\theta)$ and hence $V(z)$ to be calculated using (6.90) or (6.94), and (7.14). These can give a great deal of insight into how the contrast can be expected to vary as a function of defocus and specimen parameters.

The ray model requires many more assumptions about the contrast mechanism. It assumes that the fluid loading is light, so that the fluid causes only a small perturbation to the Rayleigh wave propagation. And any other interactions with the fluid, such as lateral wave excitation, are left out unless they are explicitly included. But when Rayleigh waves do dominate the contrast, as is the case with many materials, then the ray model gives a very useful account of what is going on, and one that is so simple that it is possible to visualize the Rayleigh waves in a way that soon becomes intuitive. In a sense the very weakness of the ray model is its strength: by restricting itself to the Rayleigh wave and the normal wave it emphasizes those very waves that, in the cases where it is applicable, do all the work. In calculating the contrast when cracks or boundaries are present (Chapter 12), a hybrid of both wave theory and ray theory is needed; the contrast from resonant structures in the surface can also be calculated (Somekh 1987).

If the first role of imaging theory is to enable the microscopist to understand the images that are seen, the second role is to show how to interpret the contrast quantitatively. By expressing the reflectance function in terms of the inverse Fourier transform of $V(z)$, it is, in principle, possible to use the Fourier theory to deduce the reflectance function from a measured $V(z)$ curve. But now the strength of the Fourier theory is its weakness. Precisely because it gives you everything, the Fourier theory is not selective. If, as is often the case, the interaction with the specimen is concentrated around the Rayleigh angle, then a ray theory that explicitly makes that assumption will, for that very reason, select the information that is most reliable. This will influence the choice of method for quantitative measurements in acoustic microscopy.

8

Experimental elastic microanalysis

> J. J. Thompson once visited the laboratory at Terling—where lenses and mirrors for optical studies were sometimes held in the desired position with a dab of sealing wax—and said 'It looks like everything is done with sealing wax and string!' What mattered to Rayleigh was not the appearance of the apparatus, but rather its calibration, in which he took great pains. (J. N. Howard 1985)

Anyone who has successfully used a microscope to image properties to which it is sensitive will sooner or later find himself wanting to be able to measure those properties with the spatial resolution that that microscope affords. Since an acoustic microscope images the elastic properties of a specimen, it must be possible to use it to measure elastic properties both as a measurement technique in its own right and also in order to interpret quantitatively the contrast in images. It emerged from contrast theory that the form of $V(z)$ could be calculated from the reflectance function of a specimen, and also that the periodicity and decay of oscillations in $V(z)$ can be directly related to the velocity and attenuation of Rayleigh waves. Both of these observations can be inverted in order to deduce elastic properties from measured $V(z)$.

8.1 Measurement of the reflectance function

8.1.1 *Fourier inversion*

The variation of signal with defocus may be written

$$V(z) = V_0 \int_0^{\theta_0} P(\theta) R(\theta) \mathrm{e}^{-\mathrm{i}2kz\cos\theta} \sin\theta \cos\theta \, \mathrm{d}\theta \tag{8.1}$$

where θ_0 is the maximum aperture of the lens and the other functions and variables are as in (7.14). Putting

$$u \equiv kz,$$

and

$$t \equiv \frac{1}{\pi} \cos\theta \tag{8.2}$$

as in (7.15) it may be rewritten

$$V(u) = V_0 \int_{1/\pi}^{\cos\theta_0/\pi} P_t(t) R_t(t)\, e^{i2\pi u t} t \, dt. \tag{8.3}$$

Since P_t is zero outside the limits of integration, the limits may be taken as $\pm\infty$, yielding a Fourier transform relationship. The Fourier transform may be inverted to yield

$$R_t(t) = \frac{1}{P_t(t)t} \int_{-\infty}^{\infty} \frac{V(u)}{V_0}\, e^{i2\pi u t}\, du. \tag{8.4}$$

Thus, by measuring $V(u)$ and inverse Fourier transforming it, the reflectance function may be deduced. Four practical constraints are immediately apparent from the theoretical formation.

1. Equation (8.4) is valid only for $1 \geq t > \cos\theta_0$; outside this range $P_t(t)$ in the denominator is zero. No information about the reflectance function can be obtained outside the aperture angle of the lens.
2. The complete Fourier transform requires measurement of $V(u)$ over an infinite range. Even though $V(u)$ may be small outside the range that can be measured, the truncation will introduce errors.
3. The inversion procedure is most straightforward when attenuation in the coupling fluid is ignored. This may present problems in high-frequency applications.
4. The inverse Fourier transform operation must be performed using complex variables. This means that both the amplitude and the phase of $V(u)$ must be measured.

Any method based on inversion of $V(z)$ actually measures the product $P(\theta)R(\theta)$, so the first step must be to measure $V(z)$ for a material with a well known and well behaved reflectance function. Suitable choices are lead or PTFE (polytetrafluoroethylene, alias Teflon), whose reflectance functions are plotted in Fig. 8.1. A mathematically ideal reflector would have a reflectance function of unity, with no phase change, for all angles of incidence. The material that approximates most closely to this ideal is lead. Its Rayleigh velocity is too slow to allow Rayleigh waves to be excited by waves in water. Even the longitudinal critical angle of 43° is beyond the angle where the lens needs to be accurately characterized in order to be able to detect the more interesting phenomena in the reflectance functions of many materials. The longitudinal impedance of polycrystalline lead is 24.5 Mrayl; the modulus of the reflection coefficient is about 0.92 up to the longitudinal critical angle and almost unity above it; the phase is almost exactly zero over the whole range of

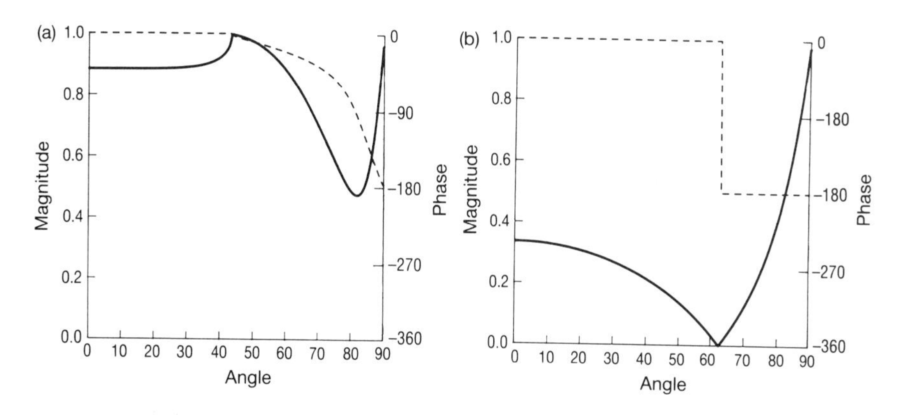

FIG. 8.1. Reflectance functions for waves in water incident on: (a) lead, $v_l = 2160\ \mathrm{m\,s^{-1}}$, $v_s = 700\ \mathrm{m\,s^{-1}}$, $\rho = 11343\ \mathrm{kg\,m^{-3}}$; (b) PTFE, $v_l = 1390\ \mathrm{m\,s^{-1}}$, $v_s = 700\ \mathrm{m\,s^{-1}}$, $\rho = 2140\ \mathrm{kg\,m^{-3}}$: —— magnitude; ---- phase. The phase shift of $-\pi$ in (b) corresponds to the amplitude becoming negative.

practical interest. But PTFE shares many of the desirable properties of lead, and it is much easier to prepare and maintain a flat and clean surface on PTFE than on lead. The phase of its reflectance function is constant over the whole range of practical interest. The only snag is that its reflectance function is more vulnerable to variations in elastic constants. If high impedance is essential, then a third possibility that does not suffer from the corrosion to which lead is vulnerable is gold (a flat surface can be prepared instantly by pressing an optically flat sapphire rod against it). By determining $P(\theta)R(\theta)$ from the measured $V(z)$ for whatever reference material is chosen, and dividing by $R(\theta)$ calculated from the known elastic constants, the pupil function $P(\theta)$ can be determined. When $P(\theta)R(\theta)$ is subsequently determined for unknown specimens, the result can be divided by the pupil function to yield the reflectance function for that material.

The need to measure $V(u)$ as a complex-valued quantity can be met by using the accurate amplitude and phase measurement system described in Chapter 5. A system like that has been used to demonstrate material characterization by inversion of $V(z)$ (Liang *et al.* 1985*b*). The frequency was about 10 MHz, so that attenuation in the water was negligible and mechanical stability could be readily achieved to a very small fraction of a wavelength. A spherical transducer was used instead of a combination of a planar transducer and a lens; this removes the $\cos\theta$ term in (8.1), but the principle of the inversion formulation remains the same. Figure 8.2(a) shows the amplitude and phase of an experimental $V(z)$ curve for fused

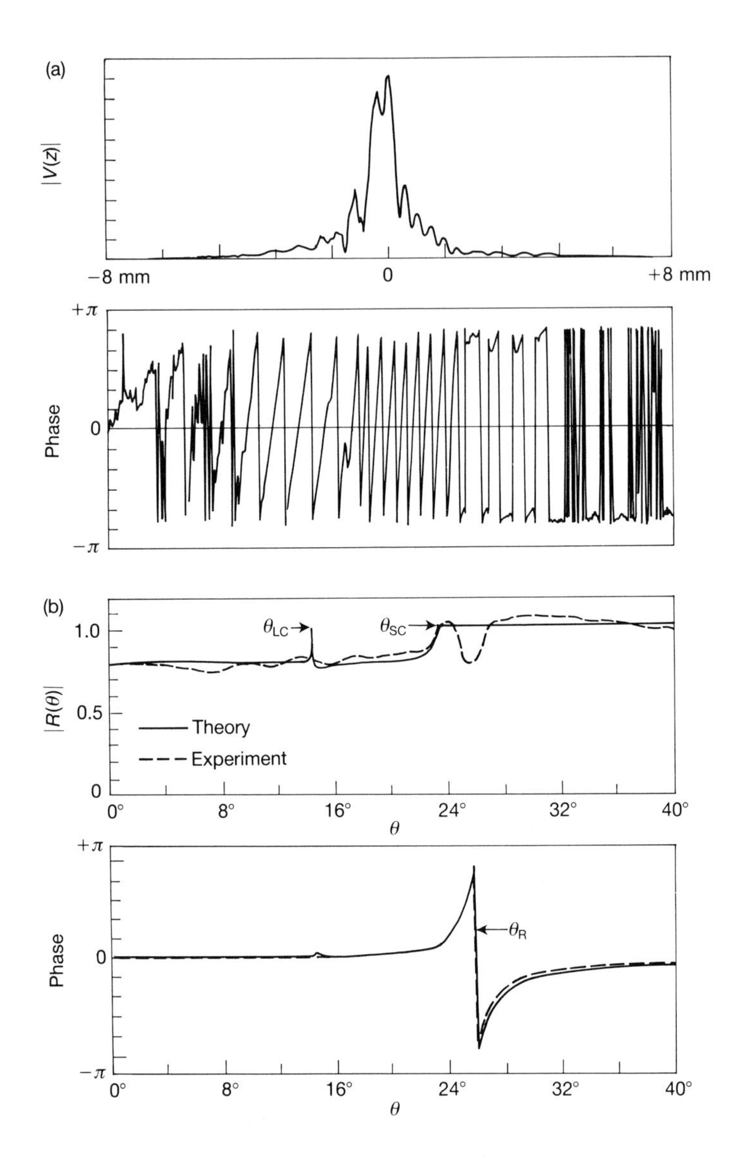

FIG. 8.2. Fused silica. (a) Magnitude and phase of an experimental $V(z)$ using a curved transducer, $\theta_0 = 45°$, 10.17 MHz. (b) Magnitude and phase of reflectance function: ---- deduced from (a) via (8.4), —— calculated from (6.90) with $v_0 = 1486\ \text{m s}^{-1}$, $v_s = 5960\ \text{m s}^{-1}$, $v_l = 3760\ \text{m s}^{-1}$, $\rho_l = 2200\ \text{kg m}^{-3}$ (Liang *et al.* 1985*b*).

silica. The phase is plotted modulo 2π with the linear term $2kz$ removed; nevertheless, fairly rapid phase variations remain, perhaps due to the focused transducer geometry (Bertoni and Somekh 1985). Figure 8.2(b) shows the modulus and phase of the reflectance function deduced using eqn (8.4), together with theoretical curves calculated from the elastic constants using (6.94). The overall agreement is impressive.

The feature in the theoretical curves that is most precisely reproduced in the experimental results is the phase change of approximately 2π at the Rayleigh angle. This enables the Rayleigh angle, and hence the Rayleigh velocity, to be measured. Here θ_R is 25.85°; taking v_0 as 1486 m s^{-1} this gives $v_R = 3408$ m s^{-1}, which compares amazingly with a theoretical value of 3409 m s^{-1} (Table 6.3). The rise to unity in the modulus of the reflectance function also appears to be reproduced, enabling a shear critical angle θ_{SC} of 23.5°, and hence a shear velocity of 3727 m s^{-1}, to be deduced. In some cases, for example aluminium, features can be seen at the longitudinal critical angle θ_{LC}, but in the experimental measurements of fused silica in Fig. 8.2(b) neither the kink in the phase nor the rise to unity in the modulus is reproduced from the experimental data. The difference between silica and aluminium is related to inequality (7.53).

The most marked deviation from the theoretical curves occurs in the modulus at the Rayleigh angle, where there is a pronounced dip in the curve deduced from the measured $V(z)$. There are various reasons why a dip may occur in the modulus of the reflectance function at the Rayleigh angle. If the material is lossy, there may be a dip associated with attenuation of the Rayleigh wave (Briggs *et al.* 1982); if there is a surface layer with shear velocity slower than the substrate there may be a dip associated with the excitation of a leaky pseudo-Sezawa wave (this will be discussed in §10.2); if it is anisotropic, there may be a dip due to phase cancellation in different directions (this will be discussed in §11.5). More than one of these effects may be combined. These effects are all genuinely present in the reflectance function of the material itself. However, there is a further reason why a dip may be present in a reflectance function calculated from a Fourier inversion procedure, and that is because of the limited extent of the scan in z that is available. The true reflectance function becomes convolved with the transform of the window, and this leads to phase cancellation, and therefore a dip, where the convolution occurs at a region of rapid phase change in $R(t)$. Since $V(u)$ is multiplied by a rectangular function, in the transform $R(t)$ becomes convolved with a sinc function. Specifically, if the range of u is u_1, then the reconstructed reflectance function will be

$$R_t(t) = \frac{\{P_t(t)R_t(t)t\} \otimes \{u_1 \operatorname{sinc}(\pi u_1 t)\}}{P_t(t)t} \tag{8.5}$$

where $\otimes$ denotes the convolution operator and $\text{sinc}(x) = \sin(x)/x$ (cf. §5.1). A corresponding effect will apply to the initial determination of the pupil function, with P and R exchanged in (8.4). It is most serious, however, when there are rapidly changing features in the function to be reconstructed. The effect of the convolution is to smear out the reconstruction in the t-domain, and also to generate side ripples. The width between zeros in the sinc function is

$$\Delta t = \frac{2}{u_1} \tag{8.6}$$

or, in the θ-domain,

$$\Delta\theta = \frac{2\pi}{u_1 \sin\theta}. \tag{8.7}$$

At the Rayleigh angle in Fig. 8.2(b), (8.7) implies a broadening of $\Delta\theta \approx 1.2°$. The phase cancellation caused by the convolution process where the phase is changing rapidly causes the dip that appears in the measured modulus. The broadening is greater the smaller the value of θ at which the rapid phase change occurs. Thus fast materials, such as ceramics, will show even greater pathologies in the reconstruction of $R(\theta)$ around the Rayleigh angle. Oscillations associated with the sinc function can be removed by applying a smooth apodization function to the $V(t)$ data, with a corresponding degradation of the resolution in the θ domain. The problems associated with the limited extent of z become more severe as the frequency is increased, because of the f^2 increase in attenuation in water and the consequent reduction in the working distance of lenses.

In order to deduce the product $P(\theta)R(\theta)$ in these experiments from a simple inversion of $V(z)$, it is necessary to measure it as a complex-valued function, i.e. with both amplitude and phase information. This can be done, though the requirements for thermal, mechanical, and electronic stability should not be underestimated. If the phase information is not available, then it must be reconstructed.

8.1.2 *Phase reconstruction*

In most acoustic microscopes the video signal is measured after detection by an envelope detector, so that the phase information has been lost. In order to attempt to reconstruct $P(\theta)R(\theta)$ from a modulus only $V(z)$ it is necessary to employ a phase retrieval algorithm (Fright *et al.* 1989). The method is based on the Gerchberg–Saxton algorithm (Gerchberg and Saxton 1972; Bates and McDonnell 1986). With the change of variables of (8.2) and with

$$Q(t) = R(t)P(t)t, \tag{8.8}$$

the expression for $V(z)$ (8.3) becomes (cf. (7.18))

$$V(u) = \mathbb{F}\{Q(t)\}. \tag{8.9}$$

$\mathbb{F}$ and $\mathbb{F}^{-1}$ representing the Fourier transform and its inverse.

The algorithm proceeds as follows. For the mth loop of the iteration, the best estimate of $Q(t)$ is Fourier transformed,

$$V'_m(u) = \mathbb{F}\{Q_{m-1}(t)\}. \tag{8.10}$$

This is forced to have the measured amplitude $|V(u)|$, while retaining the phase given by the algorithm

$$V_m(u) = |V(u)|\, \mathrm{e}^{\mathrm{i}\,\mathrm{phase}\{V'_m(u)\}}. \tag{8.11}$$

This is inverse transformed to give a new estimate of $Q(t)$

$$Q'_m(t) = \mathbb{F}^{-1}\{V_m(u)\}. \tag{8.12}$$

Finally, $Q_m(t)$ is truncated outside its allowed range by setting

$$\begin{aligned} Q_m(t) &= Q'_m(t), \qquad 0 < t \leq \frac{1}{\pi}; \\ &= 0, \qquad \text{otherwise}. \end{aligned} \tag{8.13}$$

In practice the allowed range of t may be further restricted by the pupil function of the lens. The new value of $Q(t)$ is used for the next iteration through eqns (8.10)–(8.13), and so on until satisfactory convergence is obtained.

This measured $V(z)$ may be improved by first tidying up the raw data. The autocorrelation function of $Q(t)$ is

$$QQ(t) \equiv \int_0^{t'} Q(\alpha)Q(\alpha + t)\, \mathrm{d}\alpha \tag{8.14}$$

where the limits of integration represent the range of t outside which $Q(t)$ is zero. Hence the allowed range of $QQ(t)$ is $-t' \leq t \leq t'$. Because $Q(t)$ and $V(u)$ are a Fourier transform pair,

$$QQ(t) = \mathbf{F}\{|V(u)|^2\}. \tag{8.15}$$

Thus the data can be preprocessed by Fourier transforming $|V(u)|^2$, applying a window corresponding to the extent of the pupil function of the lens, and then inverse transforming to obtain a filtered $|V(u)|$, which can then be used as the data for the Gerchberg–Saxton algorithm.

Steps in the reconstruction of $P(\theta)R(\theta)$ for duraluminium alloy are illustrated in Fig. 8.3(a). $V(z)$ was measured at 320 MHz with a range of z of $\pm 240\,\mu m$ relative to focus. Reconstructed curves of $P(\theta)R(\theta)$ after 1, 3, 10 and 30 iterations are shown, plotted as functions of θ for the sake of familiarity. For this material, with water as the coupling fluid, the Rayleigh angle is approximately 31°, and an incipient feature is present at that angle even after the first cycle. After three cycles, a phase change of 2π has developed, accompanied by a dip in the modulus. The curves after 10 and 30 iterations show little further change and thereafter

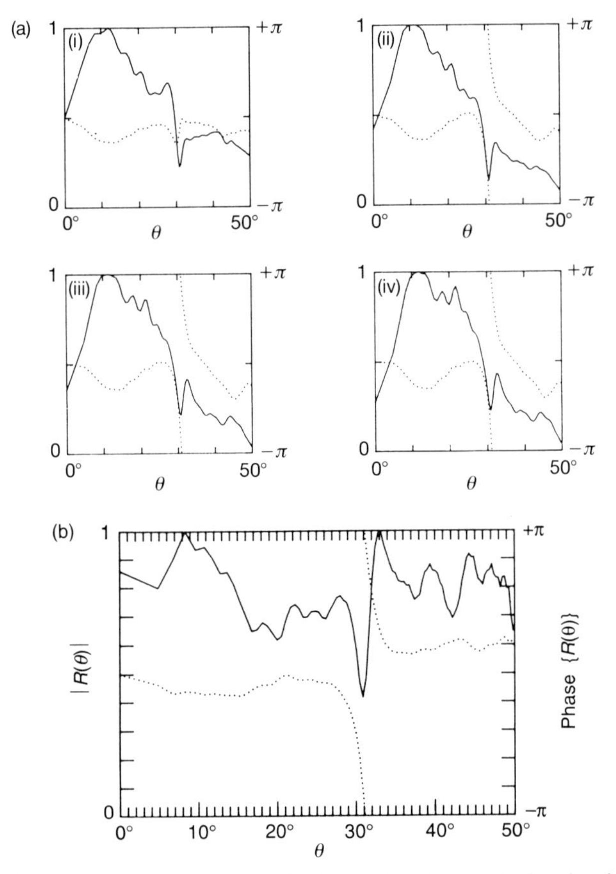

Fig. 8.3. Reconstruction of the reflectance function of duraluminium from magnitude only $V(z)$ data, measured at 320 MHz. (a) Steps in the reconstruction of $P(\theta)R(\theta)$ after: (i) 1; (ii) 3; (iii) 10; and (iv) 30 iterations of the phase retrieval algorithm. (b) Reconstructed $R(\theta)$. (Fright *et al.* 1989.)

the curves remain almost constant at this level of presentation. 100 iterations were used for the results given below.

A pupil function $P(\theta)$ was measured for the microscope using the same processing method. PTFE was used as the reference material. $V(z)$ was measured for a specimen of PTFE, and was then processed, as above, to yield $P(\theta)R(\theta)$. This was divided by the calculated $R(\theta)$ for PTFE to give a calibrated pupil function. Finally, the reconstructed $P(\theta)R(\theta)$ for the duraluminium specimen was divided by this calibrated $P(\theta)$ to give a reflectance function $R(\theta)$ for duraluminium. The result is presented in Fig. 8.3(b). The dominant feature in the result is the phase change of 2π associated with the Rayleigh angle. The angle at which this appears is 31°, in good agreement with the angle of 30.6° deduced from the data in Tables 3.2 and 6.3. There is also a dip in the modulus at this angle; once again this is due to finite range in z convolved with a region of rapid phase change.

If $R(\theta)$ in Fig. 8.3 is compared with the calculated reflectance for a similar material, it is apparent that the phase change at the Rayleigh angle is the feature that is reproduced by far the most faithfully. This must be because this feature corresponds to the strongest interaction of the acoustic waves in the acoustic microscope with the specimen itself.

8.2 Ray methods

The most prominent feature in most $V(z)$ curves, after the central maximum at focus, is the series of oscillations at negative defocus associated with Rayleigh wave excitation. It is perhaps therefore not surprising that the most accurate information in the reconstructions of $R(\theta)$ concerns the Rayleigh velocity. The period of the Rayleigh oscillations is

$$\Delta z = \frac{\lambda_0}{2(1 - \cos\theta_R)} \tag{8.16}$$

with λ_0 the water wavelength. Using Snell's law, this relationship may be inverted to give the Rayleigh velocity in terms of Δz, the frequency f, and the velocity in the coupling fluid v_0,

$$\boxed{v_R = v_0\left\{1 - \left(1 - \frac{v_0}{2f\,\Delta z}\right)^2\right\}^{-1/2}}\,. \tag{8.17}$$

Hence, by measuring the period of the Rayleigh oscillations, the Rayleigh velocity may be deduced directly. Likewise, the exponential decay of the Rayleigh oscillations is often of the form (Kushibiki *et al.* 1982)

$$\exp(\alpha z) = \exp\{2(\alpha_0 \sec \theta_R - \alpha_R \tan \theta_R)z\}, \tag{8.18}$$

where α_0 is the attenuation in the fluid. This can be immediately inverted to give the normalized Rayleigh wave attenuation

$$\boxed{\alpha_N \equiv \frac{\alpha_R \lambda_R}{2\pi} = \frac{(\alpha_0 \sec \theta_R - \alpha/2)\lambda_R}{2\pi \tan \theta_R}}. \tag{8.19}$$

Normalizing the attenuation in this way enables it to be used as an imaginary component of a factor multiplying a real wavenumber in an expression for wave propagation.

The theoretical basis of the method assumes a ray model in which the transducer signal is made up of two components (§7.2). One (denoted by subscript R) is due to the ray that is incident at the Rayleigh angle, propagates in the surface as a Rayleigh wave, and is then re-radiated symmetrically to the incident ray. The other contribution (denoted by subscript G) is that due to everything else, i.e. to rays that bounce off the surface geometrically. The model can be extended to allow for interactions with other waves that propagate along the surface, along the lines of §7.2.3. The transducer signal is thus made up of two components

$$\mathbf{V} = \mathbf{V}_G + \mathbf{V}_R. \tag{8.20}$$

Throughout this section all V are functions of z; they have been printed bold in (8.20) to emphasize that they contain phase information. They are summed as complex quantities, i.e. with respect to both their amplitude and phase. But in the usual experimental implementation, where $|V|$ is measured through a diode detector, the phase information of $\mathbf{V}$ is not available. Therefore, if the system has been calibrated to give square law detection, the measured signal may be represented as

$$|V|^2 = |V_G|^2 + |V_R|^2 + 2\,|V_G|\,|V_R| \cos \phi \tag{8.21}$$

where ϕ is the phase angle between V_G and V_R, and the z dependence of all the terms is implicit.

As with the methods of analysis based on Fourier inversion, it is necessary to characterize the lens response. This is again performed using a specimen in which waves are not excited in the surface. Lead is again suitable, and the measured curve is denoted $|V_L|$. $|V_L|$ approximates closely to $|V_G|$. Exactly how the analysis proceeds depends on lens geometry.

8.2.1 *Line-focus-beam technique*

Many materials whose elastic properties are of interest are anisotropic, so the surface wave velocity depends on the direction of propagation. In order to be able to make measurements in one direction at a time, a lens with a cylindrical surface is used (Kushibiki *et al.* 1981*b*) instead of the spherical surface needed for imaging. The cylindrical lens produces a so-called line-focus-beam, parallel to the axis of the cylindrical surface. Surface waves are excited in a direction perpendicular to the line-focus. The principle is illustrated in Fig. 8.4(a), and the dimensions of a lens for use at 225 MHz are given in Fig. 8.4(b). Cylindrical lens surfaces can be ground extremely accurately.

In the line-focus beam analysis method, two fundamental assumptions are made regarding (8.21). First, it is assumed that the Rayleigh contribution is a perturbation on the total signal, i.e. that

$$|V_R| \ll |V_G| \, . \tag{8.22}$$

This, together with the approximation $|V_L| \approx |V_G|$, means that the lead curve may be subtracted from a curve for the material being studied to give the last term on the right of (8.21), which is the term describing the interference between the Rayleigh ray and the geometrical contribution. To see this, (8.21) may be rewritten

$$|V| = |V_G| \, \{1 + (|V_R|/|V_G|)^2 + 2(|V_R|/|V_G|) \cos \phi\}^{1/2}. \tag{8.23}$$

Since $|V_R| \ll |V_G|$, the term in $|V_R|^2$ may be dropped, and a binomial expansion may be used, to give

$$|V| = |V_G| \, (1 + (|V_R|/|V_G|) \cos \phi + \ldots). \tag{8.24}$$

Since $|V_L| \approx |V_G|$, to a good approximation

$$|V| - |V_L| = |V_R| \cos \phi. \tag{8.25}$$

The second assumption that is made is that the z dependence of the phase of each of the terms V_G and V_R is linear. For V_R, this is quite accurate, the phase being

$$\phi_1 = 2kz \cos \phi_R + \pi. \tag{8.26}$$

For V_G, the situation is more complicated; the deviation from linearity is described by the $\exp\{iX\}$ term in (7.39). The phase shift for each ray will

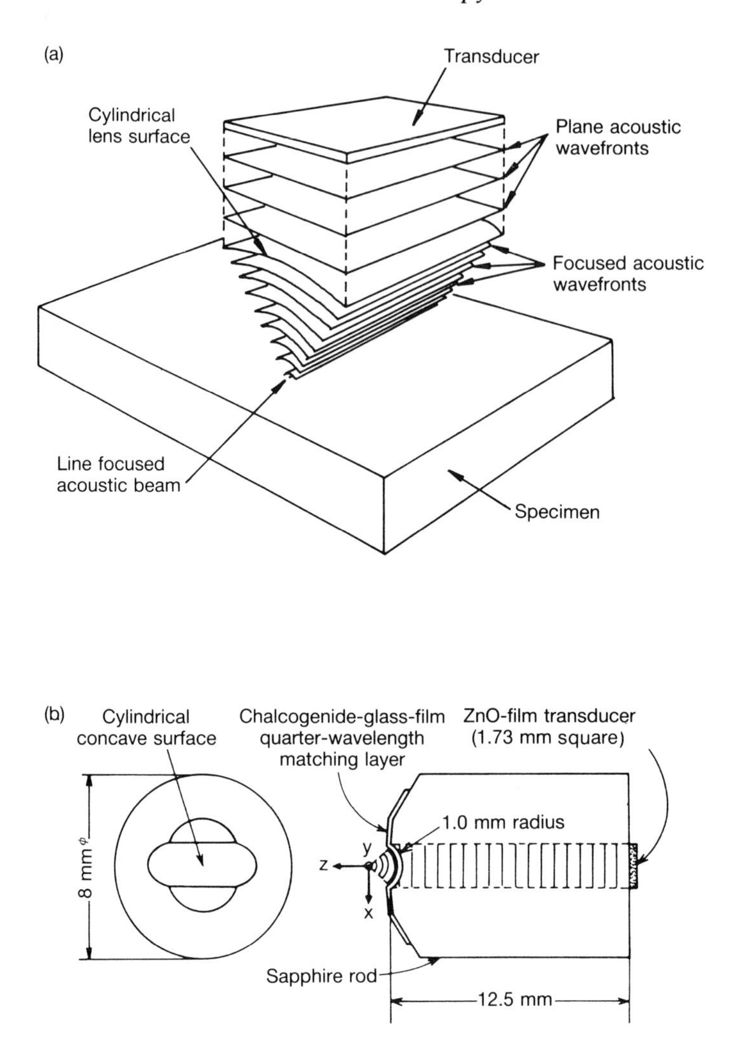

FIG. 8.4. (a) Wavefronts in a line-focus-beam microscope; (b) structure of a line-focus-beam lens with dimensions for 225 MHz (Kushibiki and Chubachi 1985).

depend on its angle of incidence on the solid. Because of phase cancellation, the contribution as a function of angle will have a form similar to a sinc function (depending in detail on the pupil function), with the width of the central maximum decreasing with increasing z. The phase of V_G will be an average of the contributions over all angles. As the defocus is increased so the phase becomes dominated by contributions closely parallel to the z-axis and, in this limit,

$$\phi_0 = -2kz. \tag{8.27}$$

The z dependence of ϕ may thus be written

$$\phi = -2kz(1 - \cos\theta_R) + \pi. \tag{8.28}$$

If the attenuation is uniform, $|V_R|$ decreases with negative z as

$$|V_R| = |V_{R0}|\,\mathrm{e}^{-2\alpha z}. \tag{8.29}$$

In the ray model, the path as a Rayleigh wave is $2z \tan\theta_R$. If α_w is the attenuation in the fluid and α_R is the Rayleigh wave attenuation, then

$$\alpha = \alpha_R \tan\theta_R - \alpha_w \sec\theta_R. \tag{8.30}$$

Fourier analysis is used to find the velocity and attenuation of surface waves. Let the range in z over which data is available be ζ. If there were no attenuation, then by the convolution theorem the Fourier transform $F(\xi)$ would be a sinc function centred at a spatial frequency

$$\xi_0 = \frac{2\pi}{\Delta z} = 2k\,(1 - \cos\theta_R). \tag{8.31}$$

The transform would have amplitude V_{R0}/π at $\xi = \xi_0$, and zero at $\xi = \xi_0 \pm 2\pi/\zeta$. When attenuation is present, however, these zeros become partially filled in. The voltage spectrum is now

$$|F(\xi)| = \left|\frac{V_{R0}}{2\pi}\int_{z_0-\zeta}^{z_0} \mathrm{e}^{\alpha z}\mathrm{e}^{\{\mathrm{i}2kz(1-\cos\theta_R)+\pi\}}\mathrm{e}^{-\mathrm{i}\zeta z}\,\mathrm{d}z\right|. \tag{8.32}$$

At $\xi = \xi_0$ this has the value

$$|F(\xi_0)| = \frac{V_{R0}}{\pi}\frac{\sinh(\alpha\zeta/2)}{\alpha} \tag{8.33}$$

and, at $\xi = \xi_0 \pm 2\pi/\zeta$ where the zeros in a sinc function would have been,

$$|F(\xi_0 \pm 2\pi/\zeta)| = \frac{V_{R0}}{\pi}\frac{\sinh(\alpha\zeta/2)}{\{(2\pi/\zeta)^2 + \alpha^2\}^{1/2}}. \tag{8.34}$$

The decay in the Rayleigh oscillations in $V(z)$ is therefore found to be

$$\alpha = \frac{2\pi}{\zeta}\frac{|F(\xi_0 \pm 2\pi/\zeta)|}{\{|F(\xi_0)|^2 - |F(\xi \pm 2\pi/\zeta)|^2\}^{1/2}}. \tag{8.35}$$

Hence the Rayleigh velocity and normalized attentuation can be calculated;

$$v_R = v_0\left\{1 - \left(1 - \frac{v_0\xi_0}{4\pi f}\right)^2\right\}^{-1/2}, \tag{8.36}$$

$$\alpha_N = \frac{\alpha \cos\theta_R + 2\alpha_0}{2k_R \sin\theta_R}. \tag{8.37}$$

Accurate measurements using the line-focus-beam technique require careful alignment of the system. If there is a misalignment δ in the parallelism between the axis of the cylindrical lens and the surface of the specimen, then because of phase cancellation the signal will be multiplied by a factor $\mathrm{sinc}(2l \tan\delta/\lambda_0)$, where l is the length of the line-focus-beam and λ_0 is the wavelength in water. For the dimensions in Fig. 8.4(b), at 225 MHz, complete phase cancellation would occur at a misalignment of 0.1°, so alignment much better than that is necessary. As with an imaging microscope, the specimen must be mounted on a levelling stage or goniometer. In order to take measurements in different propagation directions, the tilting stage must in turn be mounted on a rotating stage. An additional goniometer is therefore needed for the lens mount, in order to adjust the alignment of the lens relative to the axis of the rotation stage.

To set up a specimen, it is mounted on the stage with a suitable region under the lens and the goniometer axes parallel and perpendicular, respectively, to the axis of the line-focus-beam lens. After coupling fluid has been added and a signal obtained, z is adjusted for maximum signal, and the specimen goniometer is adjusted about the axis perpendicular to the line-focus-beam (taking care to find the maximum of the central peak in the sinc function, and not a side-lobe). The goniometer setting is noted, the stage rotated through 180°, and the maximum is again found. If this is at a different setting from the previous one then the specimen goniometer should be set halfway between the two, and the lens-mounting goniometer adjusted to find the maximum again. The specimen stage can then be rotated through 90° and a similar process repeated. A small number of careful iterations should suffice to achieve near perfect alignment provided the specimen is flat. Once the lens goniometer has been set it should not need adjusting unless the lens is changed. A thermocouple is inserted in the coupling water as close as possible to the active surface of the lens. The water droplet should be large, because cooling by evaporation from the surface can cause a temperature gradient. Provided a lead response curve, V_L, has been previously obtained for that lens, and the response of the electronic circuit has been calibrated, data for the specimen can now be acquired and analysed.

The measured $V(z)$ data may be analysed by the following procedure (Kushibiki and Chubachi 1985). The steps are illustrated for fused quartz in Fig. 8.5.

1. Convert the data to a linear scale (Fig. 8.5(a)).
2. Filter out any fast ripple of period $\lambda_w/2$ in $V(z)$, due to interference with internal reverberations in the lens (Fig. 8.5(b)). This may be achieved most simply by convolving with a rectangular function of length $\lambda_w/2$. This is known as a moving average filter; it is equivalent to a sinc filter in the Fourier domain, but is computationally somewhat more efficient. Because of its period the ripple removed at this stage is sometimes called water ripple.
3. Subtract $|V_L|$ (Fig. 8.5(c)) to yield $|V'| = |V| - |V_L|$.
4. Estimate the error in the various approximations implicit in (8.24) by applying a low-pass filter to $|V'|$, so as to remove components at and above the spatial frequency of the Rayleigh oscillations. For example a guess can be made of the period (an FFT can be used for this) and $|V'|$ can be convolved with a rectangular function of that length (i.e. another moving average filter). The resulting curve is $|\Delta V_L|$.
5. Calculate a better value of $|V''| = |V'| - |\Delta V_L|$ (Fig. 8.5(d)).
6. The record is converted to a form suitable for Fourier analysis by: (a) reducing the number of points over the measured range of z to about 15–30 per wavelength, depending on the computing power available; and (b) extending the record length to perhaps 30–50 times its original length with dummy points of value zero at each end, to give adequate resolution in the spatial frequency domain. An FFT may then be performed (Fig. 8.5(e)).
7. For enhanced accuracy, the spatial frequency ξ'' corresponding to the oscillations of interest in $|V''|$ may be measured from this first FFT, and a complex $\mathbf{V}''$ may be generated as $\mathbf{V}'' = |V''|\, e^{i\xi'' z}$. A second FFT may be performed on $\mathbf{V}''$ to give a final voltage spectrum $|F(k)|$ (Fig. 8.5(f)).
8. Using interpolation methods, the position of the peak may be found, and also $F(\xi_0)$ and $F(\xi_0 \pm 2\pi/\xi_0)$.
9. Hence v_R and α_R may be calculated from (8.36) and (8.37), using values for v_0 and α_0 at the recorded temperature from (3.11) and Tables 3.2 and 3.3.

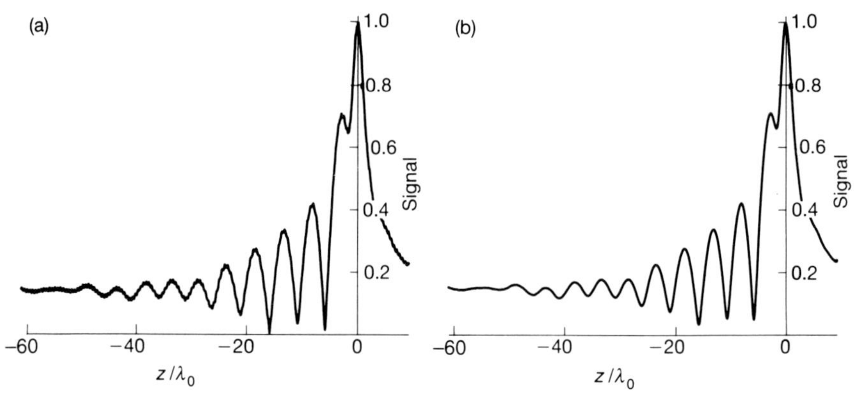

(a)
1.0
0.8
0.6
0.4
0.2
Signal
−60
−40
−20
0
z / λ0
(b)
1.0
0.8
0.6
0.4
0.2
Signal
−60
−40
−20
0
z / λ0

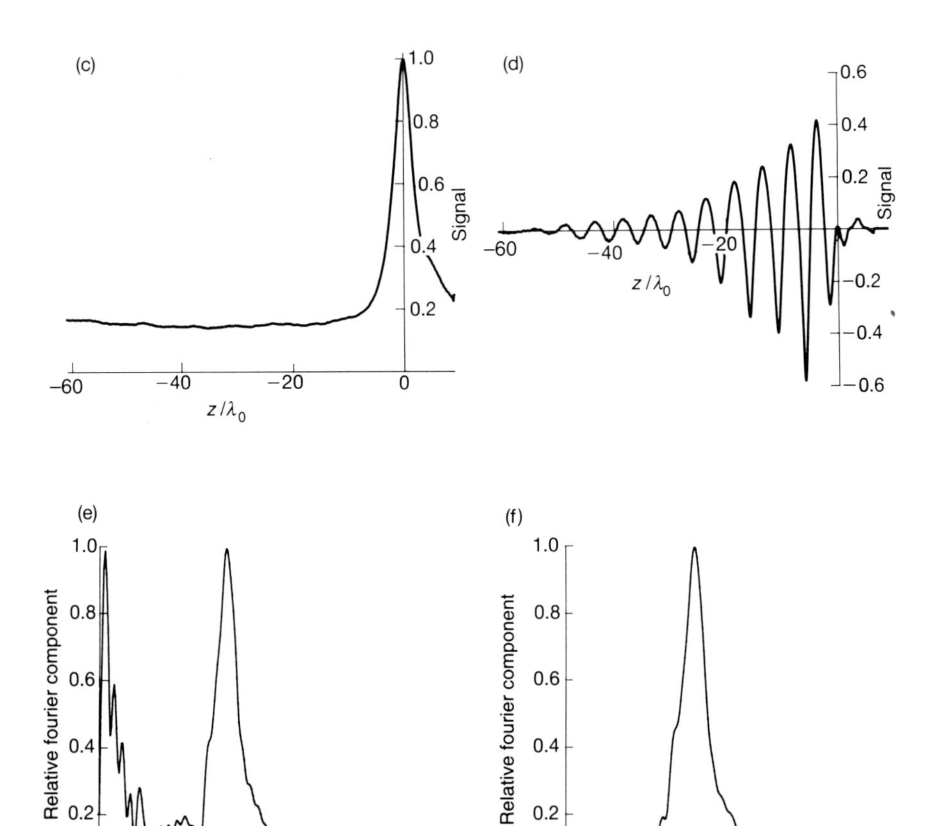

(c)
1.0
0.8
0.6
0.4
0.2
Signal
−60
−40
−20
0
z / λ0
(d)
0.6
0.4
0.2
−0.2
−0.4
−0.6
Signal
−60
−40
−20
z / λ0
(e)
1.0
0.8
0.6
0.4
0.2
0.0
Relative fourier component
0
250
500
750
1000
1250
Relative spatial frequency
(f)
1.0
0.8
0.6
0.4
0.2
0.0
Relative fourier component
0
250
500
750
1000
1250
Relative spatial frequency

10. Finally, there may be a small systematic error associated with the lens used. The origin of this is not fully understood, but it may depend on how components around the Rayleigh angle are weighted by the pupil function. Each lens may be calibrated on a well defined reference material to correct for such systematic errors.

The velocity measured in this way is the fluid-loaded velocity. The fluid may be thought of as slightly stiffening the surface, and therefore increasing the Rayleigh velocity a little (for example, the water-loaded surface wave velocity of GaAs{100} is about $7 \mathrm{~m~s}^{-1}$ faster than the free surface velocity). The attenuation may include a combination of three factors: anelastic damping (e.g. dislocation damping); elastic scattering processes (e.g. grain boundary scattering or surface roughness); and leakage of energy back into the fluid (in the case of pseudo-surface waves there may also be leakage of energy into the bulk of the solid) (Yamanaka 1982). In the examples given in the following section the first two mechanisms are negligible, and the attenuation measured may be entirely attributed to energy leaking back into the fluid.

These quantities differ slightly from what would be measured in perfect inversion technique to reconstruct $R(\theta)$. From eqn (6.90), the angle at which the phase change of π occurs, i.e. when $R(\theta) = -1$, is independent of the fluid impedance, and therefore corresponds to the free surface velocity, whereas the velocity calculated from (8.36) is the fluid-loaded Rayleigh velocity. Again, there is no dip in the modulus of the reflectance function due to radiation into the fluid, because the effect of leakage into the fluid is already fully accounted for in the phase change at the Rayleigh angle (Bertoni and Tamir 1973). Therefore, any dip in the modulus of the reflectance function at the Rayleigh angle due to attenuation would be due to attenuation in the solid alone, whether elastic or anelastic (in some experiments to measure Rayleigh angle phenomena with flat transducers, dips are found in the reflection at the Rayleigh angle that are due to the finite size of the transducers; these have nothing to do with the present discussion). In practice pathologies in the reconstruction of $R(\theta)$ due to limitations in the range of z may mask such subtleties.

FIG. 8.5. Steps in the analysis of $V(z)$ for fused quartz (Kushibiki and Chubachi 1985). (a) $V(z)$ on a linear scale; (b) $V(z)$ filtered to remove short period ripple due to lens reverberations; (c) $V(z)$ for Teflon $\approx V_{\mathrm{L}}$; (d) Best value of V'' after subtracting long period error in V_{L}; (e) Fourier transform of (d); (f) Final Fourier transform from which the Rayleigh wave velocity and attenuation are found using eqns (8.36) and (8.37). 225 MHz, $\lambda_0 = 6.6\ \mu\mathrm{m}$.

From §7.2.3, similar analysis may be used to measure the longitudinal velocity. In ray terms the mechanism may be thought of as the excitation of a longitudinal wave in the solid with an angle of refraction of 90° by a wave in the fluid incident at the longitudinal critical angle. The longitudinal wave propagates parallel to the surface, and is sometimes known as a lateral wave or as a surface skimming compressional wave (SSCW). The lateral wave in turn excites a wave in the fluid, also at the longitudinal critical angle, sometimes known as a head wave. Curiously, it does not excite a shear wave in the solid; if it did the reflectance function would not rise to unity at the longitudinal critical angle. Whereas Rayleigh waves decay exponentially as $e^{-\alpha r}$, with a corresponding decay in the amplitude of waves in a fluid generated by them, the amplitude of head waves due to excitation by lateral waves decreases algebraically. The rate of decrease depends on the dimensionality of the situation. If the lateral wave is spreading out in all directions, as though it came from a point source, the decay is as $r^{-5/2}$. If it is spreading in one direction, as if from a line source, the amplitude decays as $r^{-3/2}$ (Tew *et al.* 1988). Because in a defocused acoustic microscope the excitation is from a circle where the cone of rays at the longitudinal critical angle meet the surface of the specimen, the $r^{-3/2}$ law applies.

The longitudinal headwave interaction is not as strong as the leaky Rayleigh wave interaction, but at a sufficiently large distance an exponential decrease is always stronger than an algebraic decrease. Therefore, in general, Rayleigh wave oscillations dominate $V(z)$ up to some value of defocus, after which the longitudinal head wave oscillations may begin to take over. When both Rayleigh wave and head wave phenomena are present, it is best to analyse them one at a time, Fourier transforming $V(z)$ only over the range where the mode to be measured dominates, in order to demand less subsequently of selection in the Fourier domain. The head wave excitation mechanism is tied to the propagation of longitudinal waves in the bulk of the solid. Therefore, unlike surface waves, they do not suffer a velocity perturbation due to fluid loading, and the velocity measured is indeed the longitudinal velocity.

8.2.2 *Analysis for spherical lenses*

The line-focus-beam lens is excellent for making measurements, but it cannot be used for taking pictures. But for the spherical, or point-focus-beam, lens in an imaging microscope, the approximation of (8.25) is not valid. This is illustrated by a time-resolved measurement in Fig. 8.6, in which the geometrical and Rayleigh components are separated by defocusing (Liang *et al.* 1982; Yamanaka 1983; Weaver *et al.* 1989). The picture was made using the short pulse excitation system described in

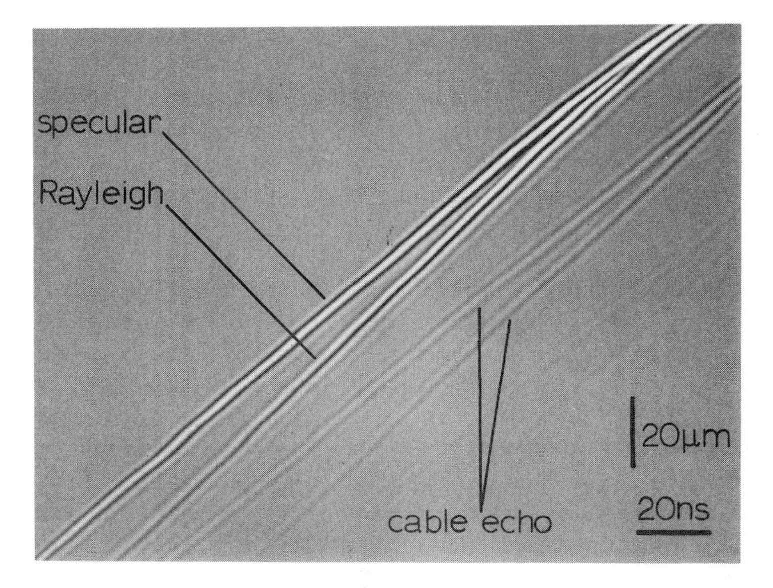

FIG. 8.6. Time-resolved measurements $S(t, z)$ separating specular (geometrical) from Rayleigh reflections; horizontal axis is time t; vertical axis is defocus z; the value of $S(t, z)$ is indicated by the intensity, with mid-grey as zero and dark and light as negative and positive values of S (Weaver *et al.* 1989).

§5.3. It is a display of $S(t, z)$. The horizontal axis is time, and each horizontal line corresponds to a trace from the sampling oscilloscope, with the instantaneous voltage modulating the brightness of the line instead of moving it up and down. The vertical axis is defocus, the top corresponding approximately to focus and the bottom to negative defocus. Across a line of given defocus, the first signal is the specular (or geometrical) reflection and the second (if they are separated) is the Rayleigh reflection. Subsequent modulations are electronic reflections of these and other signals in the cables. At focus, only one primary echo can be identified. As the lens is moved towards the specimen, there comes a defocus, rather well defined here, where the Rayleigh reflection separates out and becomes delayed relative to the specular echo. As the defocus is increased further the separation of the specular and Rayleigh reflections increases linearly with z. This provides a striking picture of the origin of the oscillations in $V(z)$ that would occur when the pulses are longer and so continue to overlap and interfere even at large negative defocus. But, for the present, the important point is that over most of the range of defocus displayed here, the two reflections are of comparable magnitude. Although at a sufficiently large defocus the exponential decay of the Rayleigh component must eventually make it smaller than the algebraically decaying geometrical component, over a range of defocus that corresponds to several oscillations in $V(z)$ the Rayleigh component is never sufficiently smaller than the specular component for the inequality (8.22) to be adequately satisfied. Therefore a method of analysis based on the full expression (8.21) must be used (Rowe 1988). Equation (8.21) may be written

$$|V_R(z)| \cos \phi(z) = \frac{|V(z)|^2 - |V_G(z)|^2 - |V_R(z)|^2}{2\,|V_G(z)|}. \tag{8.38}$$

To illustrate the ray analysis of data obtained with a spherical lens, Fig. 8.7 presents various stages in the processing of data measured at 1.5 GHz using an ELSAM microscope (Briggs *et al.* 1988). Usually a reference curve $V_L(z) \approx V_G(z)$ would be obtained with, for example, a lead specimen, but in this particular case a reference curve was calculated for a perfect reflector and a spherical lens with a Gaussian pupil function corresponding approximately to the microscope lens. The measured $|V(z)|$ was first squared and filtered. Filtering the square of the measured data is the most satisfactory way to remove interference with reverberations and other signals that do not vary with z and, provided that the lens opening angle is less than 60°, they can be completely eliminated. The reason for filtering $|V(z)|^2$ rather than $|V(z)|$ itself corresponds to truncating the autocorrelation of $Q(t)$ in §8.1.2; indeed the same variables and Fourier transform are used in order to perform the filtering. The same filtering considerations also apply when a measured $|V_0(z)|$ reference curve is used. Figure 8.7(a) is the filtered function $|V(z)|^2$ from the glass microscope slide.

At this stage the intermediate function

$$V_I(z) = \frac{|V(z)|^2 - |V_G(z)|^2}{2\,|V_G(z)|} \tag{8.39}$$

can be calculated. The final term on the top line on the right-hand side of eqn (8.38) is not known, but within the ray approximation must be a decaying exponential of the form (cf. (8.29))

$$|V_R(z)|^2 = a \exp(2\alpha z) \tag{8.40}$$

with only two unknowns, a and α. In principle, these could be found by an automatic iterative procedure, but for the results presented here various values of a and α were tried interactively until a combination was found that gave a function, corresponding to the right-hand side of eqn (8.38), that when plotted looked most nearly symmetrical along its length about the $-z$ axis. The result is shown in Fig. 8.7(b). It is by no means a sine wave with a perfect exponential decay, and there is a certain amount of operator skill in selecting the part of the curve to analyse: at the large defocus end the data may be corrupted by interference with lens

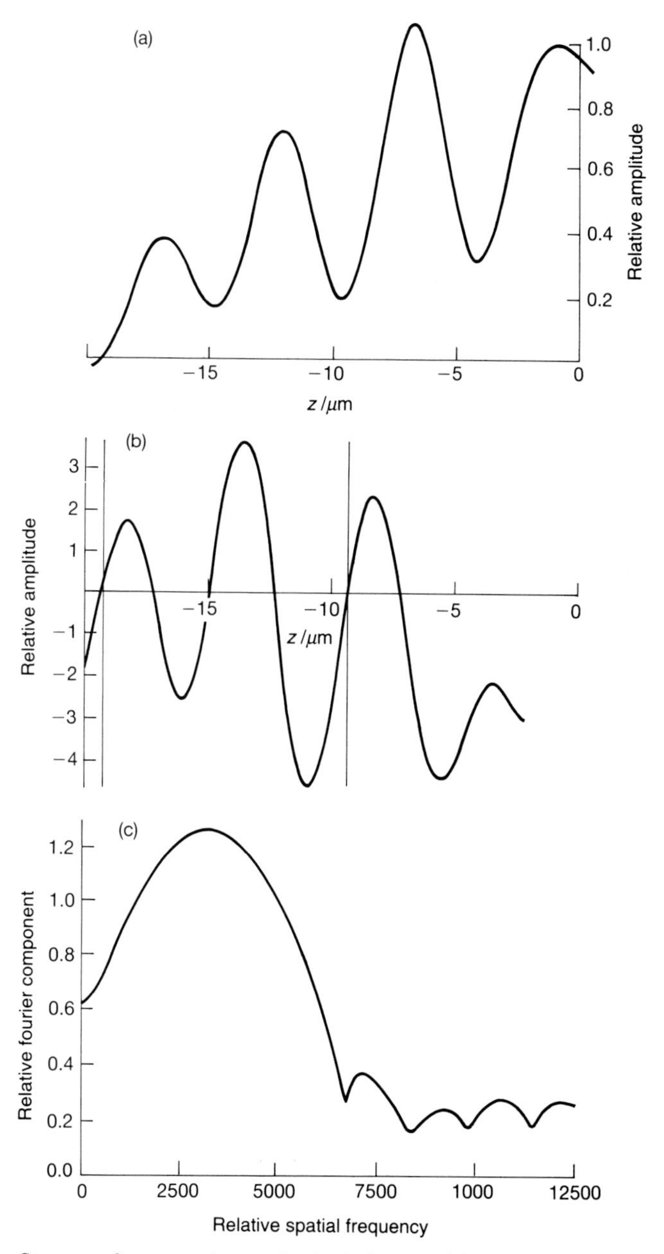

FIG. 8.7. Stages of processing spherical lens $V(z)$ data from fused silica. (a) $|V(z)|^2$, after filtering out short period ripple; (b) the intermediate function $V_{\mathrm{I}}(z)$ (8.39), with a portion of the curve selected for Fourier transforming; (c) Fourier transform of the chosen section of (b), from which the Rayleigh velocity and attenuation are deduced by finding the best fit of a decaying sinusoidal wave truncated and Fourier transformed in the same way as (b). Focus in (a) is the peak at approx. $-1\ \mu$m; it is smaller than its neighbour because of a gating effect. 1.5 GHz; ELSAM (Briggs *et al.* 1988).

reverberations and even with the second echo from the specimen, whereas close to focus there are diffraction effects outside the scope of the ray theory. If one is not selective enough these effects may cause significant errors; if one is too selective then accuracy can be lost too, especially as attenuation measurement can easily be swamped by the effects of truncation in the Fourier transform. These problems become more acute at higher frequencies. Nevertheless, because the ray analysis specifically concentrates on the information from the strong interaction with Rayleigh waves, surprisingly good results can be obtained even from imperfect data.

The part of the curve in Fig. 8.7(b) selected for subsequent analysis lies between the two vertical lines at $-8.4\ \mu m$ and $-18.4\ \mu m$. This part of the curve is Fourier transformed (using a Hamming window), and the result is shown in Fig. 8.7(c). The peak in the Fourier transform corresponding to the Rayleigh wave interference oscillations in $V(z)$ is identified and the number of points on either side of the peak that can be considered free from other effects is selected (usually mainly or even exclusively on the right-hand side because of the possible influence of interference with surface skimming longitudinal waves on the left side, especially if data with a larger amount of defocus than here is used). Using standard numerical algorithms, a best fit to these points is then found with the Fourier transform of a decaying exponential sine wave truncated in the same way as the data from Fig. 8.7(b).

From the parameters of the decaying exponential the velocity and attenuation of leaky Rayleigh waves in the specimen are deduced. In the example presented here the measured leaky Rayleigh wave velocity was found to be $v_R = 3510 \pm 25\ \mathrm{m\,s^{-1}}$, and the normalized attenuation (i.e. the fractional imaginary component of the wavevector) was $\alpha_N = 0.022 \pm 0.01$ (the errors given indicate the scatter in the results, not systematic errors). Similar measurements were made on E6 glass, with the results $v_R = 3130 \pm 25\ \mathrm{m\,s^{-1}}$ and $\alpha_N = 0.032 \pm 0.01$. Greater accuracy with a spherical lens can be obtained at lower frequencies and, because the first one or two oscillations in $V(z)$ have to be discarded, there is a disproportionate advantage in being able to use a range of z that corresponds to a greater number of wavelengths. The results here were obtained with the spatial resolution of an imaging microscope operating at 1.5 GHz.

8.2.3 *Accuracy*

The formula for determining the Rayleigh velocity v_R from the spatial period Δz of the oscillations in the $V(z)$ curve, with water velocity v_0 and frequency f, is

$$v_R = \frac{v_0}{\{1-(1-v_0/2f\,\Delta z)^2\}^{1/2}}. \tag{8.41}$$

Thus the dependence of the measured Rayleigh velocity on the three experimental quantities is

$$\frac{\partial v_R}{\partial v_0} = \frac{v_0}{2f\,\Delta z\{1-(1-v_0/2f\,\Delta z)^2\}^{3/2}}, \tag{8.42}$$

$$\frac{\partial v_R}{\partial f} = \frac{-v_0^2(1-v_0/2f\,\Delta z)}{2f^2\,\Delta z\{1-(1-v_0/2f\,\Delta z)^2\}^{3/2}}, \tag{8.43}$$

$$\frac{\partial v_R}{\partial\,\Delta z} = \frac{-v_0^2(1-v_0/2f\,\Delta z)}{2f\,\Delta z^2\{1-(1-v_0/2f\,\Delta z)^2\}^{3/2}}. \tag{8.44}$$

Putting $a \equiv (1-v_0/2f\,\Delta z)$, and $b \equiv v_0/2f\,\Delta z\{1-(1-v_0/2f\,\Delta z)^2\}$, the total derivative is

$$\frac{\mathrm{d}v_R}{v_R} = b\frac{\mathrm{d}v_0}{v_0} - ab\frac{\mathrm{d}f}{f} - ab\frac{\mathrm{d}\,\Delta z}{\Delta z}. \tag{8.45}$$

To give some idea of the behaviour, use

$$v_0/2f\Delta z = 1-\cos\theta_R, \tag{8.46}$$

where θ_R is the Rayleigh angle. Then $a=\cos\theta_R$ and $b=(1-\cos\theta_R)/\sin^2\theta_R$. As a representative value take $\theta_R = 30°$. Then $a=0.866$, $b=0.536$, and

$$\frac{\mathrm{d}v_R}{v_R} = 0.536\frac{\mathrm{d}v_0}{v_0} - 0.464\frac{\mathrm{d}f}{f} - 0.464\frac{\mathrm{d}\,\Delta z}{\Delta z}. \tag{8.47}$$

The velocity in the fluid is not usually measured directly, but deduced from the temperature. At 20°C the variation of velocity in water with temperature is $3.1\ \mathrm{m\,s^{-1}\,K^{-1}}$. This gives a fractional dependence

$$\frac{\mathrm{d}v_0}{v_0} \approx 0.0021\,\mathrm{d}T\ (\mathrm{K}). \tag{8.48}$$

For a measurement of a Rayleigh velocity in the vicinity of $3000\ \mathrm{m\,s^{-1}}$, if the fractional errors in f and Δz are $\delta f/f$ and $\delta\,\Delta z/\Delta z$ and the error in the temperature measurement is δT (measured in degrees kelvin), then the overall fractional measurement error is

$$\frac{\delta v_R}{v_R} = \sqrt{\left\{(0.0011\,\delta T)^2 + \left(0.464\frac{\delta f}{f}\right)^2 + \left(0.464\frac{\delta\,\Delta z}{\Delta z}\right)^2\right\}}. \tag{8.49}$$

Table 8.1
Experimental accuracy for Rayleigh velocity measurements

Accuracy to be achieved	Accuracy required in		
	δT (°C)	$\delta f/f$ (%)	$\delta\Delta z/\Delta z$ (%)
1 in 10^3	0.9	0.215	0.215
1 in 10^4	0.09	0.0215	0.0215

This table gives the accuracy required in each experimental parameter in order to measure a Rayleigh velocity in the vicinity of 3000 m s^{-1} from $V(z)$ with water as the coupling fluid, assuming the other parameters are exact. If each parameter contributes equal error, then from (8.47) each tolerance must be reduced by $1/\sqrt{3}$.

Table 8.1 gives upper limits in the measurement errors to give an accuracy of one part in 10^3 (the best that has ever been achieved with a standard imaging lens), and one part in 10^4 (the best that has ever been achieved with a cylindrical lens). Frequency measurement can be achieved to many orders of magnitude greater accuracy than the requirements specified here, and this is readily available using synthesizer-based r.f. electronics (§5.2). The accuracy in Δz depends primarily on the specification of the mechanical stage. The chief variable under the microscopist's direct control is the stability of the water temperature, and it is worth taking unlimited pains over this. Some idea of the extent of the problem comes from considering the old-fashioned wet and dry bulb hygrometer, in which it is quite possible to obtain a temperature difference of several degrees between the wet bulb and the dry bulb. Similar temperature differences can exist within the drop of water in the microscope. Thus, even if an accurate thermocouple is being used, there is no guarantee that the temperature measured is that of the water through which the acoustic wave is passing. It is therefore essential to use a good cabinet to reduce drafts and to allow some sort of local equilibrium vapour pressure to be approached, to have the thermocouple as close to the region of the acoustic beam as possible (*without risking damaging the lens surface,* especially at negative defocus), and to use a generous-sized water drop.

The line-focus-beam technique with its associated analysis has been more extensively used than any other method for quantitative acoustic microscopy. The results vindicate the approximations made in the theoretical model, and the method proves extremely robust (Kushibiki and Chubachi 1985). In systematic tests on isotropic materials such as fused silica and certain specified glasses the agreement between the calculated and measured Rayleigh velocity was about 0.2 per cent. If the water temperature can be stabilized then a sensitivity to change in v_R can approach one part in 10^4 (Kushibiki and Chubachi 1987). The agreement between calculated and measured Rayleigh attenuation is 2 per cent, but this, especially when α is small, may vary up to 20 per cent or more. The agreement of measured longitudinal velocity with accepted published values is about 2 per cent for the same materials; for lower velocity isotropic materials such as polystyrene, PVC, and PMMA there can be as much as 5 per cent discrepancy, but this may be due in part to genuine variation of the materials' parameters.

Anisotropic materials have been extensively measured using the line-focus-beam technique. These will be discussed further in Chapter 11, but some results are summarized in Fig. 8.8 to illustrate the range of materials over which the technique has been proven (Kushibiki and Chubachi 1985). In Fig. 8.8(a) the basic equation of the oscillations in $V(z)$ (8.17) is drawn as the continuous curve. For 31 different materials and orientations the velocity calculated using bulk elastic constants is plotted against the experimentally measured period of the oscillations

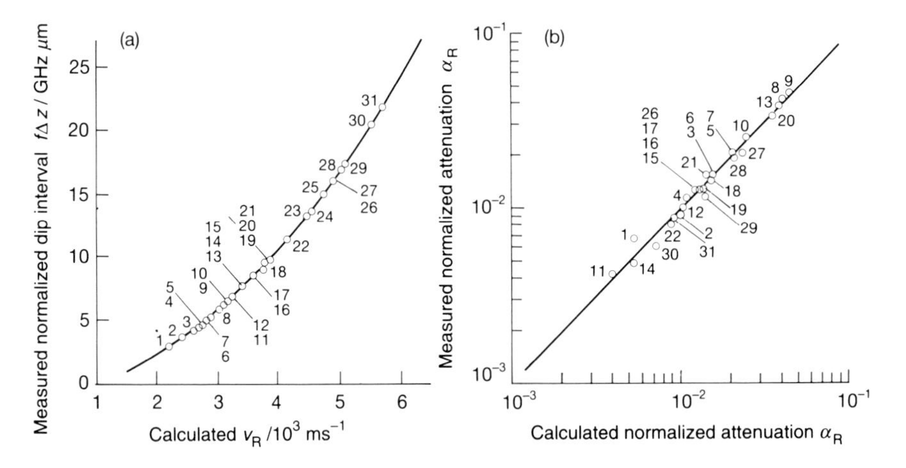

FIG. 8.8. Analysis of line-focus-beam $V(z)$ data for 31 different materials and orientations, compared with calculated values. (a) Normalized measured period of the oscillations in $V(z)$ versus calculated fluid-loaded Rayleigh wave velocity; the curve is eqn (8.17). (b) Normalized measured attenuation from (8.19) versus calculated attenuation; the line corresponds to perfect agreement. (Kushibiki and Chubachi 1985.)

Key: 1 InSb(111)[1$\bar{1}$0]†; 2 GaAs(111)[1$\bar{1}$0]; 3 GaAs(111)[11$\bar{2}$]; 4 Ge(111)[1$\bar{1}$0]; 5 GaAs(001)[010]; 6 Ge(111)[11$\bar{2}$]; 7 GaAs(001)[110]; 8 E6 glass; 9 7740 glass; 10 *XY* α-quartz; 11 GaAs(111)[1$\bar{1}$0]†; 12 GGG(111)[11$\bar{2}$]; 13 SiO_2(fused); 14 Ge(111)[1$\bar{1}$0]†; 15 *YZ* $LiNbO_3$; 16 YIG(001)[110]; 17 YIG(111)[11$\bar{2}$]; 18 *XY* $LiNbO_3$; 19 *ZX* $LiNbO_3$; 20 *YZ* α-quartz; 21 *ZY* $LiNbO_3$; 22 *XY* rutile; 23 Si(110)[$\bar{1}$10]; 24 Si(111)[1$\bar{1}$0]; 25 Si(111)[11$\bar{2}$]; 26 *XZ* rutile; 27 Si(001)[010]; 28 Si(110)[001]; 29 Si(001)[110]; 30 *ZX* sapphire; 31 *ZY* sapphire.

† Leaky pseudo-surface wave (§11.2).

in $V(z)$. This gives an indication of the reliability of the technique over the velocity range 2000–6000 m s^{-1}. A lens of semi-angle 50° with water coupling can measure velocities down to 2000 m s^{-1}; by using methanol instead of water (with a lens of reduced radius of curvature because of the higher attenuation of methanol) it is possible to measure down to 1500 m s^{-1}. For materials faster than about 6000 m s^{-1} the spacing Δz is so large that it becomes necessary to use larger lenses and lower frequencies to get enough oscillations in $V(z)$ for the Fourier analysis. In some cases where the points do not lie exactly on the curve the difference may be partly attributed to uncertainty in the constants used in the calculated velocity. Comparison between calculated and measured values of attenuation are shown in Fig. 8.8(b).

8.2.4 *Stress measurement*

If the uniqueness of acoustic microscopy lies in its ability to image and measure the mechanical properties of a specimen, then after elastic stiffness one of the properties that it would be most interesting to be able to measure is stress. This proves to be rather difficult, but not impossible. The information comes from the small changes in acoustic velocities that occur when a material is stressed, due to non-linearities in the inter-atomic or intermolecular forces (Pao *et al.* 1984). The parameters that are used to describe such effects are called third-order elastic constants, because they describe terms in the cube of the strain in the stored elastic energy, just as linear elasticity gives rise to quadratic terms in the stored energy. The third-order elastic constants are sixth rank tensors, because they relate a second rank tensor (stress) to product terms from another second rank tensor (strain). Measurement of stress ultimately means the ability to measure a tensor field, a somewhat daunting prospect! But, fortunately, a more limited objective will often do. In particular, it is often enough to be able to measure the principal stresses in two orthogonal directions in the surface. The anisotropy due to non-isotropic stress leads to different effects from ordinary elastic anisotropy. For example, the velocities of SH (shear horizontal) waves in two orthogonal directions in a surface will always be identical, no matter what the elastic anisotropy is. But if there are stress components that are different in the two directions, then the velocities will differ. Unfortunately, no-one has found a way to generate SH waves with an acoustic microscope; the predominant component in Rayleigh waves is SV (shear vertical). In successful measurements of stress by acoustic microscopy, the specimens have been elastically isotropic, and the stresses have varied either during the course of the experiment, or spatially over the surface.

With a good line-focus-beam quantitative microscope it is possible to demonstrate the variation of velocity with stress in PMMA (Shimada 1987;

Obata *et al.* 1990). In PMMA it is the excitation of lateral longitudinal waves that gives rise to oscillations in $V(z)$, as described in §7.2.3. If the specimen is mounted in a simple manually driven straining stage, preferably with a small load cell, that can be used in the microscope, then the velocity of lateral longitudinal waves can be measured as the stress is varied. The velocity of longitudinal waves parallel to the uniaxial strain axis decreases with tensile stress (denoted as positive) with a sensitivity of approximately $-1.75\,\mathrm{m\,s^{-1}\,MPa^{-1}}$; perpendicular to the strain axis the velocity of longitudinal waves is affected much less, decreasing as approximately $-0.425\,\mathrm{m\,s^{-1}\,MPa^{-1}}$. Velocity changes due to more than one stress component can be summed by linear superposition, and so biaxial stress fields can also be measured. This kind of technique, albeit using a spherical lens, has been applied to the measurement of residual stress in silicon nitride joined to metal (Narita *et al.* 1990). It was possible to correlate the conditions that produced the minimum stress in the ceramic measured this way with those that gave the best four-point bend strength.

Rather accurate measurements of stress on a microscopic scale have been measured using a lens with the combined longitudinal and shear wave transducer and double pulse electronics described in §5.2. The longitudinal wave is refracted at the lens surface and generates a focused wave in the coupling fluid in the usual way. The shear wave travels more slowly. At the lens surface it generates a wave in the fluid only where it excites motion normal to the lens–fluid interface. This means that it will not excite any axial ray, and also it will not excite waves where the surface away from the axis is tangential to the polarization. The transmitted energy is therefore concentrated towards two lobes on opposite sides of the axis, with an azimuthal distribution approximately proportional to $\cos^2\phi$, where ϕ is the azimuthal angle measured parallel to the shear wave polarization (Chou and Khuri-Yakub 1989; cf. §5.2). This field distribution almost makes it possible to have one's cake and eat it. Azimuthal resolution has been achieved for examining anisotropic surfaces, while at the same time having spatial resolution. Of course, Baron de Fourier still rules OK, and what has been achieved is a compromise. The spatial resolution of the shear wave focus is degraded somewhat in the direction where the azimuthal distribution is minimum (though in many cases in practice the degradation would scarcely be noticeable). The azimuthal resolution is adequate for detecting anisotropy with a low symmetry, but it cannot approach the angular resolution of a line-focus-beam cylindrical lens. The only way to achieve high spatial resolution and high angular resolution would be to exploit the fact that a reflection confocal microscope is effectively a folded transmission confocal microscope; therefore, in principle, two pupil functions could be used, one a full aperture to give spatial resolution, and the other a non-axially symmetric aperture to give angular resolution. This principle has been used in a reflection ultrasonic microspectrometer working in the frequency range 20–150 MHz (Tsukahara *et al.* 1990, 1991). Two transducers are used, one on a lens to focus the waves to a spot, the other on a planar buffer rod to define the angle of reflection (zenithal and azimuthal) that is measured. The whole assembly is mounted on a goniometer so that the zenithal and azimuthal angles can be scanned. The same effect could, in principle, be achieved on a single lens by depositing two transducer patterns on top of one another, one for sending and the other for receiving, with a common ground electrode between them. Experiments with so-called bow-tie transducers have shown how difficult it is to obtain the desired result, chiefly because of diffraction in the body of the lens (Davids *et al.* 1988), though some success has been achieved at 300 MHz with a bow-tie aperture near the front surface of the lens (Ishikawa *et al.* 1990). Other directional point-focus lenses have been demonstrated that give azimuthal resolution at the same time as spatial resolution better than 6 μm (Chubachi 1985; Kushibiki *et al.* 1989). This is a field of active development.

The shear wave transducer gives azimuthal discrimination while retaining most of the spatial resolution of a spherical lens. The longitudinal wave excited by the transducer gives a signal that is separated in time from the shear wave signal (because the longitudinal wave travels faster in the lens), but gives an accurate phase reference because it travels through almost the same path in the fluid. There is, therefore, compensation for fluctuations in temperature and lens height: from §7.2, the Rayleigh and axial paths differ by a fraction $1 - \cos\theta_R$ of the axial path. The double-pulse system measures the phase change $\Delta\phi$ of the returning Rayleigh reflection, from which the fractional Rayleigh velocity change can be deduced. If the measurement is made at defocus z_0, and the water wavelength and Rayleigh angle are λ_0 and θ_R as usual, then the fractional Rayleigh velocity change is

$$\frac{\Delta v_R}{v_R} = \frac{\lambda_0}{4\pi z_0 \cos\theta_R}\Delta\phi. \tag{8.50}$$

It can be seen that the accuracy of the method is increased by increasing z_0. The stress can then be deduced either from knowledge of the third-order elastic constants (Husson 1985), or by perturbation theory (Husson and Kino 1982), or from direct calibration.

As a demonstration of the ability of the technique, Fig. 8.9 shows results on a 15 μm thick layer of alumina sputtered on a glass substrate (Meeks *et al.* 1989). From the subsequent curvature of the glass, it was estimated that there was a compressive residual stress of −40 MPa in the alumina. A trench was etched in the alumina; the end of the trench can be seen in Fig. 8.9(a) and (b). The dark broad ∩-shape 50 μm wide

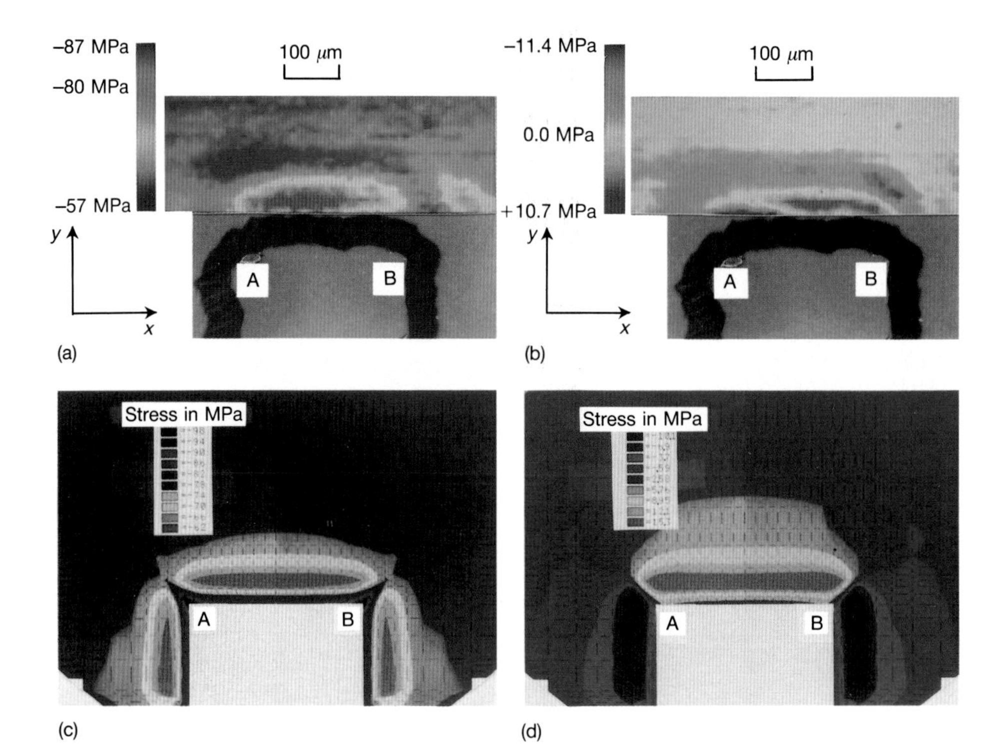

FIG. 8.9. Stress fields at the end of a trench etched in a 15 μm thick layer of sputtered alumina on a glass substrate. The trench was 15 μm deep, 0.4 mm wide, and 10 mm long. The long-range residual stress in the alumina layer measured from the curvature of the glass substrate was −40 MPa (compressive). The top two collages are photographs of one end of the trench with measurements by acoustic microscopy of (a) the sum of the stresses $\sigma_{xx} + \sigma_{yy}$ and (b) the difference of the stresses $\sigma_{yy} - \sigma_{xx}$; $f = 670$ MHz. The bottom two pictures are finite-element calculations of the same geometries, with the points AB corresponding to those in the upper pictures and the colour scales corresponding in each case to the picture above, of (c) the sum of the stresses $\sigma_{xx} + \sigma_{yy}$ and (d) the difference of the stresses $\sigma_{yy} - \sigma_{xx}$ (Meeks *et al.* 1989).

is the sloping wall of the trench, and the bottom of the trench, which is at the same depth as the top of the glass substrate, looks more or less the same as the flat surface of the alumina. Just beyond the end of the trench measurements were made of the surface stresses in the x-direction, σ_{xx}, and in the y-direction, σ_{yy}. In Fig. 8.9(a) the sum of these stresses near the trench have been represented in colour, and superimposed on to a photograph of the specimen. Away from the trench the sum of the stresses is approximately twice the residual stress, but close to the trench the stress is reduced to a value closer to the residual stress itself. More quantitative comparison is possible with theoretical results from a finite element calculation of $\sigma_{xx} + \sigma_{yy}$ in Fig. 8.9(c). The colour scales in (a) and (c) approximately correspond in the two pictures, and the points marked A and B in the two pictures roughly correspond to the same points on the specimen. The calibration of the velocity shift with stress was obtained by measuring the velocity difference in the two directions very close to the wall of the trench, and assuming that there would be a uniaxial stress of -40 MPa. The reduction in the sum of the stresses in the vicinity of the wall of trench might be supposed to be due to the relaxation of σ_{yy}, and this is confirmed by the map of the difference between the stresses, $\sigma_{yy} - \sigma_{xx}$, in Fig. 8.9(b). Far from the trench, the difference is small, but close to the wall the difference becomes positive, because both components are negative (i.e. compressive), but σ_{yy} is smaller. Comparison can be made with the finite element calculation of $\sigma_{yy} - \sigma_{xx}$ in Fig. 8.9(d).

These results are a remarkable achievement. They were obtained at 670 MHz, with a spatial resolution of 5–10 μm. The total velocity changes involved were of the order of 0.5 per cent, and the stress sensitivity achieved was 0.3 MPa.

8.3 Time-resolved techniques

If pulses can be generated and detected whose length is short compared with the time difference between reflections from the top and the bottom surfaces of a layer, then the elastic properties of the layer can be deduced from the amplitude and timing of the two echoes. The return pulses from such a situation are illustrated in Fig. 8.10. Figure 8.10(a) is an oscilloscope trace of the reference echo from the substrate at defocus z_0 and with nothing on it except the coupling fluid. We can choose to write the reference signal as

$$S_0(t) \equiv A_0 s(t - t_0) \otimes g(t, z_0). \tag{8.51}$$

In (8.51), and indeed in many of the equations in this section, the left-hand side represents what is actually measured in the experiment,

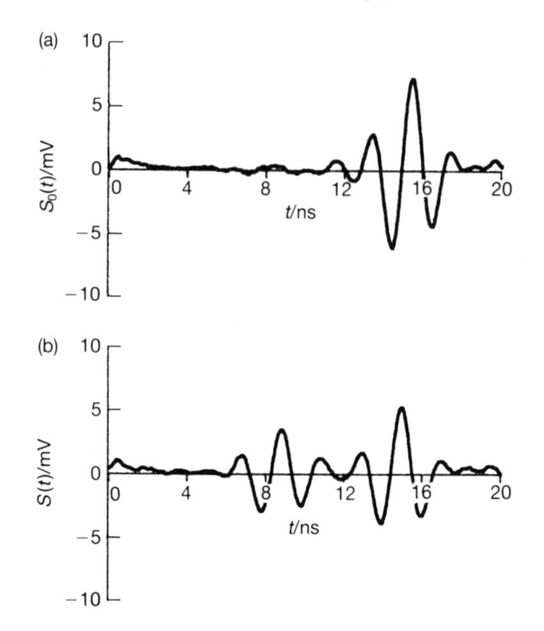

FIG. 8.10. Short pulses for time-resolved measurements; the usable bandwidth of the pulses above noise level is about 0.5 GHz, and they were digitized with an overall timing precision of 0.15 ps. (a) Reference signal reflected from a glass slide at focus with no specimen; (b) reflected signal from a cell on the glass slide, with echoes from the top of the cell and from the interface between the cell and the substrate. (Wang *et al.* 1990.)

and the right-hand side represents an interpretation in terms of the quantities of interest. A_0 is the reflection coefficient at the water–substrate interface; $s(t)$ is the two-way lens response ignoring any effects of focusing or attenuation in the coupling fluid; $g(t, z)$ describes the distortion to the waveshape due to defocus and attenuation. The symbol $\otimes$ denotes the convolution operation (5.3); it is not as frightening as it sounds, and in this context it enables the primary waveform to be described by $s(t)$, with adjustments being made by the function $g(t, z)$. Both these functions are experimentally determined. The waveform distortion function $g(t, z)$ may be arbitrarily set to unity at $z = 0$; the lens response function may then be measured with the substrate at the focus of the lens. A series of measurements of $S(t)$ can then be made for a range of values of z. Since $s(t)$ is defined to be independent of z, the function $g(t, z)$ could be found for each of the values of z by deconvolution. Deconvolution can be performed using the convolution theorem: the functions are Fourier transformed, the deconvolution is now replaced by a division to give $\bar{g}(f, z)$, and then finally $\bar{g}(f, z)$ is inverse Fourier transformed to yield $g(t, z)$. In practice, however, if the defocus is positive (i.e. $z > 0$), the shape of the waveform is almost independent of defocus over the relevant range of z, so that g becomes a function of z only. Since g is now independent of t, the convolution in eqn (8.51) reduces to a simple multiplication by a constant determined by the value of z. In principle, the functions in eqn (8.51) can be complex. The measured function $S(t)$ is real, so it is convenient to constrain $s(t - t_0)$ to be real also (this is consistent with setting $g(t, 0) = 1$). For other values of z, $g(t, z)$ would then need to be allowed to be complex. Again, however, if $z > 0$, it is usually adequate to take g to be a real variable.

If now the lens is moved to a region of the substrate that is covered by the layer of interest, two echoes will be received, one from the top surface of the layer, denoted by subscript 1, and one from the interface between the layer and the substrate, denoted by subscript 2. This is illustrated in Fig. 8.10(b) which was measured from a fibroblast cell on a polystyrene substrate. This signal can be written

$$S(t) \equiv A_1 s(t - t_1) \otimes g(t, z_1) + A_2 s(t - t_2) \otimes g(t, z_2). \qquad (8.52)$$

As in the case of the reference curve, provided the defocus is positive, it is adequate to a good approximation to constrain the function g to be independent of t and to be a real function of z only, having been previously measured for the lens and substrate being used (corrections for refraction in the layer can be made, but the approximation works best when refraction in the layer is small). The optimum value of z must be found experimentally, by scanning through z and finding the minimum positive value at which the shape of the waveform remains approximately constant as a function of z. The amplitude may vary, but that does not matter, since it is allowed for in the dependence of g upon z. It is undesirable to let the defocus be too big, since the attenuation in the coupling fluid would then reduce the signal more than necessary. Within the approximation of the independence of the waveform shape on z, eqns (8.51) and (8.52) can be written

$$S_0(t) = A_0 s(t - t_0) \times g(z_0) \qquad (8.53)$$

and

$$S(t) = A_1 s(t - t_1) \times g(z_1) + A_2 s(t - t_2) \times g(z_2). \qquad (8.54)$$

If the difference between t_1 and t_2 is greater than the length of the pulse, as is the case in Fig. 8.10, then the two signals, from the top and the bottom surfaces, can be measured by calculating the normalized correlation of $S_0(t)$ and $S(t)$

$$C(t) = \frac{\int_{-\infty}^{\infty} S(t') \times S_0(t'+t)\, \mathrm{d}t'}{\int_{-\infty}^{\infty} S_0(t')^2\, \mathrm{d}t'} \tag{8.55}$$

In practice the limits of the integrals are the range of available data. There should be two peaks in the correlation, corresponding to the optimum match between the reference signal $S_0(t)$ and the two echoes contained in the signal $S(t)$. If the two echoes are adequately separated in time it is generally best to isolate them and find their correlations separately, especially if they are of widely different magnitudes. From the height and position of each maximum, four crucial parameters can be measured, viz. $t_0 - t_1$, $t_0 - t_2$, A_1/A_0, A_2/A_0. Knowing the velocity v_0, impedance Z_0, and attenuation (taken as an average over the bandwidth) α_0 of the coupling fluid, and the impedance Z_s of the substrate, all the acoustic properties of the layer can be determined; these are denoted by subscript 1.

1. From the difference in time between the reference signal t_0 and the reflection from the top of the layer t_1, and knowing the velocity v_0 in the fluid, the thickness of the layer is

$$d = \tfrac{1}{2}(t_0 - t_1)v_0. \tag{8.56}$$

2. From the time t_2 of the echo from the interface between the layer and the substrate, and the times of the other two echoes, the acoustic velocity in the layer is

$$v_1 = v_0 \frac{t_0 - t_1}{t_2 - t_1}. \tag{8.57}$$

3. From the ratio of the magnitude of the reflection A_1 from the top of the layer to the magnitude of the reference signal A_0, and knowing the impedance Z_0 of the coupling fluid and the impedance Z_s of the substrate, the impedance of the cell is

$$Z_1 = Z_0 \frac{1 + A_1}{1 - A_1}. \tag{8.58}$$

4. From the measurements of velocity and impedance, the density is immediately

$$\rho_1 = \frac{Z_1}{v_1}. \tag{8.59}$$

5. Finally, from the amplitude A_2 of the echo from the interface between the layer and the substrate (which is described by an equation similar to (8.58)), and the amplitudes of the other two echoes, the attenuation in the cell is (in units of Napers per unit length, with attenuation taken as positive)

$$\alpha_1 = \alpha_0 + \frac{1}{2d}\log_e\left\{\frac{A_0}{A_2}\frac{Z_s - Z_1}{Z_s + Z_1}\frac{4Z_cZ_0}{(Z_c + Z_0)^2}\frac{Z_s + Z_0}{Z_s - Z_0}\right\}. \qquad (8.60)$$

It is best to measure $S_0(t)$ close to where $S(t)$ is measured, and certainly on the same substrate material. Often the time-resolved technique is chosen for low-impedance materials such as polymer coatings or biological tissue or cells. The greatest accuracy in separating two closely spaced signals is obtained when the two signals are of comparable magnitude; this will be achieved if the ratio of the substrate impedance to the layer impedance is slightly greater than the ratio of the layer impedance to the fluid impedance, so that the two terms in (8.54) are of comparable magnitude, allowing for the fact that at positive defocus $g(z_2)$ will be somewhat less than $g(z_1)$. For biological tissue this usually means choosing a polymer substrate if possible, which would probably support lateral longitudinal waves but not Rayleigh waves; for protective coatings on metals the substrate would also support Rayleigh waves. But since measurements of this type are taken at $z \geq 0$, lateral waves and surface waves are not usually a problem anyway.

When two signals are so close together that they are not adequately separated in the correlation of (8.55), then it is better to express the signals in the frequency domain and work with their Fourier transforms, denoted by a bar. Assuming that the pulse shapes are not dependent on defocus, so that the frequency dependence of $\bar{g}(z)$ can be neglected, and again letting the left-hand side correspond to what is measured and the right-hand side represent this in terms of the quantities of interest, the Fourier transforms of S_0 and S may be written

$$\bar{S}_0(f) = A_0\bar{s}(f)g(z_0)\exp(i2\pi ft_0) \qquad (8.61)$$

and

$$\bar{S}(f) = A_1\bar{s}(f)g(z_1)\exp(i2\pi ft_1) + A_2\bar{s}(f)g(z_2)\exp(i2\pi ft_2). \qquad (8.62)$$

The simplest way to sharpen up such data in the frequency domain is to use a Wiener filter (Press *et al.* 1986; Kino 1987). In the time domain, each contribution can be thought of as a Dirac delta function $\delta(t - t_m)$ with amplitude A_m convolved with the lens time-response $s(t)$; the Dirac delta function $\delta(x)$ is zero except when $x = 0$, and it has the property that the area under the function is 1. The information that is needed is the values of t_m and A_m, which could be obtained by deconvolving the lens time-response. In a perfect world the Fourier transform of the measured signal $\bar{s}(f)$ divided by the reference signal $\bar{s}_0(f)$ would give exactly what is wanted, by the convolution theorem; but because it is inevitable that noise is present this might result in sometimes dividing by zeros, and in any case it would result in wild distortions at frequencies where the reference signal is small. Therefore, both signals are multiplied by the complex conjugate of the reference signal, to ensure that the denominator is real, and then a real number is added to the denominator to ensure good behaviour when it is close to or below the noise level. In a true Wiener filter, the noise power spectrum is added in the denominator; in a simpler pseudo-inverse filter a constant M is used, chosen to be larger than the maximum value of the noise power spectrum over the frequency range of interest. The filtered signal is then, with complex conjugate denoted by *,

$$\bar{S}_2(f) = \frac{\bar{S}(f)\bar{S}_0^*(f)}{\bar{S}_0(f)\bar{S}_0^*(f) + M}. \tag{8.63}$$

The filtered signal is inverse transformed to give a shorter pulse in the time domain, and it can then be analysed by the correlation of (8.55). The reference signal must first pass through the same Weiner filter too; this is necessary because of the role of M.

A more powerful technique for analysing the Fourier transformed signals begins with cepstral filtering (Oppenheim and Schafer 1975). The logarithm is taken of the modulus of each of the two equations, and the first is subtracted from the second to give

$$\begin{aligned} \ln|\bar{S}(f)| - \ln|\bar{S}_0(f)| = \ln|&A_1\bar{s}(f)g(z_1)\exp\{i2\pi ft_1\} \\ &+ A_2\bar{s}(f)g(z_2)\exp\{i2\pi ft_2\}| \\ &- \ln|A_0\bar{s}(f)g(z_0)\exp(i2\pi ft_0)|. \end{aligned} \tag{8.64}$$

The terms 'cepstrum' and 'cepstral' come from inverting the first half of the words spectrum and spectral; they were coined because often in cepstral analysis one treats data in the frequency domain as though it were in the time domain, and vice versa. The value of cepstral analysis comes from the observation that the logarithm of the power spectrum of a signal consisting of two echoes has an additive periodic component due to the presence of the two echoes, and therefore the Fourier transform of the logarithm of the power spectrum exhibits a peak at the time interval between them. The additive component in the logarithm of the power spectrum comes from a multiplicative component in the power spectrum itself, just as the subtraction of the logarithms in eqn (8.64) corresponds to the division in eqn (8.63). For greatest accuracy in the subsequent analysis, it is best to normalize the spectra $\bar{S}_0(f)$ and $\bar{S}(f)$ so that the two logarithms on the left-hand side of (8.64) are of comparable magnitude. Taking logarithms brings analogous benefits to the factor M in (8.63). As in any filtering process, it means that information has been discarded, but the aim is to enhance the information that is of interest. Cancelling $\bar{s}(f)$ throughout the right-hand side, discarding the phase factor $\exp(\mathrm{i}2\pi f t_0)$, and putting

$$A_1' \equiv \frac{A_1 g(z_1)}{A_0 g(z_0)}, \qquad A_2' \equiv \frac{A_2 g(z_2)}{A_0 g(z_0)}, \tag{8.65}$$

(8.64) becomes

$$\begin{aligned} \ln|\bar{S}(f)| - \ln|\bar{S}_0(f)| &= \ln|A_1' \exp(\mathrm{i}2\pi f t_1) + A_2' \exp(\mathrm{i}2\pi f t_2)| \\ &= \ln|A_1'\{\cos(2\pi f t_1) + \mathrm{i}\sin(2\pi f t_1)\} \\ &\quad + A_2'\{\cos(2\pi f t_2) + \mathrm{i}\sin(2\pi f t_2)\}|. \end{aligned} \tag{8.66}$$

The modulus on the right-hand side is obtained by multiplying by the complex conjugate, giving

$$\begin{aligned} \ln|\bar{S}(f)| - \ln|\bar{S}_0(f)| = \tfrac{1}{2}\ln[2\{A_1'^2 + A_2'^2\} + A_1'A_2'\{\cos(2\pi f t_1)\cos(2\pi f t_2) \\ + \sin(2\pi f t_1)\sin(2\pi f t_2)\}]. \end{aligned} \tag{8.67}$$

This can be rewritten as

$$\ln|\bar{S}(f)| - \ln|\bar{S}_0(f)| = \tfrac{1}{2}\ln[B + C\cos\{2\pi f(t_2 - t_1)\}] \tag{8.68}$$

with

$$B = A_1'^2 + A_2'^2; \qquad C = 2A_1'A_2'. \tag{8.69}$$

If the two signals are sufficiently separated in the time domain, then a plot of $\ln|\bar{S}(f)| - \ln|\bar{S}_0(f)|$ from (8.68) contains oscillations of period $\Delta f = 1/(t_2 - t_1)$, from which $2d/v_c$ could be deduced directly (it is not possible to deduce d alone, because all information about the relation to t_0 was discarded in (8.66)). In Fig. 8.11(a) the function $\ln|\bar{S}(f)| - \ln|\bar{S}_0(f)|$ is plotted for the data in Fig. 8.10. The period of the oscillations is $\Delta f \approx 0.17\ \mathrm{ns}^{-1}$; the reciprocal of the period is 5.9 ns, which may be compared with the time interval between the two echoes in Fig. 8.11(b).

If the attenuation in the layer is frequency-dependent, as is often the case, then this is most easily incorporated into (8.64) by giving A_2 an explicit frequency dependence $A_2(f)$, and similarly $A_2'(f)$ in (8.65). These will introduce frequency-dependent shifts into both the offset and the amplitude of the oscillations of the cosine term in (8.67). How these are best determined depends on the relative magnitudes of A_1' and A_2'. If, for example, $A_1' \ll A_2'$, then the term in B would dominate, and the attenuation could be deduced from the frequency dependence of the

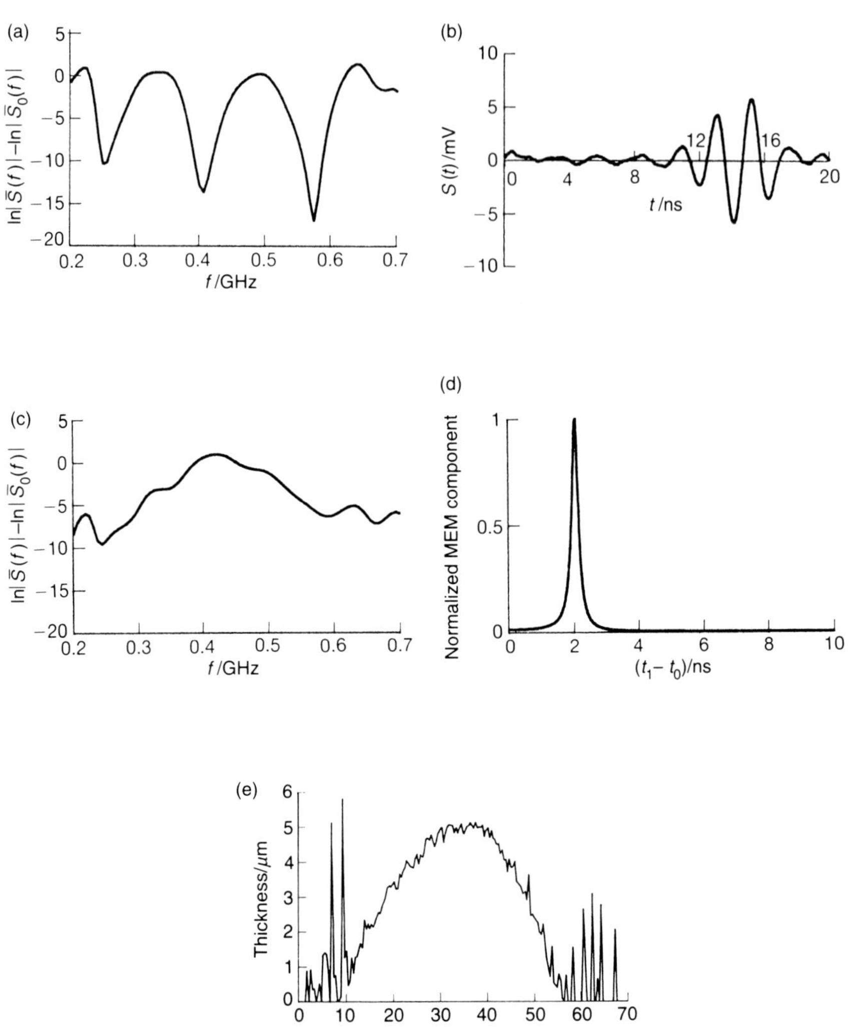

FIG. 8.11. Cepstral and maximum entropy (MEM) analysis of time-resolved signals: (a) the cepstral analysis function $\ln|\bar{S}(f)| - \ln|\bar{S}_0(f)|$ in the frequency domain for the data in Fig. 8.10, with the Fourier transforms of the signals in Fig. 8.10(a) and (b) as $\bar{S}_0(f)$ and $\bar{S}(f)$, respectively; (b) two overlapping and unresolved signals in the time domain from a very thin part of a cell on a glass substrate; (c) the cepstral analysis function $\ln|\bar{S}(f)| - \ln|\bar{S}_0(f)|$ in the frequency domain for the signal in (b); (d) the result in the time-interval domain of MEM analysis of the cepstral function in (c); (e) thickness of a cell deduced from cepstral and MEM analysis of signals measured along a line, the horizontal and vertical scales are not the same, and the thickness measurement becomes unrealiable below 1.5 μm. (Courtesy of Jun Wang.)

offset. From the frequency dependence of $A_2(f)$, the frequency dependence of the attenuation can be obtained directly by putting explicit frequency dependence in (8.60); if the velocity and therefore the impedance have negligible frequency dependence over the range of measurement, then

$$\alpha_c(f) = \alpha_0(f) + \frac{1}{2d}\log_e\left\{\frac{A_0}{A_2(f)}\frac{Z_s - Z_c}{Z_s + Z_c}\frac{4Z_cZ_0}{(Z_c + Z_0)^2}\frac{Z_s + Z_0}{Z_s - Z_0}\right\}. \quad (8.70)$$

Where several oscillations in $\ln|\bar{S}(f)| - \ln|\bar{S}_0(f)|$ occur over the frequency range of measurement, the period of the oscillations Δf can be measured most accurately by a Fourier transform into the time-interval domain. Once again, this transform cannot be related to absolute time because information relative to t_0 has been lost, but it will be possible to identify a peak corresponding to $t_2 - t_1$, yielding $2d/v_c$ as before. When, however, less than one oscillation is present in $\ln|\bar{S}(f)| - \ln|\bar{S}_0(f)|$, and even in some cases when one or more oscillations are present, the value of $t_2 - t_1$ cannot be found so simply. This corresponds to the value of $t_2 - t_1$ being less than the usable bandwidth β of the transducer, or equivalently to the pulses overlapping in time. Of course, if $t_2 - t_1 \ll 1/\beta$, then there will come a point beyond which no amount of signal processing will recover the information, but in mild cases it is possible to achieve remarkable measurements from otherwise unusable data using the maximum entropy method, or MEM (Press *et al.* 1986).

MEM, in this context, may be thought of as an alternative operation to the Fourier transform, to transform $\ln|\bar{S}(f)| - \ln|\bar{S}_0(f)|$ back into the time-interval domain. The reason that a Fourier transform does not give a very useful result when there are not well developed oscillations in $\ln|\bar{S}(f)| - \ln|\bar{S}_0(f)|$ is that information is not available over a sufficiently large range of f. MEM compensates for this by finding the transform that would correspond to the maximum possible number of different shapes of curve outside the range of f for which data is available. The term maximum entropy comes from the analogy with statistical thermodynamics, where entropy is a measure of the logarithm of the number of different arrangements that give the same overall state.

An example of the application of the power of cepstral analysis and MEM is illustrated in Fig. 8.11(b)–(d). The signal in Fig. 8.11(b) was taken from a specimen similar to the one in Fig. 8.10(b), but closer to the edge where the cell was thinner. The overlap of the signals is so great that there is no hope of separating them by eye. The application of cepstral analysis gives the curve shown in Fig. 8.11(c), and there is not even one full relevant oscillation in the frequency domain. But the maximum entropy method enables useful information to be obtained even from such poorly resolved data as this, and in the time-interval domain in Fig. 8.11(d) the MEM transform of $\ln|\bar{S}(f)| - \ln|\bar{S}_0(f)|$ has a pronounced peak from which $2d/v_l$ for the cell at that point can be determined. The time separation is about 2 ns, corresponding to a thickness of less than 2 μm. This may be a world record for acoustic distance resolution in this way.

MEM–cepstral analysis has been applied to the measurement of the thickness of biological cells (cf. §9.2.3). The peak in Fig. 8.11(d) happens to be at the reciprocal of the spectral range in Fig. 8.11(c), which looks suspiciously as though the analysis is simply measuring the bandwidth of the system. But other signals gave other values, and the results of a series of measurements along a line across a cell are plotted, expressed as cell thickness, in Fig. 8.11(e). A reasonable profile is obtained from 5 μm at the centre to the smallest reliable value of 1.5 μm near the edge.

8.4 Phew!

The last three chapters have involved a certain amount of mathematical slog. But it is worth it. These chapters have introduced the concepts governing the interaction of acoustic waves with specimens, the origin of the contrast in acoustic microscopes (especially the variation of contrast with defocus), and how the transition can be made from obtaining images to being able to interpret them quantitatively. And now the time has come to look at some of those pictures.

9
Biological tissue

Microscopy had been the principal tool of biology in the 19th century but it fell from fashion around 1900. (Sir Andrew Huxley 1990)

How are the mechanical properties related to the arrangement of macromolecular structures and their functional significance? (J. Bereiter-Hahn and N. Buhles 1987)

Provided that they are willing to let Lord Rayleigh define the limits of biology, handing over teeth, claws and bones to the materials scientists, the biologists need thus have few qualms about jumping into acoustic microscopy with scarce a thought about Rayleigh waves. (A. Howie 1987)

9.1 A soft option

Having considered the construction and operation of the acoustic microscope, and the elements of the theory of contrast, we now turn to applications of acoustic microscopy, to start to answer the question: 'Yes, but what can you actually see with an acoustic microscope?' It is natural to start with biological tissue for at least two reasons. The first is historical, in that when the scanning acoustic microscope was first developed, especially in the transmission mode, it was anticipated that the widest scope for application would lie in the fields of biology and medicine (Lemons and Quate 1979). It will be interesting to see how far that prediction is borne out, especially now that the reflection mode is used almost exclusively. The second reason for starting with biological applications is more pragmatic. In the majority of biological tissues the elastic properties are such that longitudinal waves are essentially the only waves of importance. A combination of a very low shear modulus and appreciable shear viscous damping has the consequence that shear waves propagate very slowly and with strong attenuation, so that for most purposes they may be disregarded in soft biological tissue. This makes the imaging theory very much more straightforward.

9.2 Cell cultures

In general, the acoustic properties of biological tissue show much greater variation than their optical properties. This means that acoustic images can be obtained without the need for any staining. By itself this is not a decisive advantage, because sophisticated interference contrast and other techniques are available to the light microscopist to enable him to enhance the contrast when staining is undesirable. Right from the start, therefore, we must look for the information that the acoustic microscope gives about the elastic properties of the specimen.

9.2.1 *Plane wave interference contrast*

Because there is no need for staining, living cells can be studied directly in the acoustic microscope (Hildebrand *et al.* 1981). In Fig. 9.1 a series of acoustic micrographs of endothelial cells is presented (Bereiter-Hahn 1987). The cell line was derived from *Xenopus* tadpole heart endothelia. They were stored at 27°C in amphibian culture medium, and this medium was used, at room temperature, as the coupling fluid for the acoustic microscopy. The substrate was polystyrene. It is essential that specimens adhere to the substrate and do not move around under the scanning motion of the lens; in these pictures the cells are quite well adhered, although there are some regions, for example those indicated by white arrowheads in Fig. 9.1(a), where the cell is not in contact with the substrate and therefore stronger reflection is seen from the bottom surface of the cell. No significant change could be seen in the cells while observing them for several minutes in the culture medium, so that it seemed that neither the scanning motion of the lens nor the incident ultrasonic power was affecting them. After Fig. 9.1(a) had been recorded, the cells were exposed to cytochalasin D, in the form of a dilute soluton (0.1 per cent, or $2\,\mu g\,ml^{-1}$) in the culture medium. Cytochalasin D (CD) is a drug that destroys the actin fibrillar system of the cells. Figure 9.1(b) was recorded after 75 s exposure to CD. There is a general loss of contrast in Fig. 9.1(b) compared with (a), though cell nuclei are faintly visible as indicated by the white arrowheads in Fig. 9.1(b). The dark arrows indicate regions where the central thicker cytoplasm seems to have spread into the flatter periphery. As the length of time during which the cells are exposed to CD increases, there is an increase in the formation of such protrusions, accompanied by a general reduction in the contrast. In Fig. 9.1(c), taken after 10 min exposure, the shapes of the cells have become almost unrecognizable. The disintegration of the cells in CD is not, however, completely irreversible. Figure 9.1(d) shows substantial recovery of the cell structure after 15 min back in culture medium without CD.

Morphological changes associated with the effect of a drug such as CD on cells can be studied alternatively using, for example, fluorescence labelling of the actin fibrils. In order to obtain the full benefit of the

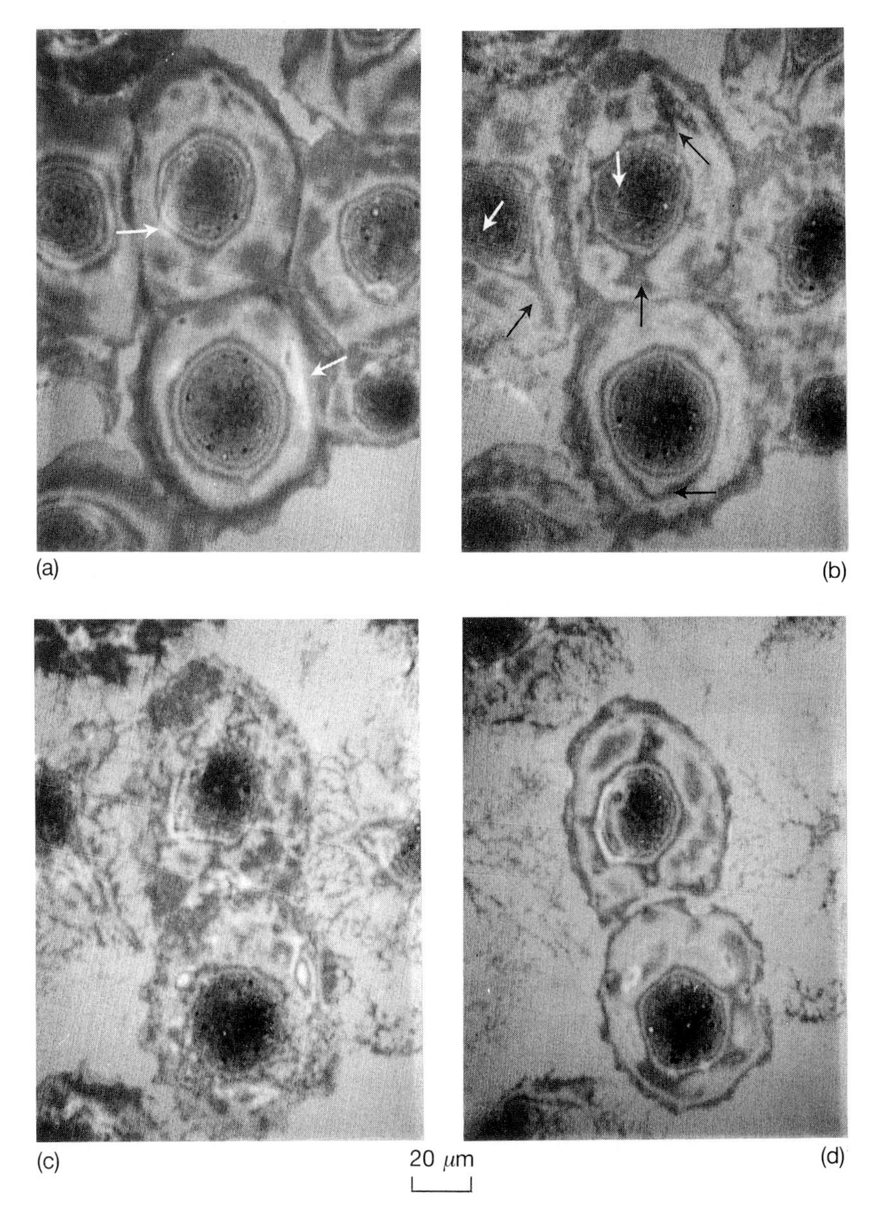

FIG. 9.1. XTH-2 cells derived from *Xenopus* tadpole heart endothelia, on polystyrene, 0.9 GHz: (a) in culture medium; (b) after 75 s in CD (cytochalasin D, 2 μg ml^{-1}); (c) after 10 min in CD; (d) after 15 min in CD-free culture medium again. In (a), the white arrows indicate bright areas where the cells are not in close contact with supporting surface; in (b) the white arrows indicate cell nuclei, which are not visible in (a), and the black arrows indicate deviations of the interference fringe towards the periphery of a cell compared with (a). (Bereiter-Hahn 1987.)

acoustic images they must be interpreted in terms of the information that they give about the elastic properties. As a starting assumption (for which some evidence will be presented later), it may be supposed that the contrast in micrographs such as those in Fig. 9.1 arises from reflections from the top and bottom of the specimen, i.e. from the interface between the coupling fluid and the cell and from the interface between the cell and the substrate. Since the pulse length in a standard acoustic microscope is rather longer than twice the thickness of a cell, the reflections from the top and bottom of the cell interfere. For normal incidence, the intensity reflection coefficient of a thin film on a substrate is

$$R = \frac{Z_1^2(Z_s - Z_0)^2 \cos^2 k_1 d + (Z_s Z_0 - Z_1^2)^2 \sin^2 k_1 d}{Z_1^2(Z_s + Z_0)^2 \cos^2 k_1 d + (Z_s Z_0 + Z_1^2)^2 \sin^2 k_1 d}, \tag{9.1}$$

where the subscripts 0, 1, s refer to the coupling fluid, the specimen, and the substrate, respectively, d is the thickness of the specimen, and k_1 is the wavenumber in it. On this basis interference fringes would be expected that indicate contours of equal thickness of the specimen (assuming constant acoustic velocity), with the spacing between the fringes corresponding to a change of thickness of half a wavelength. The strength of the fringes will depend on the relative values of the impedances of the three media. For example, if the impedance of the specimen is the geometric mean of the impedances of the coupling fluid and the substrate, then the fringes will have the greatest possible contrast, with maximum reflection whenever the specimen thickness is a multiple of half a wavelength. Variations in the strength of the fringe contrast would then indicate changes in the impedance of the specimen.

To some extent the image shown in Fig. 9.1 can be interpreted in terms of this simple interference theory (Bereiter-Hahn 1987). The shape of the cells is like a fried egg; a dome-shaped central area is surrounded by a thin peripheral cytoplasm. If the velocity is constant then the interference fringes indicate contours of equal thickness, so that the central portion has fringes that are quite closely spaced, especially towards its edge, and the relatively flat periphery has very widely spaced fringes. The bright areas in Fig. 9.1(a) that were identified with detachment from the substrate correspond to a low value of impedance (probably that of the fluid) behind the cell (i.e. Z_s in (9.1); if the impedance of the cytoplasm is intermediate between that of the water and that of the substrate, one would expect a phase reversal of the fringes where detachment has occurred). The dark areas in Fig. 9.1(b) that were identified with the spreading of the central cytoplasm out into the periphery are regions where interference between reflections from the top and the bottom interfere destructively. The loss of contrast towards the centre of the domed regions may be accounted for in terms of attenuation in the cytoplasm, which reduces the amplitude of the reflection from the bottom surface. In studies of living cells fast dynamic changes are best followed with a plastic substrate, which gives stronger changes of contrast, while a glass substrate can give stronger interference patterns for quantitative measurements in a standard microscope.

Pictures of blood cells taken in a transmission acoustic microscope were shown in Fig. 2.4. Blood cells have since been studied in a more modern reflection microscope at 0.8 GHz and 1.6 GHz (Schenk *et al.* 1988). Like the pictures of cells from the tadpole heart endothelia, the pictures showed differences in topography, density, elasticity, and absorption revealing effects of haemoglobin content and details of the cell cytoskeleton.

Attenuation can be included in (9.1) by using complex quantities for the wavenumber and the impedances, with the imaginary part corresponding to attenuation at a given frequency. By performing model calculations, it is possible to find, for given mean values of thickness and acoustic properties, frequencies at which the variation of the reflected signal will be most sensitive to changes in either velocity or attenuation (Okawai *et al.* 1987; Tanaka 1989). In this way operating conditions can be selected that give images in which the contrast is due primarily to a chosen parameter, and this can be of use in imaging sections of tissue. A more thorough analysis takes into account the confocal nature of the acoustic microscope, since focusing effects may be very important in determining which properties dominate the contrast.

9.2.2 *Focused interference contrast*

For many purposes the kind of analysis given here of the images in Fig. 9.1, based on eqn (9.1), may be adequate for qualitative work. But (9.1) is for incident plane waves only, and therefore cannot account for the phenomena associated with the focusing action of the lens. For example, if attenuation alone were responsible for the loss of contrast in the centre of the cell, then it should appear grey, with an intensity intermediate between the bright and dark fringes. The fact that it appears rather darker than that may be due to the top of the cell being out of focus, so that the signal is reduced by the $V(z)$ effect. An unapodized lens of semi-angle 60° with a perfect reflector would have the first null in $V(z)$ at a defocus of one wavelength, at which the signal would be reduced by 20 dB compared with focus. This change in defocus corresponds to about two fringes if the velocity in the cell is similar to that in water, and may help to explain why the fringes are lost towards the centre of each cell in Fig. 9.1.

A more thorough analysis of the contrast from cells and its relationship to the cellular elastic properties must take account explicitly of the focusing action of the lens (Hildebrand and Rugar 1984). The forward problem, that is the calculation of the expected signal from a structure with known parameters, is quite straightforward. First the reflectance function can be calculated using the matrix method for a layered structure, to be given in §10.2, and then that reflectance function can be used in the standard expression for $V(z)$. By means of such calculations, it is possible to investigate how the contrast at a given value of z might be expected to vary as a function of thickness, velocity, density, and attenuation in the cell. If, as is often the case, the difference between the impedance of the cell and the substrate is greater than between those of the coupling fluid and the cell, then when the bottom of the cell is at focus the reflection from that surface contributes the strongest component of the reflected signal. Hence, the contrast can be dominated by attenuation of the acoustic waves in their two-way passage through the cell. On the other hand, if the microscope is focused nearer the top surface of the cell, then the effect of $V(z)$ may to some extent compensate the different strengths of the reflections from the top and bottom surfaces of the cell, giving the fringes greater contrast. The amplitude of the fringe contrast may then give a good indication of the impedance of the cell in that region. In cases where the acoustic velocity in the cell can be assumed to be approximately constant, the elastic properties might be determined by taking two acoustic images, one with the surface of the substrate at focus, and one with it moved one or two acoustic wavelengths away from the lens. The thickness of the specimen at a point can be deduced by counting fringes, taking care to decide what thickness corresponds to the fringe at the edge of the cell (it may, for example, be the second dark fringe and not the first one), and assuming constant velocity in the cell. Knowing the thickness, the attenuation can be inferred from variations in contrast of the image obtained with the substrate at focus. Finally, the impedance can be obtained from the amplitude of the contrast in the image taken with the specimen moved away from the lens. An example of this process is shown in Fig. 9.2. The specimen was a fibroblast cell from an embryonic chicken heart ventricle (8–10 days of incubation), placed on a fused quartz substrate. A pair of images is shown with the substrate at focus and at $z = +1.8\ \mu\text{m}$, together with the profiles of thickness, attenuation, and impedance deduced from the images along the line indicated. The attenuation appeared to be roughly constant across most of this particular cell, though it could not be measured where the cell was very thin. Motile cells were also studied on both fused quartz and polystyrene substrates, and changes in impedance, such as the leading edge showing a sharp increase in impedance, have been seen.

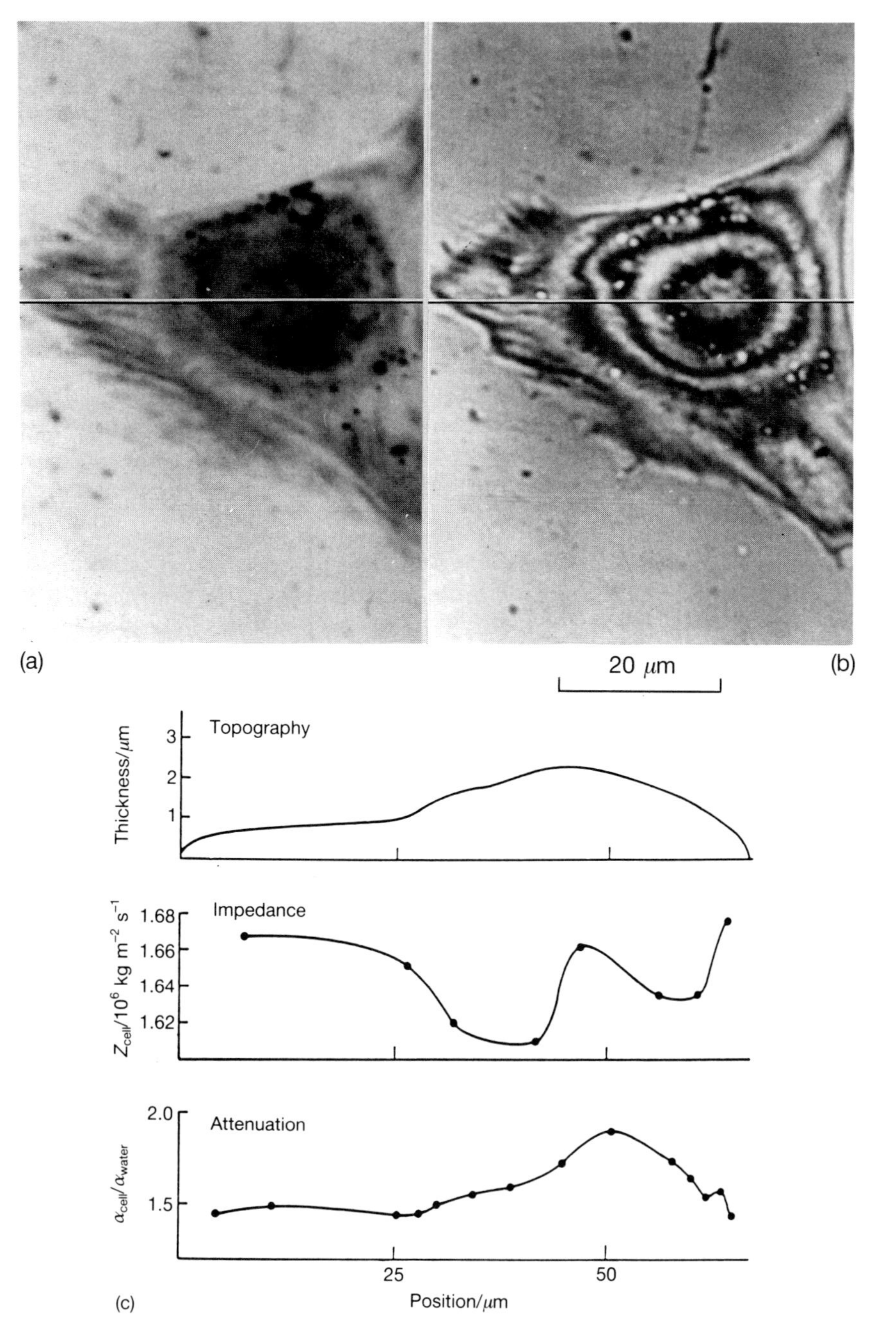

FIG. 9.2. Fibroblast cell from an embryonic chicken heart ventricle (8–10 days of incubation), on fused quartz: (a) $z = 0$; (b) $z = +1.8\,\mu$m; (c) topography, impedance and attenuation measured along the line marked; 1.7 GHz (Hildebrand and Rugar 1984).

It is possible to use a hybrid of the plane wave and focused methods to study cells, by operating at modest values of positive defocus (Litniewski and Bereiter-Hahn 1990). Provided the reflecting surfaces, starting with the top of the cell, are a few wavelengths or more away from the focus of the lens, the effect of defocus is primarily to reduce the effective aperture of the lens. This will, of course, be different for the top and bottom surfaces of the cell, and it can be allowed for by measuring the change in signal with positive defocus on a surface such as Teflon (this is the same as $V_G(z)$ in §8.2, for $z>0$). The strength of the signal from a given surface or interface at a positive defocus z' can be normalized by dividing by the value of $V_G(z')$. It may then be assumed that the effective aperture is small enough that the reflections from the top and bottom surfaces of the cell may be described by coefficients for normal incidence only. The profile of the cell is deduced simply by counting fringes, each fringe corresponding to a change in thickness of $\lambda_c/2$, where λ_c is the wavelength in the cell. Instead of considering the problem in terms of standing waves, or equivalently an infinite series of reflections, as in eqn (9.1), the reflections can be considered one at a time. The amplitude reflection is then

$$A = R_1 V_G(z_1) + R_2 T_1^2 V_G(z_2) \exp(-2\alpha_c d) \cos(4\pi d/\lambda_c) + \ldots . \qquad (9.2)$$

Instead of dividing the net signal on the left-hand side by a reference value of $V_G(z')$, in (9.2) each term on the right hand-side has been multiplied by a value of $V_G(z_n)$ appropriate to the displacement of that interface from focus. The subscripts 1 and 2 refer to the top and bottom surfaces of the cell. The terms that would describe multiple reflections $(+\ldots)$ are disregarded in this approximation. The first term describes the simple reflection from the top surface, and the second term describes the reflection from the interface between the bottom of the cell and the substrate. The exponential describes attenuation in the cell, and the cosine term describes the fact that the second echo may interfere constructively or destructively with the first reflection, depending on whether the cell is an odd or even number of quarter-wavelengths thick at that point. For a cell on a polymer or glass substrate with water or a culture medium as the coupling fluid, $R_2 \gg R_1$ and $T_1 \approx 1$. Therefore, providing the attenuation is not too great, the reflection from the interface between the cell and the substrate dominates, and A has the sign of the second term in eqn (9.2). If a bright fringe, corresponding to the second term being positive, has amplitude A_+, and an adjacent dark fringe, corresponding to the second term being negative, has amplitude

A_-, then

$$\frac{A_+ - A_-}{2V_G(z_1)} \approx R_1, \tag{9.3}$$

$$\frac{A_+ + A_-}{2V_G(z_2)} \approx R_2 T_1^2 \exp(-2\alpha_c d). \tag{9.4}$$

If the second term in (9.2) were less than the first, then tops of the fractions on the left-hand side of (9.3) and (9.4) would be exchanged. Assuming that the density of the cell is known, then by measuring the difference between the brightness of adjacent fringes, the value of R_1 can be deduced. This directly gives the acoustic impedance (§6.4) of the cell,

$$Z_c = Z_0 \frac{1 + R_1}{1 - R_1}. \tag{9.5}$$

If the density ρ_c of the cell is known, then the acoustic velocity in the cell can be immediately deduced, since $v_c = Z_c/\rho_c$. Since determination of acoustic velocity by this method depends on the measurement of relative amplitudes, the amplifiers and their gain controls must be accurately calibrated. The combination of reflection and transmission coefficients on the right-hand side of (9.4) can be expressed in terms of the acoustic impedances of the coupling fluid, the cell, and the substrate.

$$R_2 T_1^2 = \frac{Z_s - Z_c}{Z_s + Z_c} \frac{4 Z_c Z_0}{(Z_c + Z_0)^2}. \tag{9.6}$$

The thickness d can be estimated by counting the number of fringes from the edge of the cell, since each fringe corresponds to half a wavelength, and the wavelength can be deduced from the velocity in the cell and the frequency of the microscope. Hence the attenuation can be calculated from the mean brightness of adjacent fringes, via (9.4).

There are several fundamental assumptions in this analysis. It is assumed that the density of the cell is constant; this may be checked indirectly to some extent by light microscopy. The angle between the top and bottom surfaces of the cell must not be so great as to invalidate eqn (9.2). The slope can be estimated from the fringe spacing; the method is surprisingly tolerant, and a slope of up to 15° can be present before there is serious degradation in the measurement of the impedance ($\Delta Z_c > 0.02$ Mrayl). Provided that the acoustic velocity in the cell is similar to the velocity in the coupling fluid, refraction at the top surface of the cell will not be serious. In counting the fringes to find the thickness of the cell at a given point, it is important that no fringes at the periphery of the cell are missed. In some cases this can be checked by comparison with transmission electron microscopy, and it can be confirmed that at the edge the cells are much less than an acoustic wavelength thick; this means that a bright fringe at the periphery (bright because $Z_0 < Z_c < Z_s$) does indeed correspond to a vanishingly thin layer. The requirement of adequate positive defocus is not difficult to satisfy in a high-resolution acoustic microscope. For example at 1 GHz, $z = +15\ \mu m$ gives a positive defocus of 10 wavelengths. At 35°C this would give an extra attenuation in an aqueous coupling fluid of about 5 dB, which would be acceptable. Most cells have a thickness $d \leq 5\ \mu m$, so that even the top of the cell is still several wavelengths away from the focus.

This technique has been used to measure the changes in the elastic properties of cells in a healthy state, and 3 min after fixation with glutaraldehyde (Litniewski and Bereiter-Hahn 1990). It was found that fast changes were most easily followed on a plastic substrate, while glass, with its higher impedance, was best for accurate quantitative work using eqn (9.2). It was not possible to make measurements in the central 10 μm or so of the cells because the fringe pattern was disrupted by granular and vesicular cytoplasmic inclusions. But it was possible to observe an increase in attenuation, due to protein cross-linking. The change in acoustic impedance was somewhat patchy, which suggested that the tension exerted by the fibrillar system dominated any changes in stiffness due to cross-linking.

9.2.3 *Time-resolved measurements*

Information about the elastic properties of cells and tissue sections can be obtained by using pulses that are sufficiently short to allow the reflections from the top and the bottom to be separated in time. This is analogous to what is done in medical ultrasound, but the pulses can be shorter and more tightly focused. A scan along a single line can be displayed in a form that is a little unfamiliar to conventional microscopists, but which has some similarity with a medical B-scan, It is a time–distance plot of the undemodulated r.f. signal which can be designated $S(t, y)$; an example is given in Fig. 9.3. Figure 9.3(a) is a conventional acoustic image of a fibroblast; Fig. 9.3(b) is a time-resolved $S(t, y)$ along the line through the cell indicated in (a) (Wang *et al.* 1990). The vertical y-dimension corresponds to displacement of the lens, and therefore to position on the specimen, exactly as in a normal image; indeed the vertical axes of Fig. 9.3(a) and (b) are identical. But the horizontal x-dimension in (b) corresponds to time. The brightness corresponds to the signal S at that instant, with white denoting a large positive signal, black denoting a large negative signal, and mid-grey denoting no signal. Thus each horizontal line on the picture is like a single oscilloscope trace or a medical A-scan, but with the instantaneous value being indicated by brightness instead of vertical deflection as it would be on an oscilloscope screen. The complete picture is made up of a series of such time-scans for

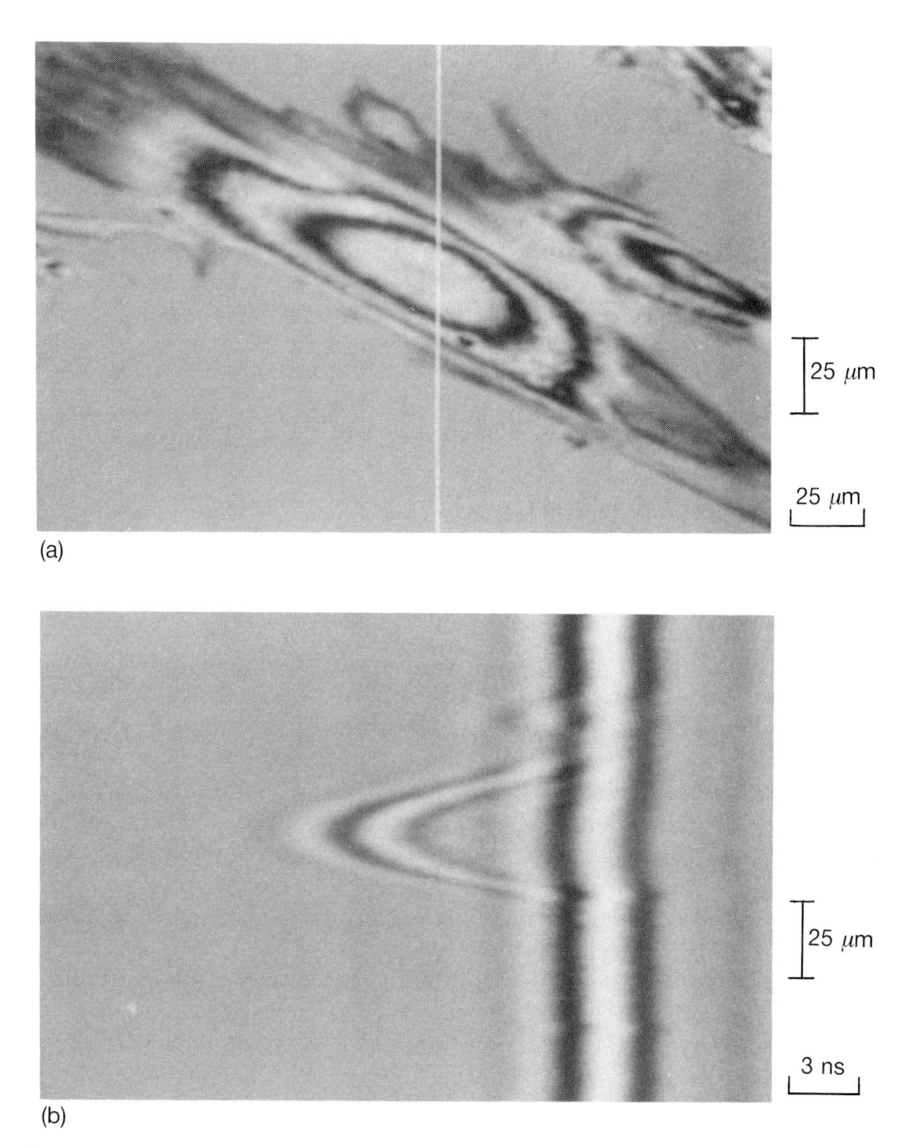

FIG. 9.3. Fibroblast on glass: (a) conventional acoustic image, 550 MHz; (b) time-resolved $S(t, y)$ along the vertical line in (a). In the $S(t, y)$ picture, the horizontal axis is time t; the vertical axis is y just as in (a), and the value of $S(t, y)$ is indicated by the intensity, with mid-grey as zero and dark and light as negative and positive values of S. (Courtesy of Jun Wang.)

each point on the y-scan. The signal at each point is denoted by S to emphasize that it is the actual radio-frequency signal detected by the transducer that is displayed, rather than the envelope-detected video signal which is used to modulate the brightness in an ordinary acoustic micrograph or is measured in a $V(z)$ curve. In principle, it would be possible to measure $S(t, x, y)$ over the whole of the area of the image in Fig. 9.3(a), but this would take a very long time. Whereas in a conventional image each r.f. pulse yields one pixel, here each r.f. pulse produces only one point in the time-scan for a point on the specimen; the data can be recorded either using repetitive sampling methods (e.g. a sampling oscilloscope) or using fast analogue-to-digital converters and an adequate computer system to handle the data. If a complete $S(t, x, y)$ is recorded it is important to think carefully how to analyse and display the data in order to obtain a clear presentation of the properties that are actually of interest.

Analysis of the time-resolved measurements does not require any a priori assumptions about the acoustic properties of the specimen, such as uniformity of velocity (Hildebrand and Rugar 1984) or uniformity of density (Litniewski and Bereiter-Hahn 1990). Self-evidently though, it does require a microscope equipped for short pulses. It also makes considerable demands on the mechanical and thermal stability of the microscope. Even small fluctuations in temperature (and therefore the velocity in the coupling fluid) or the height of the lens will cause errors in the measurements of the echo times, which are amplified when small differences are taken in eqns (8.56) and (8.57) to calculate the thickness and velocity; similarly, changes in the attenuation of the coupling fluid will cause errors in the measured amplitudes of the echoes, which may be amplified when differences are taken in eqns (8.60) and (8.70) to calculate the attenuation. The effects of fluctuations in temperature and lens height can be compensated to some extent by using a reference beam in a configuration similar to the one illustrated in Fig. 5.3.

The $S(t, y)$ recording of the fibroblast can be analysed in the way described in §8.3, specifically here using eqns (8.56)–(8.60); the results are presented in Fig. 9.4. The echo that is present at the top and bottom of Fig 9.3(a) can be used as the reference echo, to give the basic shape of the waveform $s(t)$ and the reference time t_0. The arrival time of the echo from the top surface of the cell in the central part of the picture is t_1, and from this the thickness d of the fibroblast can be deduced using eqn (8.56). For example, near the middle this echo arrives sooner (i.e. towards the left); this corresponds to the top surface of the cell being higher and therefore nearer to the lens, so that the sound does not have so far to travel and therefore gets back sooner. The calculated thickness d is plotted in Fig. 9.4(a). In $S(t, y)$ the shape of the cell appears distorted

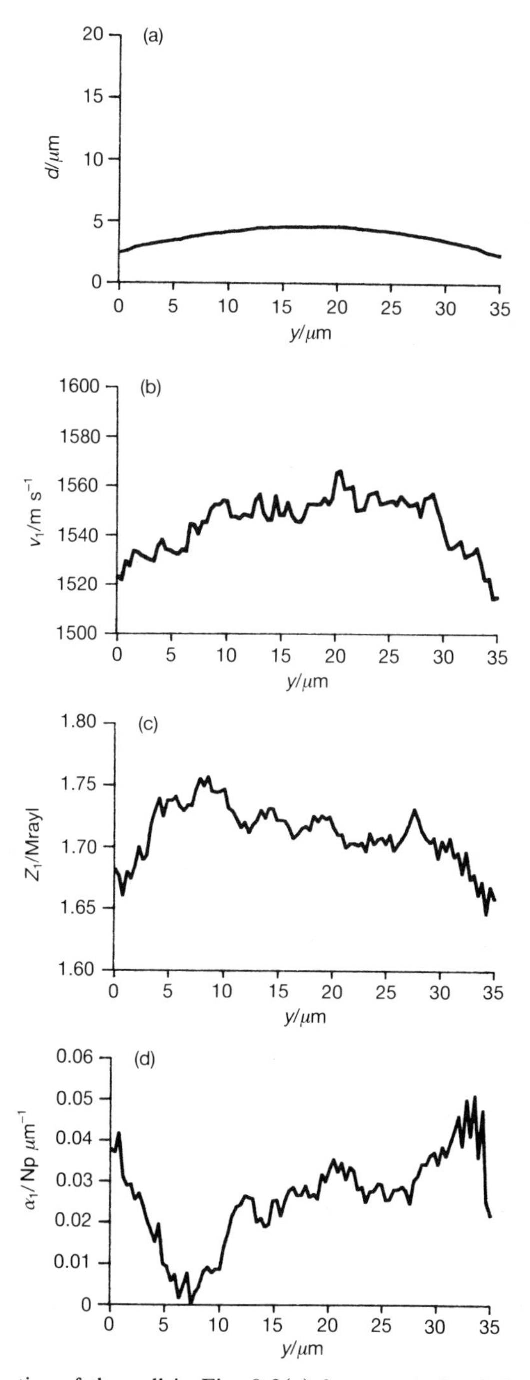

FIG. 9.4. Properties of the cell in Fig. 9.3(a) from analysis of the $S(t, y)$ data in Fig. 9.3(b): (a) thickness; (b) velocity; (c) impedance; (d) attenuation. (Courtesy of Jun Wang.)

because the two axes are not the same, so in the plot of d the axes have deliberately been given the same scale in order to indicate the true shape of the cell. Knowing the thickness, the velocity can be deduced from the arrival time t_2 of the echo from the interface between the cell and the substrate. Equation (8.57) can be rewritten

$$v_1 = v_0 + v_0 \frac{t_0 - t_2}{t_2 - t_1}. \tag{9.7}$$

It is the small amount by which the interface echo arrives earlier than the reference echo that corresponds to the change in velocity. In the $S(t, y)$ of the fibroblast in Fig. 9.3(b), the interface echo is slightly to the left relative to the reference echo above and below it, indicating that the signal travels slightly faster in the cell than it does in the culture fluid. But the difference is only small, and, therefore, in the plot of the velocity in Fig. 9.4(b) the range of velocity is not great. If, as here, $t_0 - t_2$ is close to the sampling interval, especial care must be taken over the handling of the data to yield smooth and accurate variations of the velocity.

The first two quantities calculated from $S(t, y)$ depended on measurements of times; the second two depend on measurements of amplitudes. From the amplitude of the reflection from the top surface of the cell the impedance, and hence the density, can be found using eqns (8.58) and (8.59). The impedance is plotted in Fig. 9.4(c). Finally, knowing the thickness and the impedance of the cell, the attenuation can be deduced from the amplitude of the echo from the interface between the cell and the substrate; the weaker this echo, the greater the attenuation. The attenuation calculated from (8.60), neglecting frequency dependence, is plotted in Fig. 9.4(d). It is also possible to calculate the frequency dependence of the attenuation using (8.70) (Daft *et al.* 1989).

9.2.4 *Beam damage*

At sufficiently high intensities an acoustic beam can interact harmfully with biological specimens (Suslik 1988; cf. Briggs 1990*b*). This has long been a concern in medical ultrasound, especially because at the stage in the use of X-rays in medicine corresponding to the current use of ultrasound in medicine, the harmful effects of X-rays had not been epidemiologically identified. Acoustic waves can cause cavitation *in vivo* at relatively modest powers. For continuous waves the threshold intensity can be a little over 10^{-6} mW μm^{-2} (Ter Haar *et al.* 1986), although for pulsed waves the peak intensity to cause cavitation can be much greater. Cultured cells can begin to demonstrate adverse effects due to prolonged exposure to pulsed acoustic beams of somewhat higher intensity (Maeda *et al.* 1986). In experiments to measure this, pulses at 1, 2, and 4 MHz up to 10 μs long were repeated at a rate of 1 kHz, and the cells were exposed for 1 h. Damage was assessed by measuring the culture growth after 4 and 7 days, and comparing it with a control culture. Growth suppression appeared after exposure to peak intensities above 0.8×10^{-3} mW μm^{-2}. The mechanism of the inhibition of cell proliferation above this intensity is not known, nor is the relationship between the dependence on mean power and the dependence on peak power.

These intensity levels are modest compared with the intensity of the beam at the focus of an acoustic microscope. From §3.1.2, the peak power in the beam at focus can be −10 dBm. At 2 GHz the focused beam area is about 1 μm^2, giving a peak intensity of 0.1 mW μm^{-2}. But the pulse length is usually less than 1 per cent of the interval between the pulses, and any one spot on the specimen experiences the acoustic beam for less than 1/250 000 of the total scan time (for a 512×512 scan). The temperature rise due to the acoustic power will usually be less than 1°C, and in most cases very much less than that (Maev and Maslov 1991). It is not clear how to relate these parameters predictively to the experiments on the effects of ultrasound, but in any case no adverse effects have been observed in an acoustic microscope.

9.3 Histological sections

Sections of biological tissue can give beautiful contrast in transmission acoustic microscopy (Kolosov *et al.* 1987). Figure 9.5 shows acoustic (a) and optical (b) micrographs of a 7 μm section through human skin tissue. The specimens were not fixed or stained; they were cut frozen. The acoustic image was obtained using a frequency of 450 MHz. The bright region on the right of the acoustic image is water; almost nothing is more acoustically transparent than that. The first dark layer is keratin, a dense fibrillar protein, though it is difficult to tell from this picture alone whether all the contrast is due to the high attenuation associated with collagenous tissue, or whether some of it may be due to phase cancellation associated with abrupt velocity changes within the acoustic beam. The homogeneous and relatively transparent layer contains epithelial cells, but very little fibrillar protein. The rest of the area imaged is dermis, which has the most heterogeneous structure of all. The elastic properties of tissue show much greater variations than its (unstained) optical properties, and this is also true of the elastic anisotropy. Figure 9.6 shows acoustic and optical pictures of two 10 μm sections of schlera cut in perpendicular directions. In (a) the view is along the general direction of the collagen fibres (though they are by no means straight), so it is difficult to discern much structure. In (b) the fibres lie more in the plane of the pictures, and it is easier to identify bundles of fibres. It is even possible to use differential phase

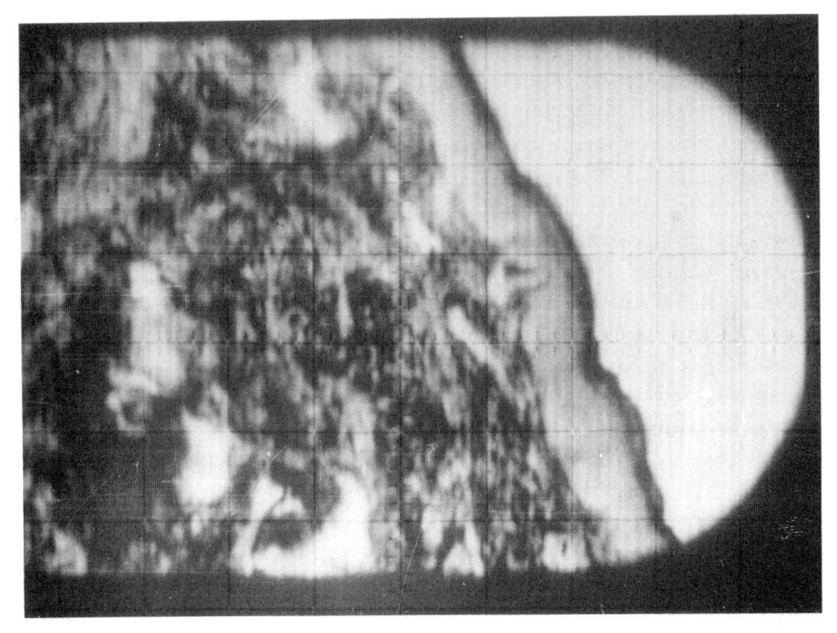
(a)

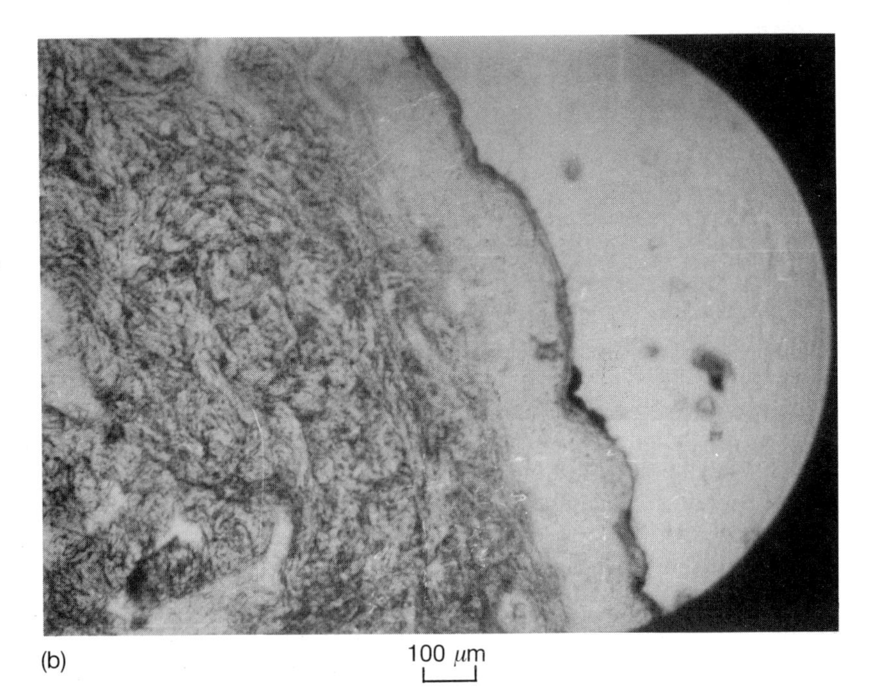
(b)

Fig. 9.5. Section of human skin, 7 μm thick, unstained and unfixed: (a) acoustic, 450 MHz; (b) optical. (Courtesy of Roman Maev.)

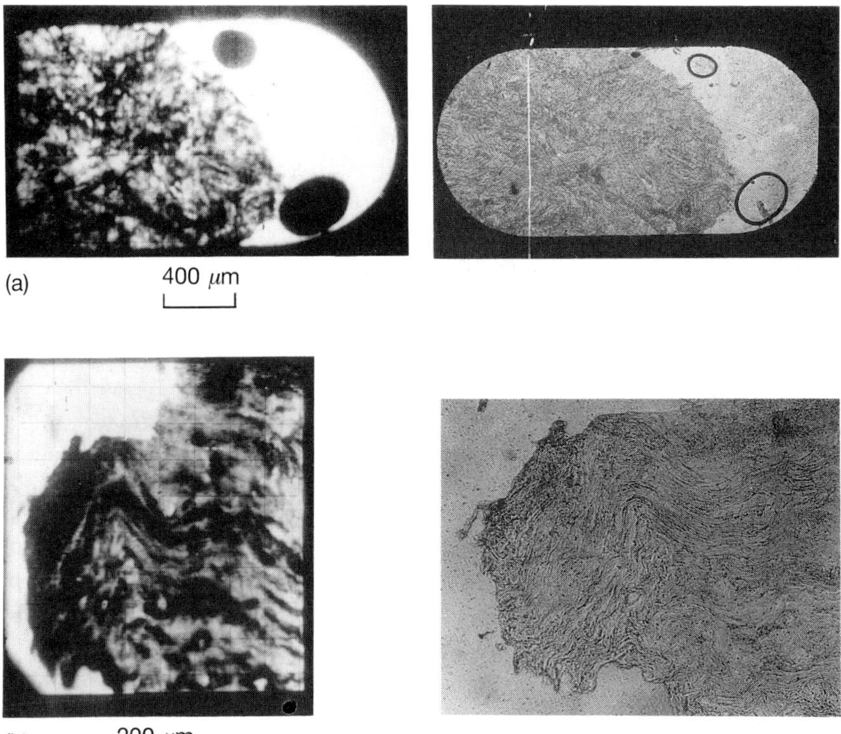

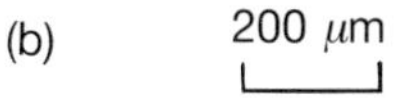

FIG. 9.6. Section of human schlera, 10 μm thick, unstained and unfixed. (a) Cut parallel to the surface: (left) acoustic; (right) optical. (b) Cut perpendicular to the surface: (left) acoustic; (right) optical; 450 MHz. (Courtesy of Roman Maev.)

recording to observe changes in a specimen between pictures (Bennett and Ash 1981), though this requires exceptional thermal stability. Backscatter acoustic microscopy has been developed up to 100 MHz (Sherar *et al.* 1987; Sherar and Foster 1988), and images like medical ultrasound B-scans are obtained. Multicellular tumours and human ocular tissue have been imaged in this way, with a penetration of 300 μm into the tissue. In these pictures, bright contrast comes from scattering of the acoustic waves by elastic homogeneities in the depth of the specimen. The elastic microstructure of botanical sections has also been studied (Kolodziejczyk *et al.* 1988).

The contrast from histological sections in reflection depends not only on the type of tissue, but also on the frequency and the focusing conditions. Sections that contain little structural protein generally have both impedance and attenuation quite similar to the coupling fluid. The acoustic image is then primarily dependent on the signal that passes through the specimen, is reflected by the substrate, and comes back through the specimen. The resulting contrast is generally dominated by variations in attenuation in the specimen. But, when the amount of structural protein in the section is greater, the picture can be quite different. The impedance of tissue containing structural protein such as collagen is significantly greater than that of water, so the reflection from the top surface is strengthened at the expense of transmission through the section; at the bottom surface the opposite happens, and transmission into the substrate increases at the expense of reflection. Moreover, the attenuation can become so great that the contrast due to variations in attenuation can become lost because the signal reflected from the bottom surface of the specimen is too small. The contrast is then dominated by the reflection from the top surface and therefore by variations in impedance. Since attenuation in general increases with frequency, these effects may be expected to become all the more pronounced at higher frequencies. But these generalizations are also tempered by the focusing conditions that are used. The enhanced depth discrimination of a confocal system is becoming familiar through confocal light microscopy (Wilson and Sheppard 1984; Boyde 1987; Shotton 1989; Sheppard and Cogswell 1990). With a wavelength λ_0 and a lens of semi-angle θ_0, the 3 dB (or *half-intensity*) focal range is (Liang *et al.* 1985*b*)

$$\Delta z_{3\mathrm{dB}} = \frac{0.45\lambda_0}{1 - \cos\theta_0}. \tag{9.8}$$

The effects of frequency and defocus are illustrated by Fig. 9.7, which contains four pictures of a section through muscle at two different frequencies and in two different focus conditions (Daft and Briggs 1989*b*). In Fig. 9.7(a) the frequency is 240 MHz, and the surface of the substrate is in focus. The lighter tissue at the top and bottom contains muscle cells, the darker tissue across the lower centre is collagenous. The specimen had been cut in a freeze-microtome to a nominal thickness of 14 μm. Figure 9.7(b) shows the image obtained when the substrate surface was moved 14 μm away from focus, so that notionally the top surface of the specimen was at focus. While there are some changes in the contrast from the muscle tissue (bearing in mind that the gain settings of the microscope may have been changed so that the absolute contrast levels may not be the same), the most noticeable effects are in the collagenous tissue. More structure is seen, and in some parts the contrast has reverses from dark to bright. It must be emphasized that this contrast reversal is for the reasons described in the last paragraph, and has nothing whatever to do with the contrast reversal due to Rayleigh wave interference that will be described in §9.4. The effects are even more marked at higher frequency. Figure 9.7(c) shows the same area with the

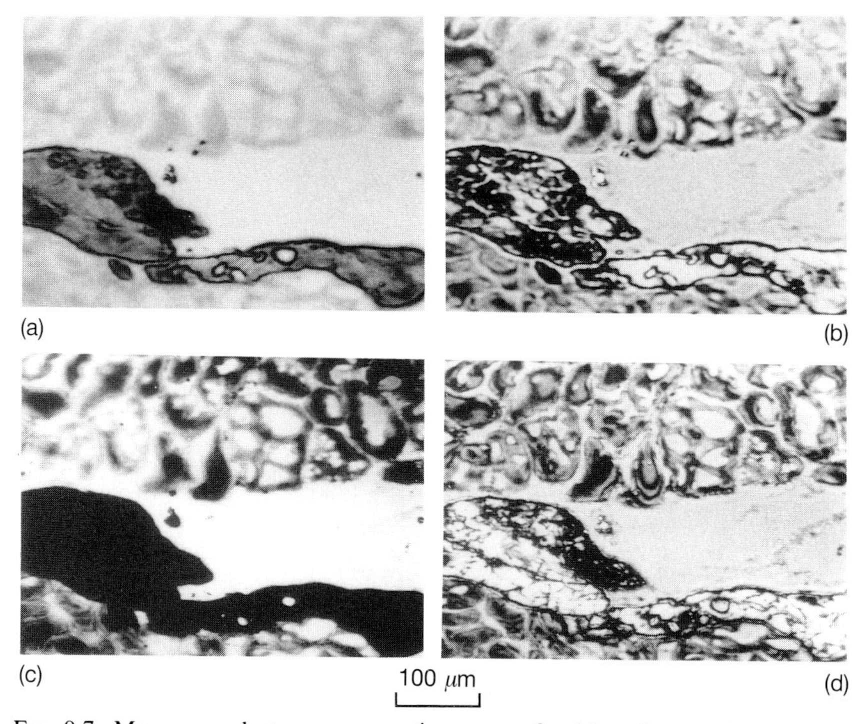

FIG. 9.7. Mouse muscle transverse section, cut unfixed in a freeze-microtome to a nominal thickness of 14 μm, on a glass substrate: (a) $z = 0$, 240 MHz; (b) $z = +14\ \mu$m, 240 MHz; (c) $z = 0$, 425 MHz; (d) $z = +14\ \mu$m, 425 MHz. The datum of z is the top surface of the substrate (Daft and Briggs 1989*b*).

substrate surface again at focus but at 425 MHz. Now the collagenous tissue looks quite black, because the acoustic waves have been so strongly attenuated in their passage through it. But there is plenty of detail in the contrast from the muscle tissue. When the substrate is moved 14 μm away from focus, so that the top of the specimen is again at focus, shown in Fig. 9.7(d), there is a wealth of contrast in the collagen. This contrast depends primarily on variations in the acoustic impedance, and the detail is finer than in Fig. 9.7(b) because the higher frequency gives better resolution.

Histological sections of tissue can be measured using the time-resolved technique (Daft and Briggs 1989*a*,*b*). Indeed, this is almost the only way to study them quantitatively if their thickness is not known for certain, because even in thin sections they do not give the nice fringe patterns that enable the profiles of cells to be deduced. Figure 9.8 shows images of a transverse section of mouse muscle. A lower leg muscle of a mouse was

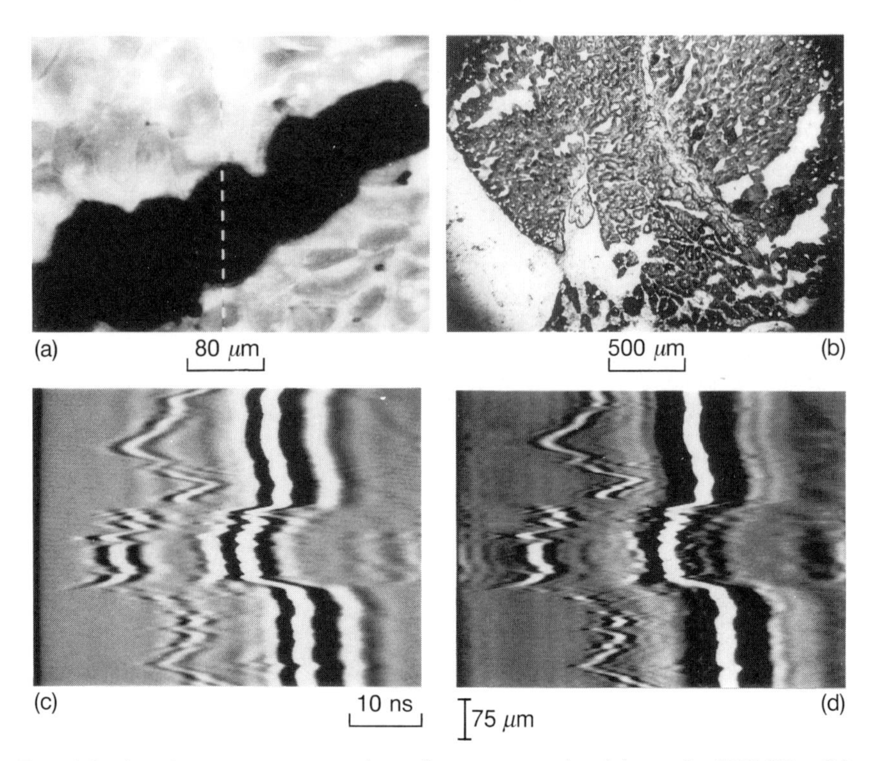

FIG. 9.8. Another transverse section of mouse muscle: (a) $z = 0$, 425 MHz; (b) optical; (c) time-resolved $S(t, y)$ along a line 600 μm long just to the left of the dashed line on (a); (d) $S(t, y)$ after Wiener filtering the data in (c) (Daft and Briggs 1989*b*).

dissected and frozen in liquid nitrogen. A section nominally 14 μm thick was cut using a cryo-microtome, and immediately deposited on a glass microscope slide without using any special adhesive. An acoustic micrograph taken with conventional gated continuous wave excitation and with the surface of the glass substrate at focus is shown in Fig. 9.8(a). Muscle tissue is visible in the upper part of the picture and in the lower right corner. The black region in the middle is connective tissue, which has a high concentration of collagen fibres. Another section was cut and stained with toluidine blue; an optical micrograph is shown in Fig. 9.8(b) to help relate the acoustic images to pictures that may be somewhat more familiar. Figure 9.8(c) shows an $S(t, y)$ scan, taken along a line just to the left of the dashed line on Fig. 9.8(a) and covering a length somewhat greater than the height of the picture. The data of Fig. 9.8(c) has been sharpened up by Wiener filtering (§8.3; eqn (8.63)) in Fig. 9.8(d). The effect of the Wiener filter is to shorten the main part of the pulse, while at the same time giving more prominence to ripples before and after it due primarily to the reference pulse having a different spectral content from the returning signal.

The variation of several properties can be seen simply by inspection of $S(t, y)$ in Fig. 9.8(c) and (d). First, from the wild variations of the arrival time of the echo from the top surface of the tissue section it is apparent that the top is by no means flat. The bottom surface of the section seemed to be in good contact with the glass substrate, so there must have been large variations in the thickness of the section. Second, the top surface echo is considerably stronger over the collagenous tissue, indicating that it has higher impedance than the muscle tissue. Third, the bottom surface echo occurs at an earlier time in the collagenous region, indicating a higher velocity. It is not easy to say much about the density and the attenuation from the pictures alone; for those properties more exact analysis is needed.

Quantitative analysis of the data is presented in Fig. 9.9, in which the thickness (a), velocity (b), impedance (c), and frequency-averaged

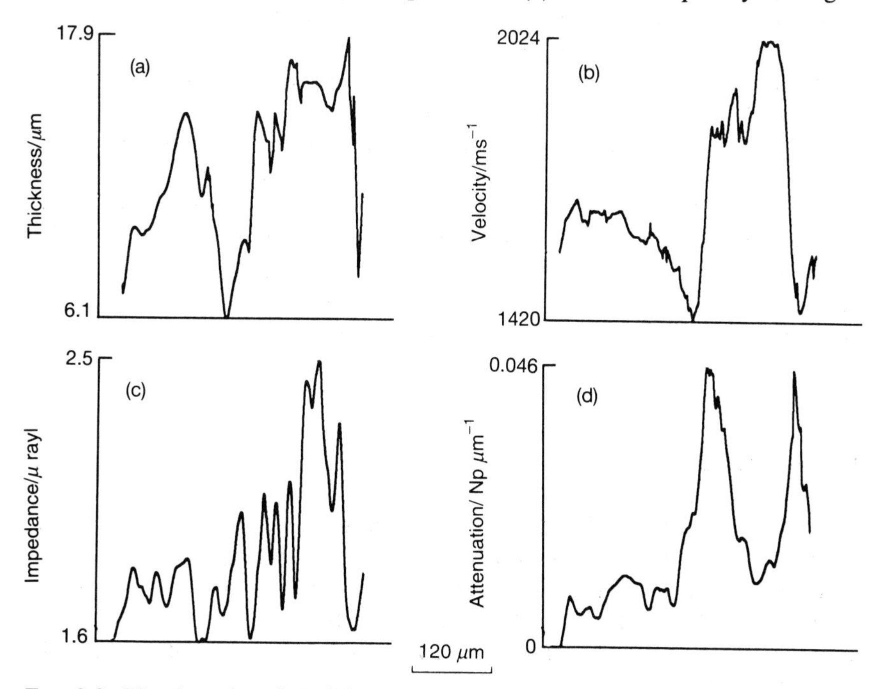

FIG. 9.9. Elastic properties of the transverse section of mouse muscle in Fig. 9.8. The horizontal axis (from left to right) of each plot corresponds to 600 μm along the vertical axis (from top to bottom) in Fig. 9.8(c), (d). (a) Thickness; (b) velocity; (c) impedance; (d) frequency-averaged attenuation (Daft and Briggs 1989*b*).

attenuation (d), deduced from the $S(t, y)$ data are given. The horizontal axis (from left to right) of each figure corresponds to the vertical axis (from top to bottom) of Fig. 9.8(c) and (d). The increased thickness, velocity, and impedance of the collagenous tissue that were inferred qualitatively are nicely brought out in the quantitative analysis. In the attenuation plot it seems that there is strong attenuation (400 dB mm^{-1}) at the boundaries between the muscles and the connective tissue. This may not be genuine; it may be rather that where the acoustic beam passes through a strongly inhomogeneous region, destructive interference effects occur and give the appearance of attenuation in the specimen. Dark lines can be seen at some of the interfaces between muscle and collagen in Fig. 9.7, and probably have the same explanation. Frequency analysis of the attenuation showed remarkable differences between the muscle and the collagenous tissue (Daft *et al.* 1989). The signals contained useful information over the frequency range 120–480 MHz. In the muscle tissue the attenuation was roughly proportional to the frequency over this range. But in the connective tissue the attenuation showed a minimum at about 240 MHz, above which the attenuation rose quite rapidly. It has been speculated that this might correspond to a resonance associated with the diameter of collagen fibres; other samples of collagenous material have also shown minima in attenuation, sometimes at even higher frequencies. It is probable that all the properties of the tissue described here are anisotropic, and that a section cut with a different orientation would yield different results containing additional information. It would also be nice to be able to measure the non-linear parameters of tissue, with the wealth of information that they provide (Bjørnø and Lewin 1986), and this might be possible using a single toneburst technique demonstrated at lower frequencies (Nikoonahad and Liu 1990).

9.4 Stiff tissue

All the studies described so far in this chapter have been on tissue whose density and elastic properties are not very different from those of water. This means that refractive indices and impedance mismatches are not too large (especially at the interface between the coupling fluid and the specimen). Moreover, shear waves have such a small velocity (generally less than 50 m s^{-1}), and such a high attenuation, that they play very little role in the tissue. Only longitudinal waves need be considered, and their behaviour has much in common with the behaviour of light in a confocal microscope (and to some extent in a conventional light microscope too). Therefore, a biologist may be able work to a large extent with concepts that have become intuitive to him from his previous experience of light microscopy (Howie 1987). With stiff tissue, however, the situation is quite different. Impedances are higher, the refractive indices can be several times bigger than they could ever be with light, and shear wave excitation is important. Much of the theory described in Chapters 6–8 needs to be understood, and, in particular, a key role is played by the Rayleigh waves with which so much acoustic microscopy of materials is obsessed.

9.4.1 *Lesions in tooth enamel*

A section through a human tooth is shown in Fig. 9.10. The tooth had been extracted from a patient (for orthodontal reasons, not just for this experiment). It was then exposed to an acid environment, to simulate the process of demineralization that leads to the formation of a caries lesion. Finally, it was cut in two longitudinally (i.e. roughly parallel to the growth direction), and one of the exposed surfaces was carefully polished. The area of the lesion was too big to be scanned in one go, and so four images have been mounted in a collage. In Fig. 9.10(a) the polished surface lies at the focus of the lens. The anatomical surface is towards the top of the picture; the region above that is simply mounting resin. The contrast is disappointing. It is possible to make out the interface between the enamel and the dentine, and also a hazy darkish region in the enamel, but not much else. Figure 9.10(b) is the same area at the same magnification; the difference is that the specimen has been

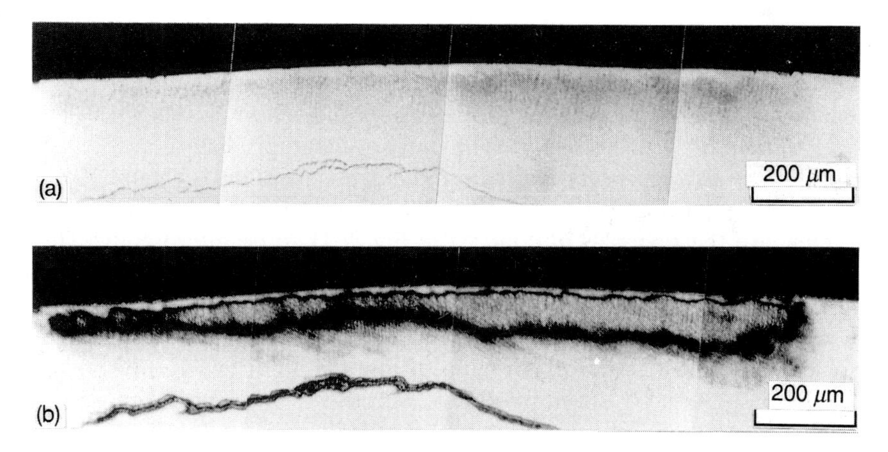

FIG. 9.10. Artificial caries lesion in human tooth enamel by the ten Cate method: (a) $z = 0$; (b) $z = -14\ \mu\text{m}$; 370 MHz. (Peck and Briggs 1986.)

moved 14 μm towards the lens. Now a great deal more of the structure of the lesion is apparent. At the advancing front of the lesion (the part nearest the dentine), there is a gradation in properties. On the side nearest the anatomical surface the changes are much more abrupt; indeed, there is a region about 20 μm wide that shows uniform contrast similar to that of the healthy enamel away from the lesion. When lesions of this type occur naturally they have a slightly milky appearance due to the way that light is scattered in them (Angmar-Månsson and ten Bosch 1987), so they are often called white spot lesions. Provided the thin layer of healthy enamel at the surface remains intact, it is possible for the lesion to recover naturally. It is, therefore, of great interest to study the changes in mechanical properties associated with the formation of the lesion.

The variation of contrast with defocus can be illustrated by a series of snapshots from the examination of another caries lesion, this time one that had occurred naturally in a tooth (Peck and Briggs 1987). The sequence in Fig. 9.11 shows what would be seen over the shoulder of the microscopist as he varied the defocus. In the first picture the surface of the section is at focus, and in each subsequent picture the specimen was moved 6 μm towards the lens. At focus (Fig. 9.11(a)), $z = 0$, no contrast is visible, except for the anatomical surface at the left of the picture; the lesion in this specimen may well have been at an earlier stage than the more severe artificial lesion in Fig. 9.10. At modest defocus (Fig. 9.11(b)), $z = -6\,\mu$m, the lesion appears lighter than the rest of the enamel; some detail can also be seen of the prismatic structure in the healthy enamel around the lesion. As the defocus is increased (Fig. 9.11(c)), $z = -12\,\mu$m, the contrast reverses, and the lesion appears dark relative to the surrounding enamel. At further defocus still (d), $z = -18\,\mu$m, the centre of the lesion becomes bright again, and the dark area extends beyond what appeared at $-12\,\mu$m defocus; there is also some evidence of the intact surface zone. For comparison, a polarized light micrograph and a microradiograph of the same lesion are shown in Fig. 9.11(e) and (f), respectively. The microradiograph has a particularly straightforward interpretation: where the exposure is darker more X-rays have penetrated, indicating that there is less mineral in the section.

How are the reversals of contrast in Fig. 9.11 to be interpreted? They are related to the changes in density revealed by polarized light microscopy or by microradiography (Peck and Briggs 1986), but clearly it is not a simple relationship. We cannot say simply that a region of light (or dark) contrast indicates a region of low (or high) density, or of high (or low) stiffness or elastic modulus, because a given area can appear light and dark in different pictures, depending on the defocus. So in order to interpret the acoustic images we must ask how the contrast varies with defocus. The effect of defocus on contrast is best thought of

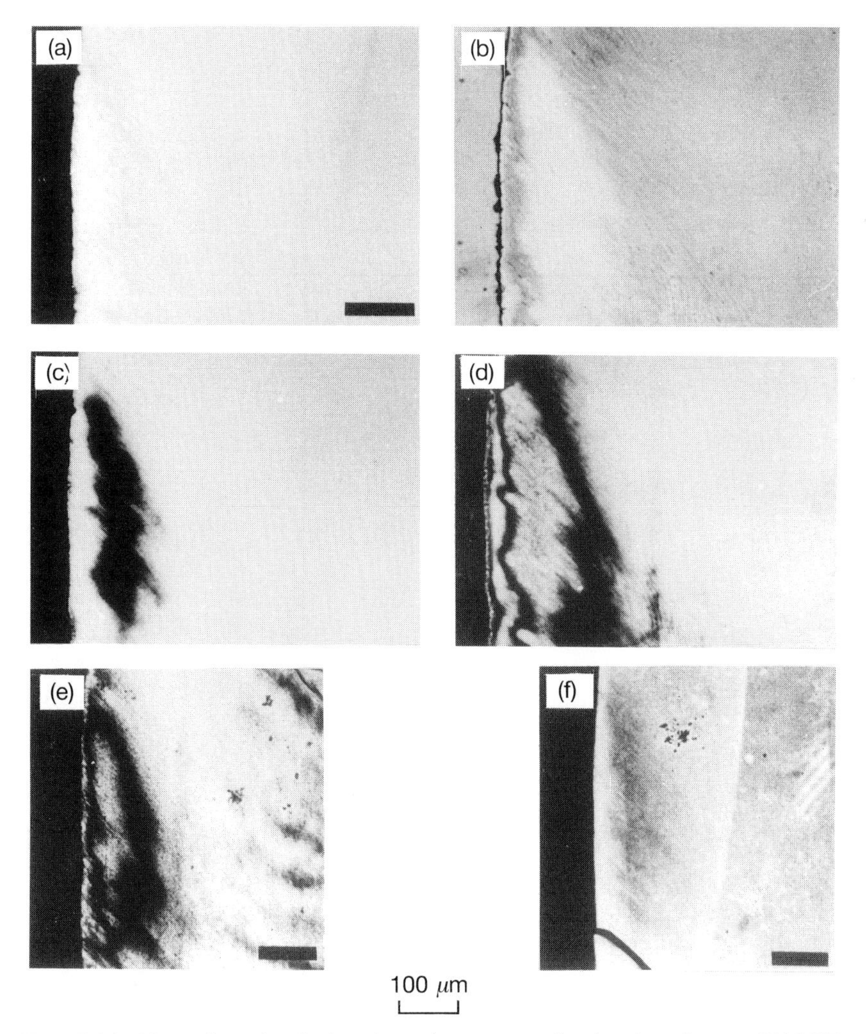

FIG. 9.11. Natural caries lesion in a human tooth: focal series at 370 MHz. (a) $z = 0$; (b) $z = -6\ \mu m$; (c) $z = -12\ \mu m$; (d) $z = -18\ \mu m$; (e) polarized light (water imbibed); (f) microradiograph. (Peck and Briggs 1987.)

in terms of a $V(z)$ curve, which portrays the variation of the video (or envelope-detected) signal V as a function of the defocus z. The video signal is what determines the brightness of the image at any point. The defocus z is defined as the distance of the surface of the specimen beyond the focus of the lens, a negative value of z (which usually gives the most interesting contrast) indicating defocus towards the lens.

In order to illustrate this, Fig. 9.12 shows three pictures of a specimen

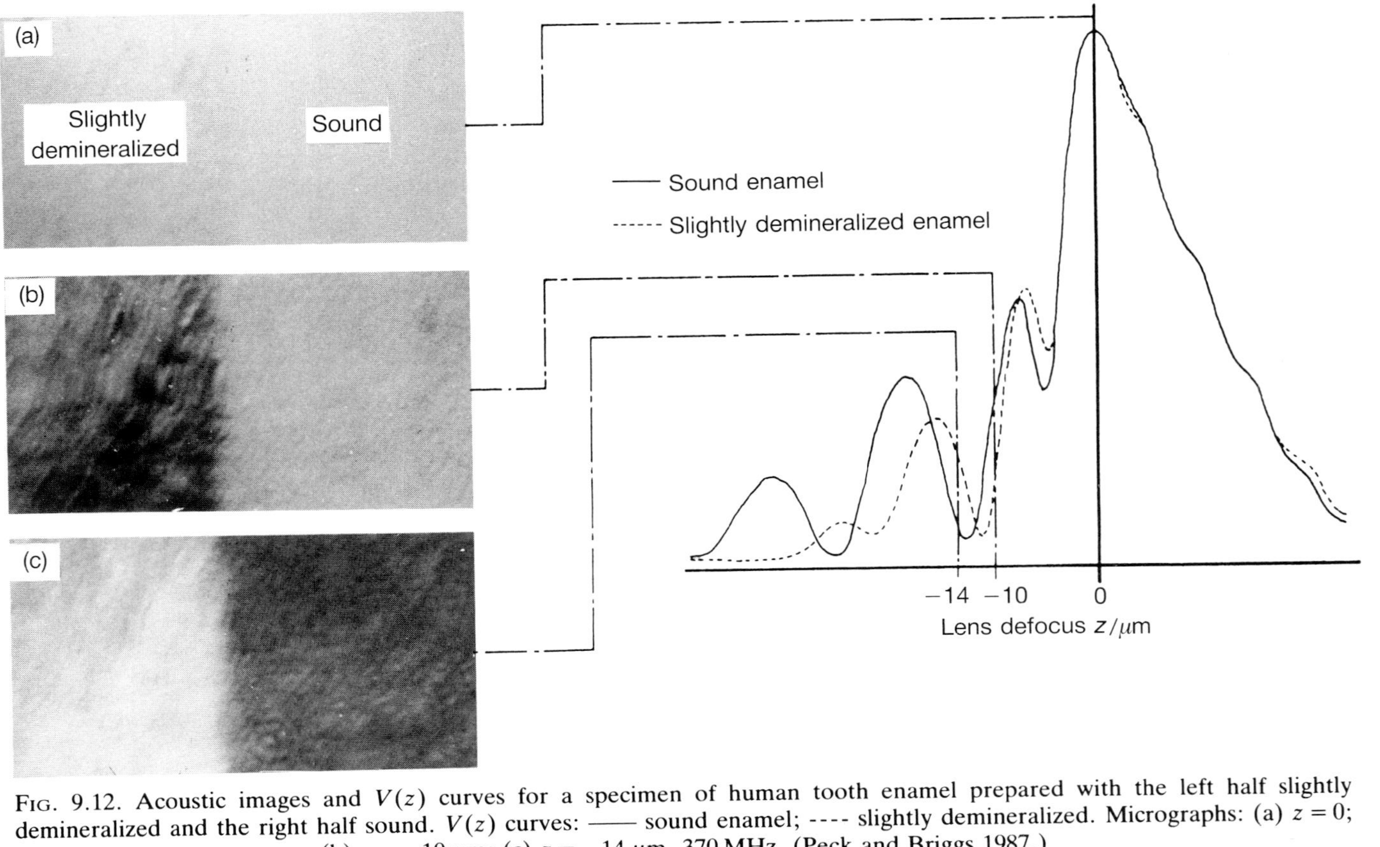

FIG. 9.12. Acoustic images and $V(z)$ curves for a specimen of human tooth enamel prepared with the left half slightly demineralized and the right half sound. $V(z)$ curves: —— sound enamel; ---- slightly demineralized. Micrographs: (a) $z = 0$; (b) $z = -10\,\mu$m; (c) $z = -14\,\mu$m. 370 MHz. (Peck and Briggs 1987.)

of enamel that was specially prepared so that the left half was slightly demineralized and the right half was sound. $V(z)$ curves were measured over each half of the specimen, and these are shown as a continuous curve for the sound enamel and a broken curve for the demineralized enamel. At focus (and indeed for a fair range of z either side of focus, especially positive defocus) the two $V(z)$ curves lie almost on top of each other. This implies that at focus the two halves of the specimen will give almost identical contrast, and this is what is seen in the top picture. By a defocus of $z = -10\,\mu m$, the two curves have separated, with the sound enamel curve lying above the demineralized curve. This corresponds to the demineralized enamel appearing darker than the sound enamel, which is what the second picture shows. As the defocus is increased to $z = -14\,\mu m$, the two curves swap over, so that the demineralized curve lies above the sound one, as indeed it did briefly around $z = -5\,\mu m$. This means that there should be a contrast reversal, with the demineralized enamel now appearing bright relative to the sound enamel, which is precisely what is seen in the third picture.

Because of the great sensitivity of the contrast to the distance between the lens and the point being imaged at any moment in the scan, it is essential that the surface of the specimen must be flat to rather better than a wavelength. This requirement is much more severe than simply the half-intensity focal range Δz_{3dB} described by eqn (9.8). As can be seen from the two $V(z)$ curves, the most interesting contrast often occurs when the gradient of $V(z)$ is steep, and this exacerbates the need to get the surface flat in order that contrast from topography should not mask the much more significant contrast from the changes in the elastic properties within the surface. The specimens of tooth described here were prepared by sectioning, and then grinding with successively finer grades of silicon carbide paper ending with 1200-grade. For the final polishing it was essential to avoid relief polishing (for example, removing the weaker enamel in the lesion to a greater depth than the sound enamel). They were therefore mounted in a precision polishing jig, and lapped against a solder-faced lapping wheel impregnated with oil-soluble $1\,\mu m$ diamond paste. The flatness of the finished surface can be checked by interference microscopy; a flatness better than 300 nm was obtained over the area examined.

When studying any specimen in which contrast reversals are seen as a function of defocus, it is invaluable to keep mind the relevant $V(z)$ curves such as the two shown in Fig. 9.12, and to relate the effects of defocus to them. The curves in Fig. 9.12 are purely experimental, and no theory has been presupposed in presenting them or relating them to the contrast variations in the pictures. But, but, but, . . . (before too many sighs of relief are breathed at having escaped theory altogether), if they can be related to theory, then it is possible to extract a great deal of additional information from them. The oscillations in the two $V(z)$ curves in Fig. 9.12 are due to the mechanism described in §7.2.1, namely, interference between the geometrically reflected waves and the ubiquitous Rayleigh waves that are excited in the surface of the enamel. The path lengths of the two contributions to the signal change at different rates with defocus, and so the components alternate between being in phase and out of phase. Peaks in $V(z)$ correspond to constructive interference, and dips to destructive interference. The propagation of Rayleigh waves in the enamel is affected by the demineralization, so the $V(z)$ curve measured over the slightly demineralized enamel has oscillations whose periodicity and amplitude are different from those of the curve measured over the sound enamel. The period is related to the Rayleigh angle, and hence the Rayleigh velocity, by the formulae given in eqns (7.7) and (7.8).

The way the $V(z)$ curves change over enamel containing a white spot lesion is illustrated in Fig. 9.13 (Peck *et al.* 1989). At each of the points labelled by letters along the line across the centre of the picture in Fig. 9.13(i), measurements of $V(z)$ were made. Four of these are presented in Fig. 9.13(ii); the letters correspond to the points in Fig. 9.13(i). Curve (a) comes from the intact surface zone. There are three well pronounced maxima apart from the focal maximum, with a fourth trying to appear. The period of the oscillations is about 13 μm. Curve (f) comes from the core of the lesion. The attenuation is much more severe here, and so the heights of the maxima fall off much more rapidly away from focus. The period is also smaller; it is reduced to about 10 μm, corresponding to a considerably lower velocity. Curve (k) is taken from the advancing front of the lesion. The heights of the peaks have returned to something close to those in the first curve, and the period is similar to that curve too. Finally, in curve (m), taken from healthy enamel beyond the front of the lesion, there is little change in the attenuation (as indicated by the decay in the heights of the peaks), but there is some increase in the period, indicating greater Rayleigh velocity.

A qualitative discussion such as this is useful in seeing how to relate the key features of a $V(z)$ curve, namely the period of the oscillations and their rate of decay, to velocity and attenuation of Rayleigh waves in the surface of the specimen. But it is also possible to analyse the $V(z)$ measurements more accurately. The $V(z)$ measurements for each of the lettered points in Fig. 9.13(i), not just the four shown in Fig. 9.13(ii), were analysed using the method described in §8.2.2. The Rayleigh wave velocity and attenuation are plotted as filled and open circles in Fig. 9.13(iii); the curves have no significance except to indicate the trends.

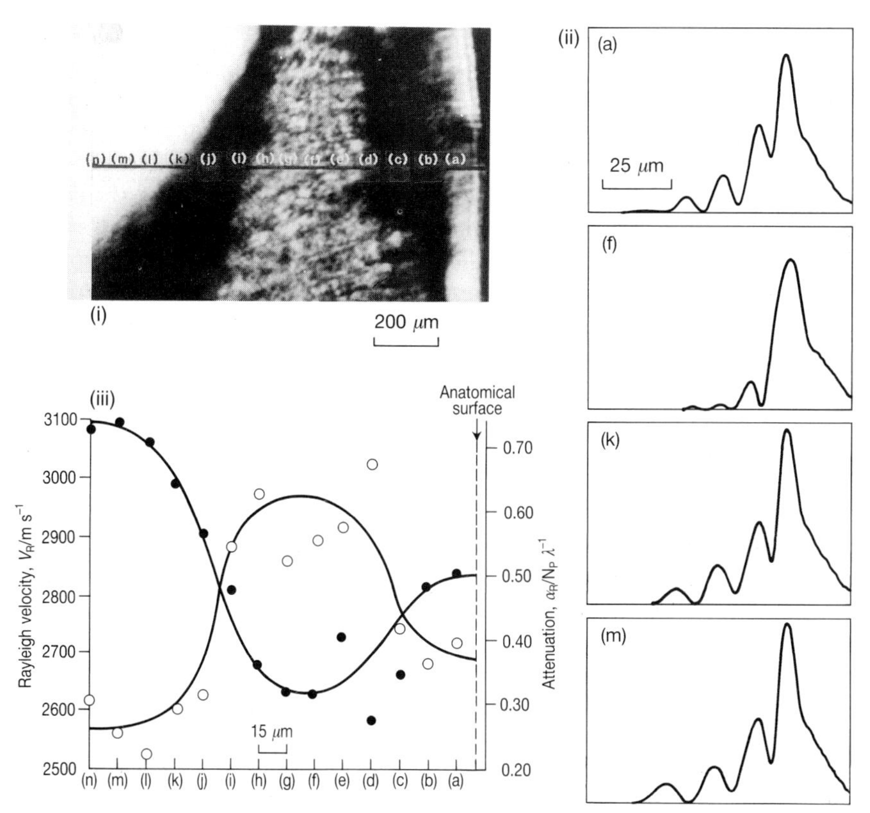

FIG. 9.13. Quantitative analysis of the contrast from a white spot lesion in human tooth enamel: (i) micrograph, 370 MHz; (ii) $V(z)$ curves of selected points along the line in the micrograph; (iii) Rayleigh velocity and attenuation calculated from $V(z)$ measured at each of the points on the line in (i). (Peck *et al.* 1989.)

There is considerable scatter, but the overall pattern can be discerned. The velocity drops from about $2800\,\mathrm{m\,s^{-1}}$ near the anatomical surface to a value $200\,\mathrm{m\,s^{-1}}$ lower in the core of the lesion. Then it gradually increases through the advancing front to nearly $3100\,\mathrm{m\,s^{-1}}$ in the sound enamel beyond the lesion. The attenuation has almost the opposite behaviour. It is relatively low near the anatomical surface, it is much greater in the core of the lesion, and in the sound enamel it decreases to less than half the value in the lesion. The increase of attenuation in the lesion is probably due to additional scattering within the enamel as a result of the demineralization, though there may also be an increase in the leaking into the coupling water due to decrease in both velocity and density. The decrease in velocity in the lesion is particularly interesting. The reduction in density in the core of a lesion as a result of demineralization can be 25 per cent or more (Darling 1956; cf. the radiograph in Fig. 9.11(f)). Since acoustic velocity varies inversely as the square root of density, then if the loss in density were the sole effect of demineralization an increase in velocity of 15 per cent or so would be expected. But what is observed is just the opposite, a decrease in Rayleigh velocity in the core of the lesion of 15 per cent or so compared with the sound enamel. Since acoustic velocity increases with the square root of the elastic stiffness, this implies that in the lesion there is a reduction in stiffness of about twice the magnitude of the decrease in density that is deduced from other techniques.

9.4.2 *Anisotropic properties of tooth enamel*

The measurements on tooth enamel described above were analysed on the assumption that the enamel is, over the scale of each measurement, an isotropic homogeneous medium. The radius of the maximum area sampled in a measurement is

$$r_{\max} = -z_{\max} \tan \theta_{\mathrm{R}} \tag{9.9}$$

with $\theta_{\mathrm{R}} = \sin^{-1}(v_0/v_{\mathrm{R}})$ by Snell's law (7.8); $z_{\max}$ is the maximum negative defocus from which data is used in the analysis. Of course, the sample is heavily weighted towards the centre of this area, because at small defocus only a smaller circle in the centre is measured and, even at larger defocus, the Rayleigh wave amplitude is concentrated towards the centre (§3.6). In the results in Fig. 9.13, $-z_{\max}$ was about 45 μm, corresponding to a maximum sampled radius of about 25 μm. Tooth enamel is neither isotropic nor homogeneous over this kind of distance scale (Boyde 1987). The inhomogeneity can be studied only by using higher acoustic frequencies, up to the practical limit of 2 GHz. The anisotropy can be measured using the line-focus-beam technique with a cylindrical lens which was described in §8.2.1 (Kushibiki and Chubachi 1985).

Figure 9.14 shows measurements that were made on a buccal surface of a human molar tooth (the buccal surface is parallel to the cheek). The surface was prepared by polishing it in the way described in §9.4.1 to give a flat area large enough for the line-focus of the cylindrical lens (i.e. greater than 1.73 mm in every direction). It was then mounted in a line-focus-beam microscope with a rotating stage, and measurements of $V(z)$ were made over 180° at 10° intervals. The sketches are at the angles at which the Rayleigh wave propagation would have been roughly perpendicular and parallel to the long axis of the tooth. The Rayleigh velocity is marked by filled circles, and attenuation by open

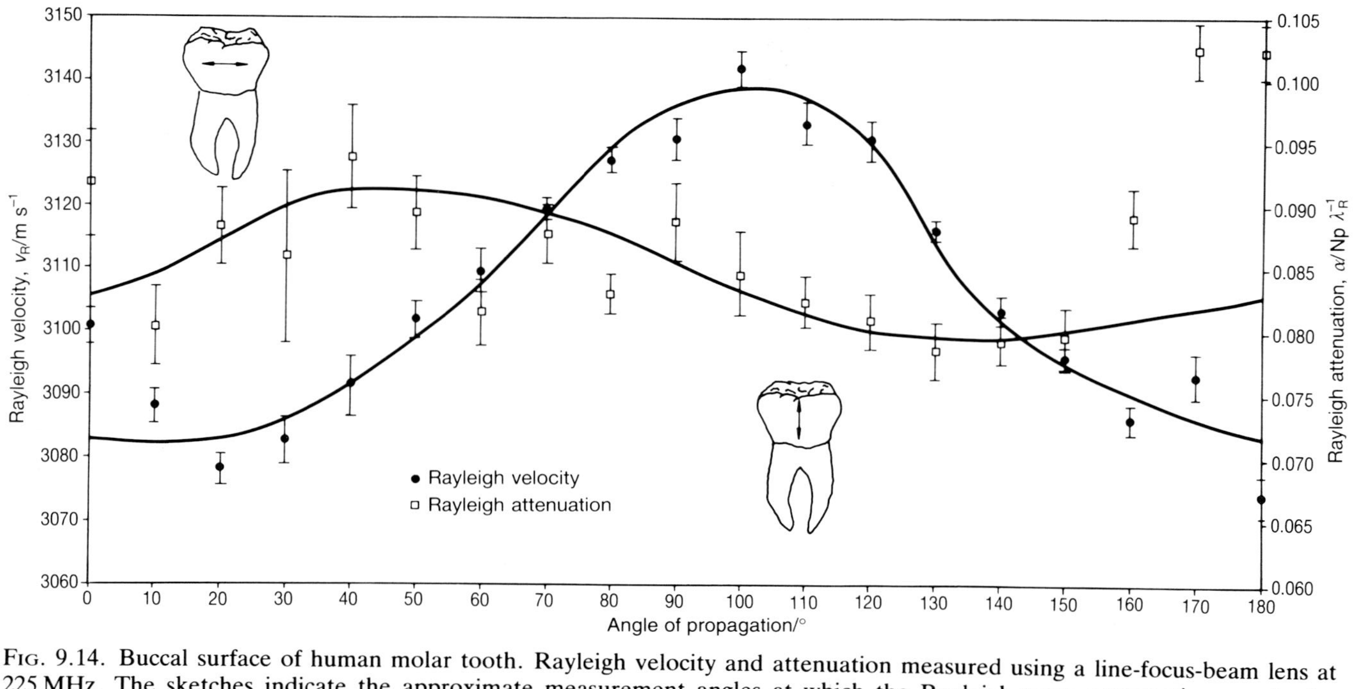

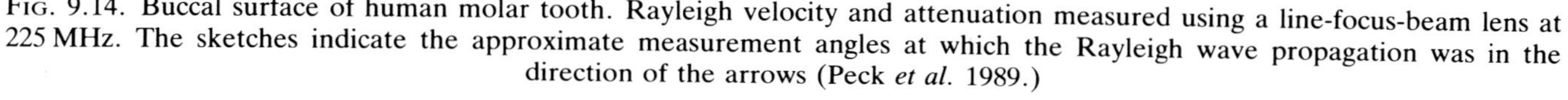
Fig. 9.14. Buccal surface of human molar tooth. Rayleigh velocity and attenuation measured using a line-focus-beam lens at 225 MHz. The sketches indicate the approximate measurement angles at which the Rayleigh wave propagation was in the direction of the arrows (Peck *et al.* 1989.)

squares. Once again the curves have no significance except to indicate trends. The velocity is least perpendicular to the long axis of the tooth, and greatest parallel to it, and the difference between the two is about $60\,\mathrm{m\,s^{-1}}$. It is difficult to be confident about any systematic trend in the attenuation. In an independent study (Kushibiki *et al.* 1987), line-focus-beam measurements on the front surface of a human incisor yield a range of Rayleigh velocities from about $3106\,\mathrm{m\,s^{-1}}$ to $3153\,\mathrm{m\,s^{-1}}$; the greatest velocity again occurring parallel to the tooth axis. The attenuation showed somewhat clearer trends than in Fig. 9.14; in particular, the attenuation was least at the angle of maximum velocity and greatest at the angle of minimum velocity.

It is more difficult to find a suitable longitudinal section for line-focus-beam measurements, because of the requirement that the enamel must be thick enough to yield a flat area of adequate size. The most likely place is at the side of the tooth, as far from the root as possible but avoiding the cusp region where the orientation of the apatite crystallites changes rapidly. Measurements on such a region are plotted in Fig. 9.15. The scales in this figure are different from those in Fig. 9.14. $V(z)$ data were again measured at 10° intervals, but this time over a full 360°. The measurements should repeat exactly over the second 180°; the reason that they do not may in part be that the axis of rotation did not coincide exactly with the centre of the cylindrical lens, so that slightly different regions were being sampled. The two sketches of the simplified keyhole structure of enamel (Meckel *et al.* 1965) mark the angles where the Rayleigh wave propagation was approximately parallel to the estimated mean direction of the long axes of the prisms. The variation in Rayleigh velocity is much greater than in the buccal surface, ranging from $3124 \pm 10\,\mathrm{m\,s^{-1}}$ parallel to the axes to $2943 \pm 10\,\mathrm{m\,s^{-1}}$ in the perpendicular direction. The variation in attenuation is large enough to allow some confidence in it in this case. It shows the opposite trend to the velocity, being smallest in the parallel direction, $0.087 \pm 0.05\,\mathrm{Np}\,\lambda_{\mathrm{R}}^{-1}$, and greatest in the perpendicular direction, $0.381 \pm 0.05\,\mathrm{Np}\,\lambda_{\mathrm{R}}^{-1}$.

The apatite crystallites in human tooth enamel are bundled into prisms a few micrometres in size. In cross-section these prisms have a keyhole shape, so that they interlock in the way illustrated by the sketches in Fig. 9.15, although the boundary of the part of the prism corresponding to the bottom wedge-shaped part of the keyhole is much less well defined than the head of the prism, corresponding to the round part of the keyhole (Boyde 1987). The majority of the apatite crystallites within the prism heads are aligned with their c-axes parallel to the long axes of the prisms, but the embryological crystallites between the prisms may be quite different. The boundary regions between prisms can also have higher porosity. These two factors would both contribute to the higher attenua-

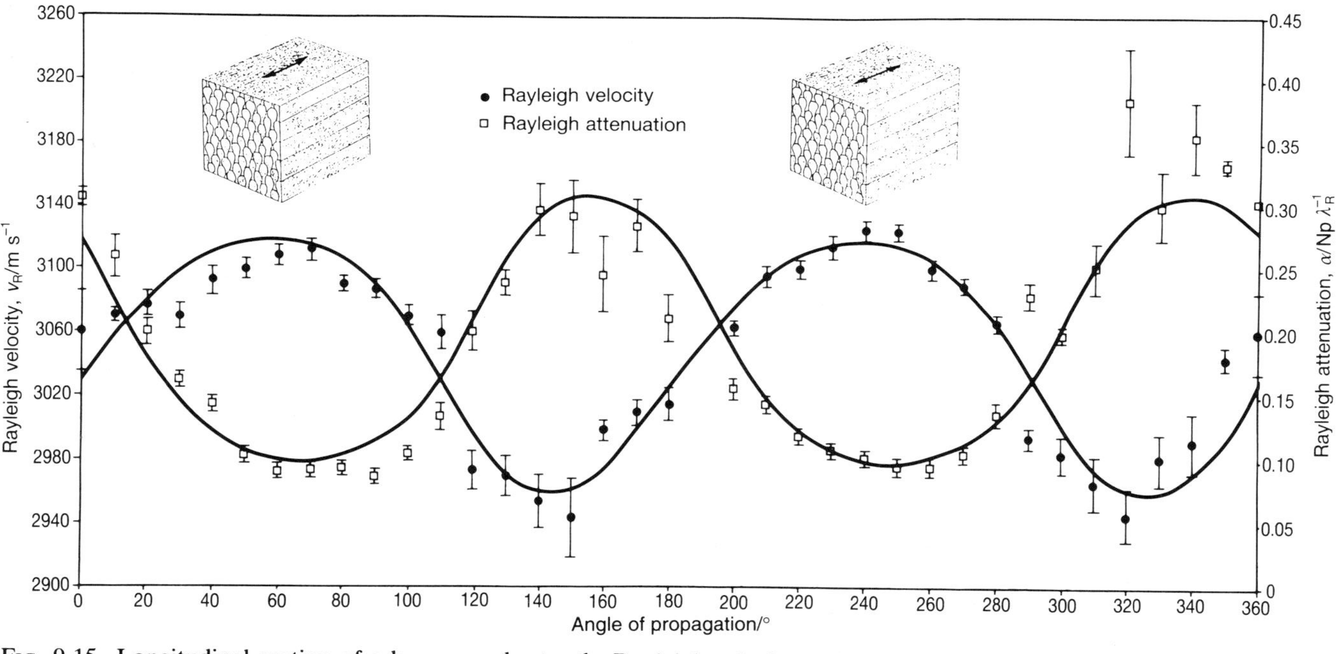

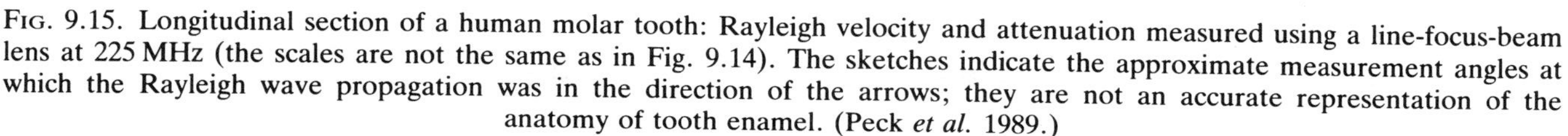

FIG. 9.15. Longitudinal section of a human molar tooth: Rayleigh velocity and attenuation measured using a line-focus-beam lens at 225 MHz (the scales are not the same as in Fig. 9.14). The sketches indicate the approximate measurement angles at which the Rayleigh wave propagation was in the direction of the arrows; they are not an accurate representation of the anatomy of tooth enamel. (Peck *et al.* 1989.)

tion that is found when Rayleigh waves are propagating perpendicular to the prism axes. The greater velocity in the parallel direction probably corresponds to higher apatite crystal stiffness constants for Rayleigh wave propagation parallel to the c-axis. But because there is some variation in the crystallite orientation, and indeed in the prism orientation, even in the parallel direction the velocity would be expected to be less, and the attenuation greater, than in a perfect single crystal. In cross-section, the dominant alignment is within the tails of the keyholes, where the crystallites tend to have a preferred orientation with their c-axes at approximately 20° to the plane of the section, and roughly pointing towards the middle of the head of the prism. This could to some extent account for the higher velocity parallel to the long axis of the tooth that is found in the buccal surface. But in the transverse section more or less the same amount of interprismatic material is encountered in propagation in all directions, so that there is little or no variation in attenuation with direction (though this does not explain why the attenuation is comparable with the attenuation in the parallel direction in the longitudinal surface).

When studying a complex material such as tooth enamel it is invaluable to have accurate data on single crystals of the apatite of which it comprised. Figure 9.16(a) is a $V(z)$ curve measured using a line-focus-beam microscope at 225 MHz, on the biggest crystal of hydroxyapatite in the world! At least 12 oscillations can be counted. The quality of the data is far better than the $V(z)$ curves in Fig. 9.13(b), partly because the specimen is more homogeneous, and partly because a line-focus-beam lens has been used. $V(z)$ curves were measured over a complete range of azimuthal angles from 0 to 360°, and analysed to find the Rayleigh velocity. The results are plotted in Fig. 9.16(b). In Fig. 9.16(c) results of similar measurements on a crystal of fluorapatite are plotted. The fluorapatite shows a greater variation of velocity with direction, and a lower mean velocity. In each case the anisotropic Rayleigh velocity is also plotted, calculated using published elastic constants and densities (Yoon and Newnham 1969; Katz 1971; Katz and Ukraincik 1971; Lees and Rollins 1972). The calculated hydroxyapatite curve has considerably lower values than the measured curve; this may be partly because of the way that the crystal elastic constants for hydroxyapatite were found, viz. scaling the fluorapatite crystal constants by the ratio of the stiffness of polycrystalline hydroxyapatite to polycrystalline fluorapatite (Katz and Ukraincik 1971). The agreement between the measured and calculated fluorapatite curves is somewhat better. The theory of waves in anisotropic surfaces will be described further in Chapter 11.

Both the range of velocity and its peak value in the tooth sections (Figs 9.14 and 15) are considerably less than in the single crystals. The range of values in the buccal surface would be expected to be smaller, since the

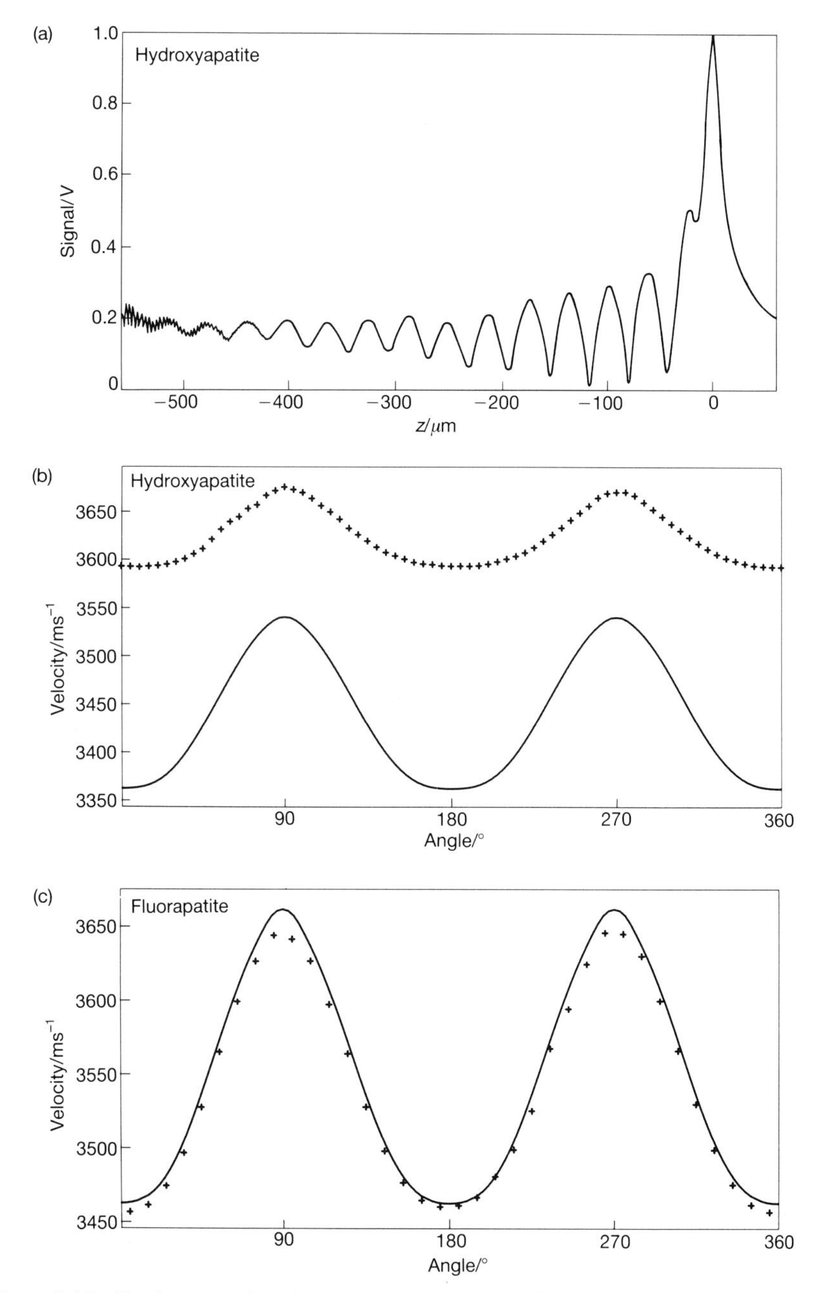

FIG. 9.16. Single crystals of apatite compounds: (a) $V(z)$ measured with a line-focus-beam lens at 225 MHz; (b) measured and calculated Rayleigh velocity in the surface of a hydroxyapatite crystal; (c) measured and calculated Rayleigh velocity in the surface of a fluorapatite crystal; + measured, —— calculated, propagation is in a median plane, and the angle is relative to the c-axis. (Courtesy of Trevor Gardner.)

crystallites are more nearly end on. The smaller range in the longitudinal section could be explained, at least in part, by variations in the orientation of the crystallites. The lower overall mean velocities in the tooth material are not so easy to account for in that way, but no doubt neither the composition nor the elastic properties of the components of the enamel correspond exactly to either of the pure crystals. The measurements on the enamel illustrate how much harder it is to measure inhomogeneous materials than nice uniform single crystals.

9.5 Bone

Bone falls rather heavily between two stools in acoustic microscopy. Soft tissue can be cut into thin sections. It can be imaged with a standard toneburst microscope, focused either on the substrate to let attenuation dominate the contrast, or on the top surface of the section to let impedance dominate (§9.3). Quantitative measurements of thickness, velocity, impedance, and attenuation can be made using time resolved techniques to separate the echoes from the top and bottom of the section. Structural tissue with a high collagenous content has both a higher velocity and a higher attenuation than other soft tissue. Tooth enamel is at the other end of the stiffness spectrum for biological materials (§9.4). It supports Rayleigh waves very nicely, with a velocity around $3000\,\mathrm{m\,s^{-1}}$, corresponding to a Rayleigh angle $\theta_R \approx 30°$, and modest attenuation. Therefore, all the beautiful Rayleigh wave contrast mechanisms that are used so successfully in other materials such as metals, semiconductors, and ceramics can be exploited for looking at tooth enamel, and probably dentine as well. But bone lies somewhere in the middle. It has a high attenuation because of its porosity. Because it has a higher velocity than soft tissue, a thicker section would have to be used for time-resolution measurements, which would create problems with attenuation; the alternative of using a higher bandwidth would exacerbate the problems of attenuation too. On the other hand, it has not proved possible to obtain useful Rayleigh contrast on a polished thick specimen; perhaps partly because of the attenuation, certainly also because the low Rayleigh velocity puts the Rayleigh angle outside the useful aperture angle of a water-coupled lens. But bone is such an important material, and so much is to be gained from understanding the micro-biomechanics, that it is worth persevering. In transmission at relatively low frequency it is possible to map the change in mechanical properties associated with local adaptive remodelling of the bone in the vicinity of a prosthetic hip implant (Zimmerman *et al.* 1990). With care, at higher resolution in reflection, lateral longitudinal waves can be made to perform a role very similar to that of the Rayleigh waves in stiffer materials (as they do in PMMA; cf. §7.2.3).

The term bone does not designate one single material, rather it describes a whole spectrum of heterogeneous anisotropic composites; indeed the properties vary greatly even within a single bone. The value of

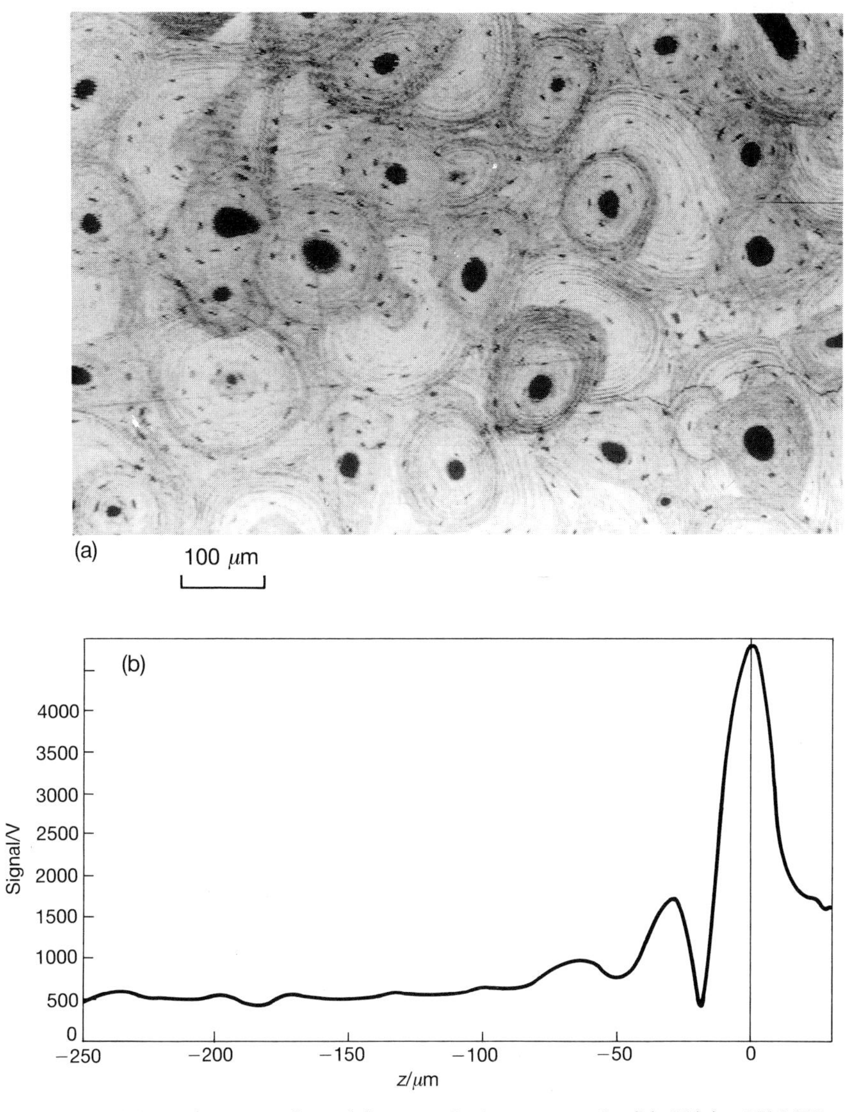

FIG. 9.17. Horse bone section: (a) acoustic image, $z = 0$; (b) $V(z)$; 650 MHz. (Courtesy of Trevor Gardner.)

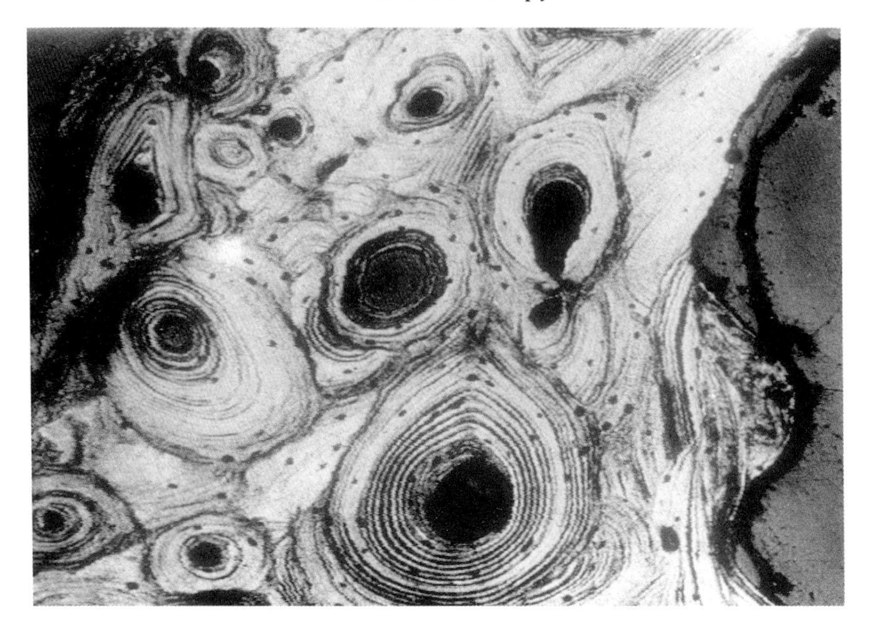

FIG. 9.18. Human toe bone section (5th metatarsal); 650 MHz, $z = 0$. (Courtesy of Trevor Gardner and Roger Gundle.)

the acoustic micrographs of bone is that they reveal the elastic nature of those inhomogeneities. Figure 9.17(a) is an acoustic image of a section through a horse bone. Great care was taken in the preparation of this specimen to eliminate all relief polish. Evidence of the excitation of lateral longitudinal waves comes from a $V(z)$ curve in Fig. 9.17(b), which was selected from a whole family of curves measured along a line on a surface like the one in Fig. 9.17(a). While it would be a foolhardy to attempt to make an accurate measurement from such a mediocre $V(z)$ curve as this one, a simple analysis of the dip spacing indicates a velocity in the vicinity of 3610 m s^{-1} which is not unreasonable for bone (Meunier *et al.* 1988).

A section through a human toe bone (fifth metatarsal, amputated by Roger Gundle who took the pictures in §1.3) is shown in Fig. 9.18. The circular patterns relate to the Haversian system responsible for blood flow in the bone. The regions around the holes are osteons. The osteons appear with different contrast in this picture. As always this relates to different mechanical properties. In this case it enables you to distinguish the different ages of osteons, because the variation in contrast is related to different degrees of mineralization.

10
Layered structures

'Yes, I have a pair of eyes,' replied Sam, 'and that's just it. If they wos a pair o' patent double million magnifyin' gas microscopes of hextra power, p'raps I might be able to see through a flight o' stairs and a deal door; but bein' only eyes, you see, my wision's limited.' (*The Pickwick Papers,* Charles Dickens 1836–7)

There is a long and distinguished history of the analysis of elastic waves in plates (Auld 1985), both isotropic (Cauchy 1828; Poisson 1829) and anisotropic (Cauchy 1829). In a free isotropic plate two kinds of waves can propagate: Love waves, which are polarized parallel to the plate surfaces, and Lamb waves, which are superpositions of longitudinal waves and shear waves polarized in a plane normal to the surfaces. Rayleigh waves may be thought of as a special case of Lamb waves in the limit of a plate many wavelengths thick (Rayleigh 1889; Auld 1973).

Acoustic microscopy of layered structures can be divided into two categories: those in which it is possible to distinguish reflections from adjacent surfaces, and those in which the layer affects the propagation of surface waves.

10.1 Subsurface imaging

Acoustic microscopy offers at least three advantages over conventional microscopy for imaging defects below the surface. The first is that many materials that are opaque to light are transparent to acoustic waves. In polycrystalline material the penetration is generally limited by attenuation due to grain scattering (Papadakis 1968; Stanke and Kino 1984) so that the resolution is subject to the requirement that the grain size be much less than the wavelength. But in many single crystals, especially semiconductors and insulators, the attenuation (which generally follows an f^2 law) can be negligible in acoustic microscopy even at 1 GHz or more (Auld 1973); for example, in silicon at 1 GHz the attenuation for longitudinal waves propagating in a $\langle 100 \rangle$ direction is only 1 dB mm^{-1}.

The second advantage is that, because the velocity of acoustic waves is five orders of magnitude slower than the velocity of light, it is possible to use time-resolved techniques to separate echoes from different surfaces. The capabilities that applied to the measurement of thin biological sections apply also to thin layers of solids, except that in solids the velocity is generally three or four times faster than in tissue, so that the limits on layer thickness are that much larger, and also the impedance is higher, so that the echo from a subsurface interface will be smaller relative to the echo from the top of the specimen. These two properties of acoustic waves have long been exploited in ultrasonic non-destructive testing (n.d.t.). Indeed, the use of acoustic microscopy techniques to inspect small components using an intermediate frequency range (say 20–200 MHz) is an important area of application (Burton *et al.* 1985; Smith *et al.* 1985; Gilmore *et al.* 1986; Vetters *et al.* 1989), especially in the examination of semiconductor device packaging (Howard 1990; Kulik *et al.* 1990; Revay *et al.* 1990; Briggs and Hoppe 1991). Figure 10.1 shows an area of a disbond in an electronic device imaged at 60 MHz.

The third advantage has nothing to do with acoustic properties of waves, but derives from the confocal configuration of scanning acoustic microscopes. As in a scanning optical microscope, this gives rise to greatly enhanced depth discrimination. In many discussions of the $V(z)$ curve the role of the reflectance function in the integral is of interest, but in depth discrimination the pupil function plays a primary role. The central peak in $V(z)$ is quite strong so that, as a reflecting surface moves away from the focal plane, the signal decreases. In the simplest case of an unapodized lens of semi-angle θ_0, in the paraxial approximation (i.e. when θ_0 is small), with an ideal reflector with no surface wave activity (Sheppard and Wilson 1981),

$$V(z) = \mathrm{sinc}\{2\pi z(1 - \cos\theta_0)/\lambda_0\} \tag{10.1}$$

where $\mathrm{sinc}(x) \equiv \sin(x)/x$. $V(z)$ would become zero when $z = \pm 0.5\lambda_0/(1 - \cos\theta_0)$, and the next maximum would occur at $z \approx \pm 0.72\lambda_0/(1 - \cos\theta_0)$. As the lens aperture increases the simple expression of (10.1) becomes less accurate, and the troughs in $V(z)$ become filled out (especially if the lens illumination decreases towards its edge), but even for a 60° lens the signal falls to -20 dB at a defocus of one wavelength, and is still -13 dB down at the next maximum at $z = \pm 1.44\lambda_0$.

The loss of signal away from focus has been extensively studied for scanning optical microscopy (Wilson and Sheppard 1984; Boyde 1987; Shotton 1989; Forsgren 1990; Sheppard and Cogswell 1990), and similar analysis can be applied to scanning acoustic microscopy (Heygster *et al.* 1990; Matthaei *et al.* 1990). Indeed, the confocal optical microscope is closely connected to the acoustic microscope, and both arose out of joint activities at Stanford and Oxford Universities. In conventional microscopes an illuminated point reflector that is out of focus appears dimmer because the light from it is spread over a larger area of the image, but, if a whole reflecting plane is out of focus, then, provided the illumination does not change, it will contribute the same intensity to the image whatever the defocus. In a confocal microscope, the contribution to the brightness of an image from a plane reflector diminishes dramatically away from focus. This effect can be used to enhance the signal from subsurface planes. Figure 10.2 shows a series of images at relatively low frequency, focusing down through the layers of a glass-fibre-reinforced composite. The material had been subject to impact damage, and it is well-known that this can seriously affect the structural properties of the composite even when the damage is scarcely visible on the surface. As the microscope focuses from one ply to the next, the damage follows the orientation of the ply. The contrast from each layer is almost completely unclouded by scattering from the layers that precede it, and it is alarming to see how much the damage increases with depth.

Just as ultrasonic techniques at conventional n.d.t. frequencies are invaluable for the inspection of mechanical joins in large structures, so acoustic microscopy can be used to examine bonding to substrates and heat sinks in electronic and optoelectronic structures. Figure 10.3 shows two acoustic images of an InP-based laser that was supposed to be gold-bonded to a ceramic substrate. Figure 10.3(a) was taken with the top surface at the focal plane, $z = 0$. The large defects at the edge and near the middle are due to handling and the removal of the ball-bonded

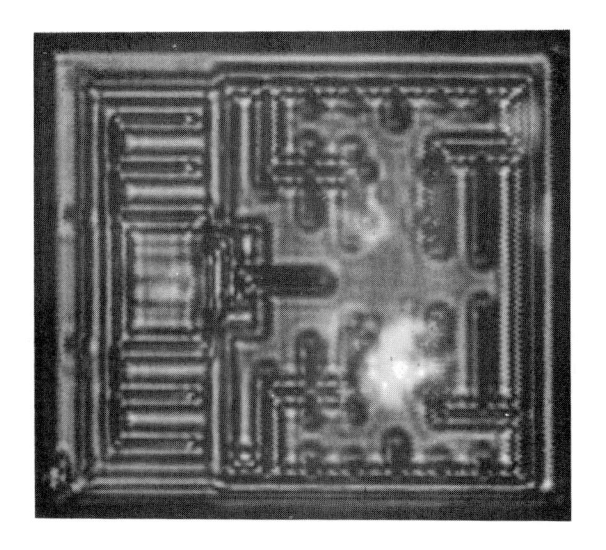

Fig. 10.1. Delamination in a semiconductor device; 60 MHz, $z = -50\,\mu$m. (Courtesy of Leica, Wetzlar.)

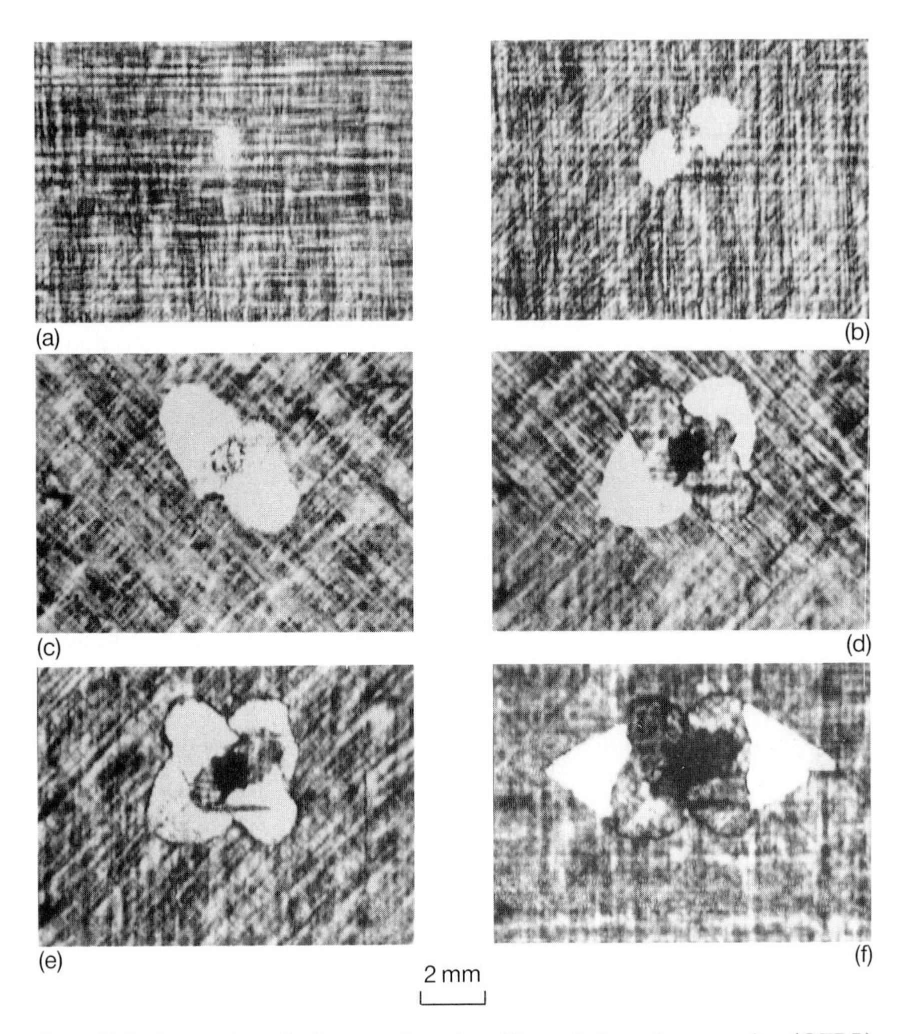

FIG. 10.2. Successive ply layers of a glass-fibre-reinforced composite (GFRP) damaged by impact of a missile from an air-gun; 50 MHz. (Sample from Kyoto University; courtesy of Olympus.)

gold wire that was connected to the top surface. There are some slightly brighter regions in the specimen, but the contrast is not easy to see in this picture. Figure 10.3(b) was taken at $z = -180\,\mu$m. The specimen was about $80\,\mu$m thick, so this defocus probably corresponds to shear waves excited in the solid being focused on the interface. Since shear waves have roughly half the velocity of longitudinal waves, they are the first to be focused on an interior plane as the specimen is moved towards the lens, and they also have half the wavelength so that the resolution is correspondingly better. In the defocused image the faint patches in the previous picture show up as strongly reflecting, corresponding to serious disbonds. This specimen was taken from a batch that were giving widely varying lifetimes, some unacceptably short. The disbonds to the heat sink would more than account for the failures, especially as high currents are needed to achieve population inversion of the electrons and the material was known to be liable to thermal degradation. Fascinating pictures of the bonds in so-called 'flip-chip' microelectronic modules have been obtained, through 0.5 mm or so of silicon (Weglein 1983; Tsai and Lee 1987). All these images were obtained without using time resolved techniques. With short pulses, a broad band transducer, and time-resolved detection, even better discrimination is possible.

In polymers, surface wave modes are much less strongly excited than in stiffer materials such as metals, semiconductors and ceramics. In most polymers the Rayleigh velocity is either slower than the velocity in water, or at least too slow to permit excitation by a 60° lens. Besides, the low impedance means that the surface waves would be strongly attenuated. These problems can be overcome to some extent by using methanol instead of water as the coupling fluid, providing that is does not dissolve the specimen. If water is used as the coupling fluid, then in many cases longitudinal lateral waves can be generated, and these can play a similar role to Rayleigh waves (§7.2.3). But the low impedance of polymers enables a greater proportion of power to be transmitted through the surface into bulk waves, thus making interior imaging easier (Hollis *et al.* 1984). In a study of a glass fibre reinforced polycarbonate specimen (Hoppe and Bereiter-Hahn 1985), where optical microscopy showed only the surface topography, acoustic microscopy at 1 GHz with a 50° lens angle also gave contrast from near-surface glass fibres, because of their different elastic properties. Acoustic microscopy at the same frequency, but with a lens whose angle was less than the longitudinal critical angle of 24°, eliminated all lateral wave phenomena in the matrix and caused a higher proportion of the incident energy to be transmitted into the bulk (Nikoonahad *et al.* 1983). This gives much better contrast from the matrix-fibre interfaces. Interference between these reflections and the reflection from the suface gives rise to fringes which are contours of equal depth, spaced roughly at intervals of $\lambda_l/2$ (i.e. about 1 μm at 1 GHz). Information can be obtained from at least 5 μm depth at 1 GHz in this kind of specimen, due not to perturbation of waves propagating parallel to the surface, but to waves propagating straight down into the solid and being reflected. The chief limitation on imaging through polymers is that they generally have rather high attenuation. But good methods exist for

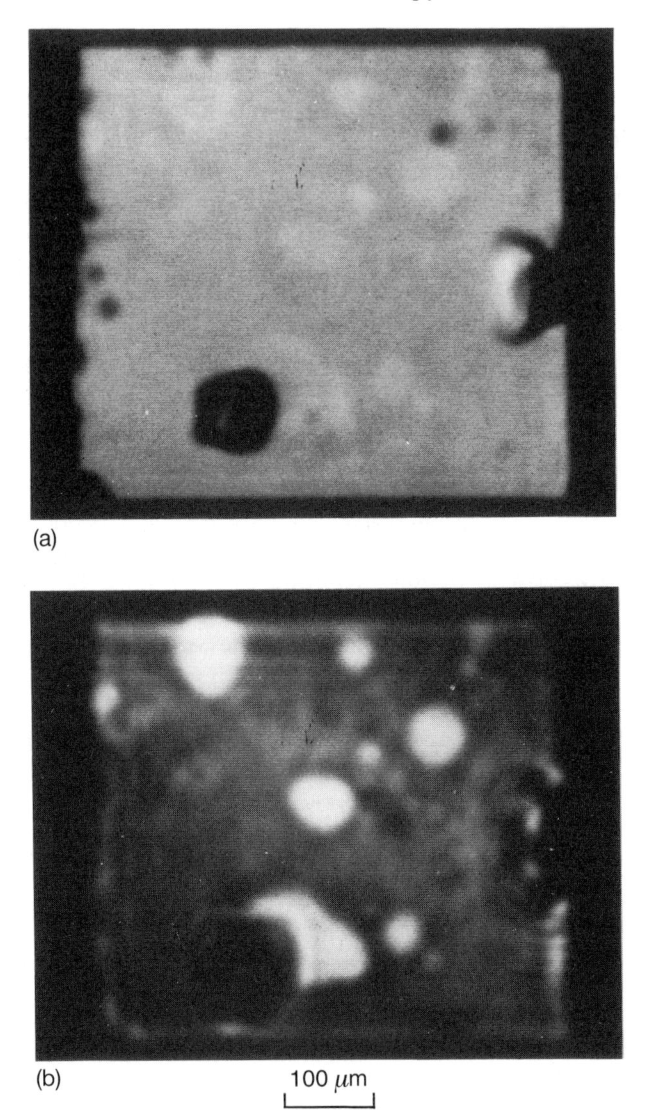

FIG. 10.3. An InP-based laser gold-bonded to a ceramic substrate: (a) $z = 0$; (b) $z = -180\ \mu$m; 400 MHz.

measuring attenuation up to 100 MHz (Tsukahara and Ohira 1989), and these can be used to design subsurface techniques for given materials. The attenuation of most polymers decreases with temperature, so there is benefit in using lower temperatures. Methanol is the preferred liquid, because it has the lowest attenuation of any organic solvent, it has a high enough impedance to ensure adequate acoustic coupling into polymers, and its freezing point, −94°C, is low enough for many applications. If methanol would dissolve the specimen too strongly, ethanol is the next best choice. A low-temperature microscope has been used to study a 0.5 mm thick epoxy bond between two 1 mm PMMA sheets at 40 MHz at −50°C (Yamanaka *et al.* 1989). Even at room temperature, useful studies on polymers have been achieved. Images and quantitative measurements have been made on two kinds of spherulites in isothermally crystallized isotactic polypropylene at 775 MHz (Duquesne *et al.* 1989), and composites, tapes, and two-phase polymers have been studied (Fagan *et al.* 1989). Figure 10.4 is a picture of a polymer containing glass beads, which had been uniaxially strained. The tiny bright dumb-bells are where the beads have come away from the polymer. It had not been possible to see that optically.

Acoustic microscopy of polymers is much more tolerant of topography than is the case for faster materials, where the defocus is often chosen for maximum sensitivity to surface waves at a region of steep slope in $V(z)$. Moreover, the refractive index is closer to unity, so that for a given roughness the distortion due to refraction is less. The requirements for

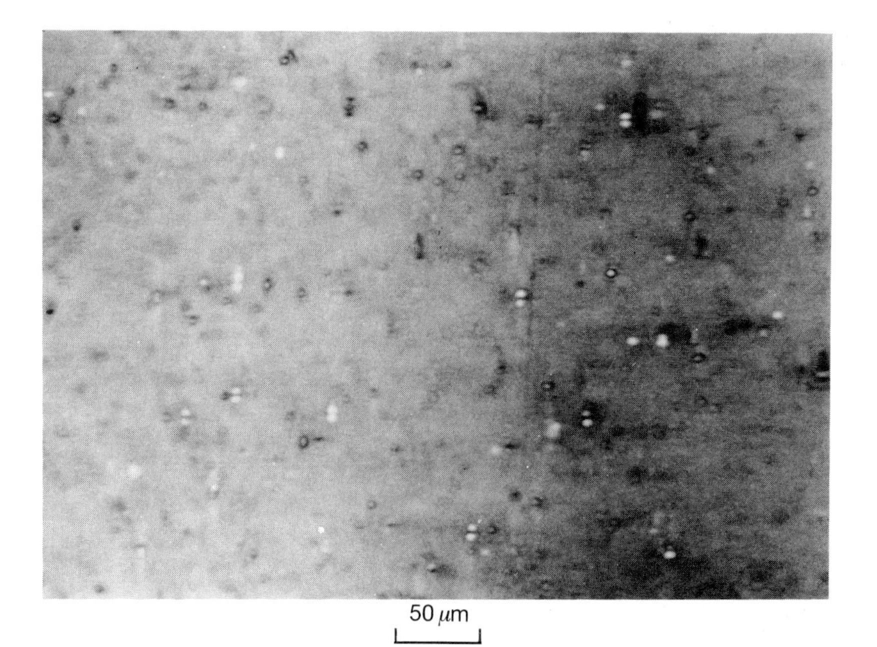

FIG. 10.4. Polymer containing glass bead filler particles that had been strained in the vertical direction of the picture; 370 MHz (Fagan *et al.* 1989).

the surface finish scale with the acoustic wavelength used. Images of flaws in composites (Khuri-Yakub *et al.* 1985) showed that the images were tolerant of surface roughness up to $\frac{1}{5}\lambda_0$ (the water wavelength). Excellent pictures can be obtained of defects in specimens of carbon–carbon and carbon–epoxy laminates with surfaces of comparable roughness. Other experiments at 3–10 MHz (Reinholdtsen and Khuri-Yakub 1986, 1991) showed that the effects of larger surface undulations can be subtracted out by measuring them at focus and then using $V(z)$ theory to calculate their contribution to a defocused image. In order to do this amplitude and phase information must be recorded for all signals, so that they can be manipulated as complex quantities. To implement this at higher frequencies would require exceptionally good mechanical, thermal, and electronic stability.

Polymer coatings on stiffer substrates can be measured by time-resolved techniques (Sinton *et al.* 1989). Often in these cases it is not convenient to measure a direct reflection from an uncoated part of the substrate at more or less the same time, and anyway the substrate may not be flat, but this may not matter if it can be assumed that *either* the thickness *or* the longitudinal velocity of the coating does not vary. The time interval between the echoes from the top and bottom surfaces of the coating can then be used to determine the unknown quantity. An example of the kind of signal that can be obtained is shown in Fig. 10.5. The specimen was a coating of PET (polyethylene tetraphthalate) 15 μm thick on a stone-finish rolled steel substrate. Although there is some overlap of the two echoes, there is no difficulty in distinguishing them in order to measure the time difference. Echoes that were more closely spaced could be distinguished using the MEM–cepstrum technique

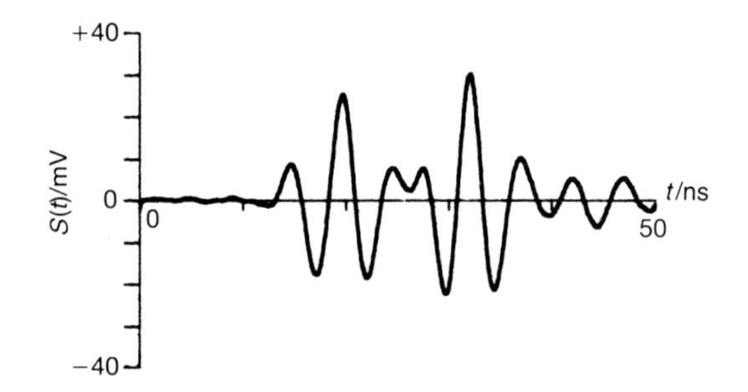

FIG. 10.5. Signals reflected from the top and bottom surfaces of a 15 μm thick coating of polyethylene teraphthalate on a stone-finish rolled steel substrate, using a short pulse of centre frequency 230 MHz and half-power bandwidth 110 MHz; $z = +40\ \mu$m (with the top surface of the polymer as datum) (Sinton *et al.* 1989).

described in §8.3. Indeed, the signals could be completely analysed using the methods of §8.3 (if either a reference signal could be obtained, or one of the other unknowns eliminated). But in some cases that much detail may not be needed.

Since the shapes of the two pulses in Fig. 10.5 are so similar, a simple pattern-matching algorithm may be used with great advantages in computational speed over Fourier domain techniques such as correlation or cepstral analysis. After suitable filtering, the turning points (i.e. the maxima and minima) of the waveform are found, and all the sets of three adjacent points are identified satisfying the criteria of each point exceeding a minimum magnitude, and each adjacent pair exceeding a minimum separation in time and having second derivatives of opposite sign. For each set of three turning points, indicated by suffixes a, b, c, at time t with size s, a figure of merit is calculated in order to find the set with best symmetry in time and size. The time asymmetry is defined as

$$A_t \equiv \frac{(t_c - t_b) - (t_b - t_a)}{t_c - t_a}, \tag{10.2}$$

and the size asymmetry is defined as

$$A_s \equiv \frac{s_c - s_a}{s_b}. \tag{10.3}$$

A figure of merit may then be defined as

$$F_A \equiv |s_b|\,(1 - \gamma\,|A_t| - \delta\,|A_s|), \tag{10.4}$$

in which γ and δ may be chosen empirically to give the best reliability of pulse identification; from experience the values $\gamma = 1$ and $\delta = 2$ work quite well (Sinton *et al.* 1989). The positions of the pulses are determined from the two sets of three points with the greatest figures of merit. Hence the time difference can be measured; if required the amplitudes of the two reflections can be determined also.

This method not only gives a computationally fast method of finding the thickness, it also enables disbonds between the coating and the substrate to be readily detected. In Fig. 10.6, signals from two films of polypropylene, each 40 μm thick, are shown. The first (Fig. 10.6(a)) was adhered to a steel substrate, while the second (Fig. 10.6(b)) was freely supported with air backing. Not only is the bottom surface reflection in the unbonded case larger than in the bonded case, but it has suffered a

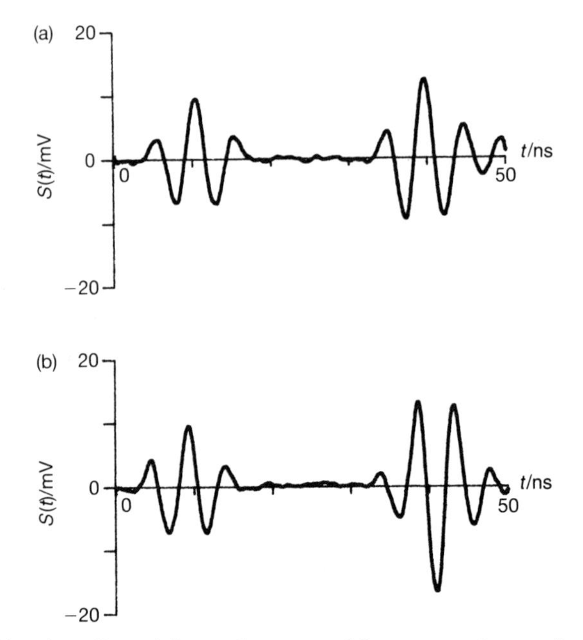

FIG. 10.6. Signals reflected from the top and bottom surfaces of a 40 μm thick layer of polypropylene, with the same acoustic parameters as in Fig. 10.4: (a) bonded to steel; (b) unbonded, with air backing (Sinton *et al.* 1989).

reverse in polarity. This is readily detected by the pattern matching algorithm, from the sign of s_b for the second echo. Of course, for imaging purposes changes in the condition of the bond between a layer and a substrate can also be detected by the change in the reflected signal without time resolution. This has been exploited in some fascinating *in situ* studies of cathodic disbonding and corrosion beneath protective coatings in electrochemical cells, at frequencies up to 1 GHz (Addison *et al.* 1989; Vetters *et al.* 1989; Honda and Ohashi 1990).

Time-resolved techniques are very powerful for examining structures where the useful information is contained in the normally reflected signal. Frequency analysis of a reflected broadband signal can also be used for film characterization (Wang and Tsai 1984; Lee *et al.* 1985). But in many other problems, especially in materials science, there is a great deal of information contained in the way that the coefficient of reflection changes with angle of incidence. It is therefore important to understand the behaviour of the reflectance function $R(\theta)$ of a layered structure.

10.2 Waves in layers

If a substrate has one or more surface layers that are too thin to allow the echoes from different interfaces to be discriminated, then the contrast may be analysed in terms of the reflection of plane waves from the layered surface. The reflection of waves in a fluid from a layered structure may be calculated by a matrix method (Brekhovskikh 1980; Brekhovskikh and Godin 1990): only the result will be presented here. The substrate is taken as semi-infinite, and is designated by the subscript n, and there are $n-1$ layers on it designated $n-1$, $n-2, \ldots, 1$, with the fluid finally designated by the subscript 0. There is a longitudinal wave in the fluid incident at an angle θ_0, which excites four waves in each layer, one longitudinal and one shear in each direction, except in layer n which has only one wave of each kind, propagating away from the interface. The tangential component of wavevector is β as before (§6.3–4), and is conserved throughout the problem. The boundary conditions at each interface are continuity of displacement and traction and, for isotropic materials, the problem is confined to a plane containing the incident ray and the surface normal, so that these reduce to four equations at each interface. The displacements and stresses at the upper boundary of the ith layer can be related to the stresses at its lower boundary by a matrix

$$\begin{pmatrix} u_{ix} \\ u_{iz} \\ \sigma_{izz} \\ \sigma_{ixz} \end{pmatrix} = \begin{pmatrix} a_{11} & a_{12} & a_{13} & a_{14} \\ a_{21} & a_{22} & a_{23} & a_{24} \\ a_{31} & a_{32} & a_{33} & a_{34} \\ a_{41} & a_{42} & a_{43} & a_{44} \end{pmatrix} \begin{pmatrix} u_{(i+1)x} \\ u_{(i+1)z} \\ \sigma_{(i+1)zz} \\ \sigma_{(i+1)xz} \end{pmatrix}. \tag{10.5}$$

The matrix elements a_{jk} are given in Table 10.1. In each layer the angles of the longitudinal and shear waves are given by Snell's law as in (6.87). When $(v_m/v_0)\sin\theta_0 > 1$ for a particular mode m in a particular layer, then for that mode the angle of refraction θ_m is complex. This can be seen by defining $\chi \equiv (v_m/v_0)\sin\theta_0$, and let $\theta_m = \pi/2 + \mathrm{i}\varepsilon$. Then from $\sin(\pi/2 + \mathrm{i}\varepsilon) = \chi$,

$$\varepsilon = \ln\left(\chi + \surd(\chi^2 - 1\right); \tag{10.6}$$

and

$$\theta_m = \frac{\pi}{2} + \mathrm{i}\ln(\chi + \surd\chi^2 - 1)). \tag{10.7}$$

The effect of a series of layers may be obtained by matrix multiplication, so that if the matrix describing layer i is denoted $\mathbf{a}_i$, and if the product of the matrices for all the layers is written

$$\mathbf{A} = \prod_{i=1}^{n-1} \mathbf{a}_i \equiv \mathbf{a}_1\mathbf{a}_2 \ldots \mathbf{a}_{n-1}, \tag{10.8}$$

Table 10.1
Multilayer matrix elements

$a_{11} = 2\sin^2\theta_s \cos P + \cos 2\theta_s \cos Q$

$a_{12} = \mathrm{i}(\tan\theta_l \cos 2\theta_s \sin P - \sin 2\theta_s \sin Q)$

$a_{13} = -\mathrm{i}(\beta/\rho\omega^2)(\cos P - \cos Q)$

$a_{14} = (\beta/\rho\omega^2)(\tan\theta_l \sin P + \cot\theta_s \sin Q)$

$a_{21} = \mathrm{i}(2\cot\theta_l \sin^2\theta_s \sin P - \tan\theta_s \cos 2\theta_s \sin Q)$

$a_{22} = \cos 2\theta_s \cos P + 2\sin^2\theta_s \cos Q$

$a_{23} = (\beta/\rho\omega^2)(\cot\theta_l \sin P + \tan\theta_s \sin Q)$

$a_{24} = a_{13}$

$a_{31} = 2\mathrm{i}(\rho\omega^2/\beta)\sin^2\theta_s \cos 2\theta_s(\cos P - \cos Q)$

$a_{32} = -(\rho\omega^2/\beta)(\tan\theta_l \cos^2 2\theta_l \sin P + 4\sin^3\theta_s \cos\theta_s \sin Q)$

$a_{33} = a_{22}$

$a_{34} = a_{12}$

$a_{41} = -(\rho\omega^2/\beta)(4\sin^4\theta_s \cot\theta_l \sin P + \tan\theta_s \cos^2 2\theta_s \sin Q)$

$a_{42} = a_{31}$

$a_{43} = a_{21}$

$a_{44} = a_{11}$

Matrix elements a_{jk} for the ith elastic layer (Brekhovskikh 1980, with unpublished corrections by Tsukahara; cf. Brekhovskikh and Godin 1990). The variables are defined: ρ is the density of the layer; d is the thickness of the layer; v_l, v_s are the velocities of the longitudinal and shear waves in the layer; k_l, k_s are their wave-vectors with components α_l, α_s normal to the interfaces between layers; and their angles θ_l, θ_s are related to the angle of the incident wave by Snell's law in the usual way,

$$\frac{\sin\theta_0}{v_0} = \frac{\sin\theta_l}{v_l} = \frac{\sin\theta_s}{v_s}; \qquad \frac{\alpha_l}{k_l} = \cos\theta_l; \qquad \frac{\alpha_s}{k_s} = \cos\theta_s.$$

The angular frequency is $\omega \equiv 2\pi f$. For convenience the notation $P \equiv \alpha_l d$, $Q \equiv \alpha_s d$ is introduced.

then

$$\begin{pmatrix} u_{1x} \\ u_{1z} \\ \sigma_{1zz} \\ \sigma_{1xz} \end{pmatrix} = \begin{pmatrix} A_{11} & A_{12} & A_{13} & A_{14} \\ A_{21} & A_{22} & A_{23} & A_{24} \\ A_{31} & A_{32} & A_{33} & A_{34} \\ A_{41} & A_{42} & A_{43} & A_{44} \end{pmatrix} \begin{pmatrix} u_{nx} \\ u_{nz} \\ \sigma_{nzz} \\ \sigma_{nxz} \end{pmatrix}. \tag{10.9}$$

With the further definitions

$$M_{jk} \equiv A_{jk} - A_{j1}A_{4k}/A_{41}, \qquad j, k = 1, 2, 3, 4, \tag{10.10}$$

and

$$B = \frac{-\beta A_{41} + \alpha_l A_{42} - i\rho\omega^2\{\cos 2\theta_s A_{43} - (v_s/v_l)^2 \sin 2\theta_l A_{44}\}}{\alpha_s A_{41} + \beta A_{42} - i\rho\omega^2\{\sin 2\theta_s A_{43} + \cos 2\theta_s A_{44}\}}, \quad (10.11)$$

the input impedance of the set of layers is

$$Z_{\text{in}} = \frac{i}{\omega} \frac{\alpha_l M_{32} - i\rho\omega^2[\cos 2\theta_s M_{33} - (v_s/v_l)^2 \sin 2\theta_l M_{34}] - B\{\beta M_{32} - i\rho\omega^2[\sin 2\theta_s M_{33} + \cos 2\theta_s M_{34}]\}}{\alpha_l M_{22} - i\rho\omega^2[\cos 2\theta_s M_{23} - (v_s/v_l)^2 \sin 2\theta_l M_{24}] - B\{\beta M_{22} - i\rho\omega^2[\sin 2\theta_s M_{23} - \cos 2\theta_s M_{24}]\}}. \quad (10.12)$$

In (10.12) all the parameters except the M_{ik} refer to the substrate. With the normal impedance of the fluid defined as $Z_0 \equiv \rho_0 v_0/\cos\theta_0$ (§6.4.1), the reflectance function has the familiar form

$$R(\theta) = \frac{Z_{\text{in}} - Z_0}{Z_{\text{in}} + Z_0}. \quad (10.13)$$

This solution may be readily implemented on a computer, the entire calculation being performed with complex variables. Attenuation in the media can be allowed for by using complex wavenumbers and velocities. An alternative method of calculation uses a single $(4n-1) \times (4n-1)$ matrix (Chimenti *et al.* 1982), but for more than one layer on the substrate this becomes extremely tedious. Any method for calculations of layered media is vulnerable to instabilities if any of the layers is a large number of wavelengths thick. In each layer there will be both positive and negative exponents and, although these are well behaved mathematically, special precautions may be necessary to prevent them from becoming ill conditioned numerically.

From the reflectance function, $V(z)$ can be directly calculated. As usual, the dominant effect in $V(z)$ will arise from surface waves. If the depth of the first interface is greater than about one Rayleigh wavelength in the top layer (rather more for low-impedance materials, less for high-impedance materials; Atalar 1985), it will have little effect on surface wave propagation, and the only hope is to try to use subsurface focusing techniques, which can give sensitive information about the presence of defects and their depth below the surface, especially if a reduced aperture lens is used (Wang and Tsai 1984; cf. §4.1). But for surface layers less than a wavelength thick there may be perturbation of the Rayleigh velocity (Weglein 1979*b*) and also the development of one or more new modes.

The difference between a specimen that is effectively a semi-infinite half-space and one that has a boundary at a finite depth is illustrated in the $V(z)$ curves of two glass specimens in Fig. 10.7. The dashed curve is $V(z)$ for a microscope slide 2 mm thick. The solid curve is $V(z)$ for a glass cover-slip 0.11 mm thick. In a thin plate such as a cover-slip Lamb waves can be excited (Auld 1973). At low frequencies the lowest-order symmetric mode is like a longitudinal wave, except that, because the plate is thin, the modulus governing the velocity is $c_{11} - c_{12}^2/c_{11}$ (Table 6.1). If the frequency is increased until the shear wavelength becomes comparable with twice the plate thickness, then the velocity begins to decrease, eventually approaching the Rayleigh velocity. The lowest-order antisymmetric mode has a vanishingly small velocity at low frequency, where the motion can be thought of as the plate flapping up and down. As the frequency increases a small longitudinal component of motion is introduced and at high frequency the velocity also approaches the Rayleigh velocity; indeed a Rayleigh wave can be considered as a summation in the high-frequency limit of a pair of degenerate symmetric and antisymmetric waves. If $V(z)$ is measured at a frequency for a plate whose thickness is less than half the shear wavelength then the two lowest-order modes are the ones present; $V(z)$ contains two oscillations, each with exponential decay, corresponding to these two modes (Kushibiki and Chubachi 1985). If the frequency is increased, more Lamb waves appear. A new symmetric mode appears for each increase in thickness of

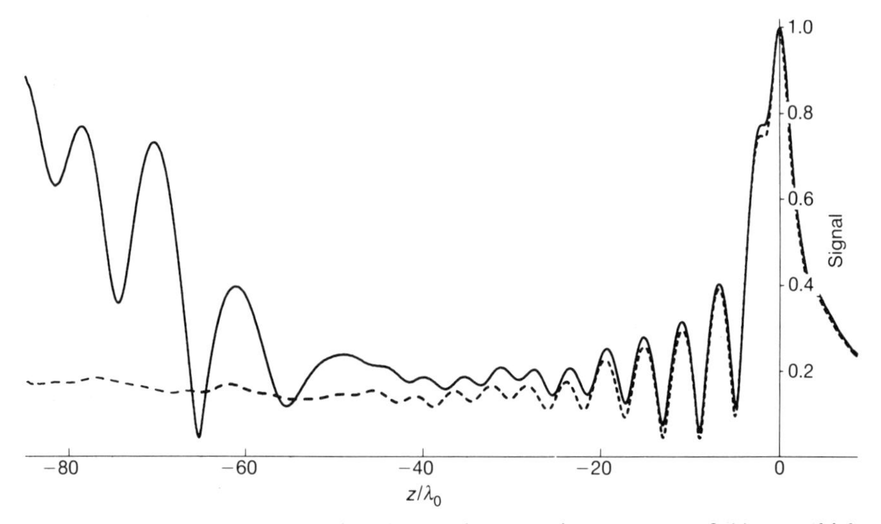

FIG. 10.7. $V(z)$ curves measured on glass specimens: —— 0.11 mm thick cover-slip; ---- 2 mm thick microscope slide (in this context effectively a semi-infinite half-space solid); line-focus-beam, 225 MHz.

half a longitudinal wavelength, and a new antisymmetric mode for each half shear wavelength. Each new mode starts with an infinite phase velocity, which decreases towards the shear velocity with increasing plate thickness. When there are more than a very small number of modes present, it becomes unrealistic to try to identify the contributions of each Lamb wave. Perhaps it is best to say simply that Lamb wave modes become important in $V(z)$ at sufficient defocus, and the values of z for paraxial focus of shear or longitudinal waves on the lower surface indicate the defocus at which the higher Lamb wave modes begin to be significant. This is analogous to the simple ray argument that, in the absence of scattering, Rayleigh waves contribute to $V(z)$ only at negative defocus (§7.2.1).

In the curves in Fig. 10.7, at small defocus both curves show oscillations of period $4.1\lambda_0$, corresponding to a velocity of $3100\,\mathrm{m\,s^{-1}}$. This is a respectable value for the Rayleigh velocity in crown glass (Table 6.3). With a wavelength $\lambda_0 = 6.6\,\mu\mathrm{m}$ and a refractive index $n = 0.43$ for shear waves, the first paraxial focus on the lower surface of the cover-slip will occur at $z = -40\lambda_0$. Beyond this defocus, oscillations develop in the $V(z)$ curve of the cover-slip with a period of $9.4\lambda_0$, indicating a velocity of $4600\,\mathrm{m\,s^{-1}}$. This is too slow to be a lateral longitudinal velocity (and if it were, why are the same oscillations not present in the dashed curve?), so it must relate to a mode that involves shear waves and reflection from the bottom surface of the cover slip. There is no objection to the phase velocity ω/k, which is what the $V(z)$ technique measures, being greater than the velocity of the constituent waves; it is the group velocity $\partial\omega/\partial k$, which would be measured by time-resolved techniques, that cannot be greater. For specimens of greater thickness relative to the wavelength than the cover-slip in Fig. 10.7. the Lamb modes form almost a continuum, and the concept of focusing on the lower surface becomes more applicable. A similar approach can be applied to a multilayered structure. If it is desired, not to image a particular interface, but rather to search for defects at a given depth below a surface, then it may not be necessary to consider Lamb modes at all. Instead, the confocal depth discrimination can be exploited to enhance the strength of the signal from a sub-surface scatterer that is in focus. At sufficient depth below the surface it may be possible also to use a time-resolved technique. But the scope for subsurface focusing (as opposed to Rayleigh wave imaging) in stiff materials becomes severely limited at high frequencies, because of the short working distance imposed by the attenuation in the coupling fluid.

If the specimen has a single surface layer on a substrate, then leaky generalized Lamb waves can be excited (Auld 1973). The character of these waves depends on the ratio of the shear velocities of the layer v_{s1}

and the substrate v_{s2}. In each case there is a fundamental mode, which in the limit of a very thin surface layer approximates to a Rayleigh wave with a slightly shifted velocity which may be calculated by treating the surface layer as a perturbation. If $v_{s1} \geq v_{s2}$, then this is the only wave mode that is bound to the surface, and it exists at all frequencies for which its phase velocity is less than v_{s2}. In many cases of interest, for example, a polymer coating on a metal substrate, or a metal film on a semiconductor, $v_{s1} < v_{s2}$. In this case the perturbed Rayleigh wave is again the lowest-order mode and, in general, it exists for all thicknesses, becoming a Rayleigh wave in the top coating when the layer becomes much more than a wavelength thick. When the layer is very thin this mode is the only one to be trapped at the surface; as the thickness is increased at a given frequency the next mode to become trapped is the Sezawa wave.

Dispersion relationships of Rayleigh and Sezawa waves for a layer of gold on 42-alloy are plotted in Fig. 10.8 (Tsukahara *et al.* 1989*a*). The formation of a Sezawa wave can be followed by considering the propagation of an essentially longitudinal wave parallel to the surface with a layer that in the first instance is very thin (Farnell and Adler 1972). The layer perturbs the wave, so that it is slower than the substrate longitudinal wave velocity. Coupling between longitudinal and shear waves can occur only at surfaces and interfaces; if by magic this could be switched off, then the slowed-down longitudinal wave would be trapped at the surface. But because it does couple into shear waves there can be a leakage into bulk shear waves in the substrate. At this stage the wave

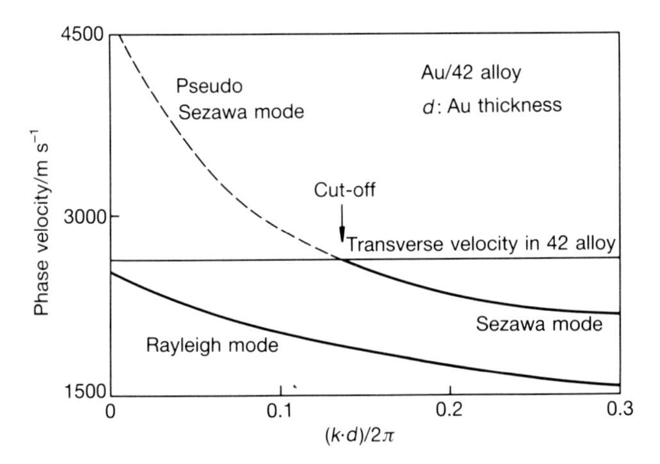

FIG. 10.8. Dispersion relationships for surface waves excited in a half space of 42%Ni–Fe alloy with a layer of gold of thickness d, and with water loading; $k = v/\omega$ is the propagation number along the specimen surface (Tsukahara *et al.* 1989*a*).

may be called a pseudo-Sezawa wave. At the layer thickness is increased, the motion in the layer becomes increasingly shear in character, and the velocity of the wave becomes even slower, until at a critical layer thickness it is slower than the substrate shear velocity, at which point the wave becomes a true Sezawa wave because it can no longer leak energy into the substrate. In the limit of a very thick surface layer the Sezawa wave becomes a shear wave in the layer (in certain special combinations of material parameters with $v_{s1} \approx v_{s2}$ it becomes instead a Stoneley wave trapped at the interface between the layer and the substrate). Both true Sezawa waves and pseudo-Sezawa waves can couple strongly to waves in a fluid, and therefore cause phase changes in the reflectance function. In addition, the pseudo-Sezawa wave causes a dip in the magnitude of the reflectance function because it serves as an effective means of coupling energy to shear waves in the substrate. If the film is lossy (as a polymer coating might well be), then the Sezawa wave can also cause a dip in the reflectance function: in ray terms this corresponds to a ray propagating along the surface, rather like the Rayleigh ray in Fig. 7.6, and dissipating energy in the film as it goes. Direct comparison has been made between measurements of polymer coatings by looking for Sezawa wave activity and time-resolved measurements of the same specimens (Tsukahara *et al.* 1989*b*; Sinton *et al.* 1989).

The dispersive reflectance function for a layer of gold on 42-alloy is shown in Fig. 10.9 (Tsukahara *et al.* 1989*a*). The reflection coefficient (Fig. 10.9(a)) has been plotted upside down in order to display high coupling into the substrate as peaks in $1 - |R(\theta)|$. At zero normalized thickness $R(\theta)$ has the familiar form of a fluid-loaded half space. As the thickness is increased so the pseudo-Sezawa wave can be seen to develop, with decreasing velocity (given by Snell's law as $v_0/\sin\theta$), until the critical angle for shear waves in the substrate, when it becomes a true Sezawa wave and ceases to couple any energy into the substrate, manifesting itself only in the phase of $R(\theta)$. Higher modes can also be seen beginning to develop at greater layer thicknesses. Just as in a uniform half space the interaction between waves in the fluid and the solid does not stop at the shear critical angle, but is manifested in phase change of 2π, so also there is a phase change associated with Sezawa waves. In the plot of the phase of the dispersive reflectance function (Fig. 10.9(b)), the 2π phase change is well pronounced, especially in the region where the reflectance function is unity. The analogy with Rayleigh waves is a close one; the 2π phase change is associated with the excitation of Sezawa waves which cannot radiate energy into the solid but can couple weakly into waves in the fluid. Pseudo-Sezawa waves have been observed in $V(z)$ data in acoustic microscopy (Tsukahara *et al.* 1984), and they have also been identified in a reconstructed $R(\theta)$ by the inversion of $V(z)$ (Liang *et al.* 1985*b*; cf. §8.1). Figure 10.10 shows calculated and measured $R(\theta)$

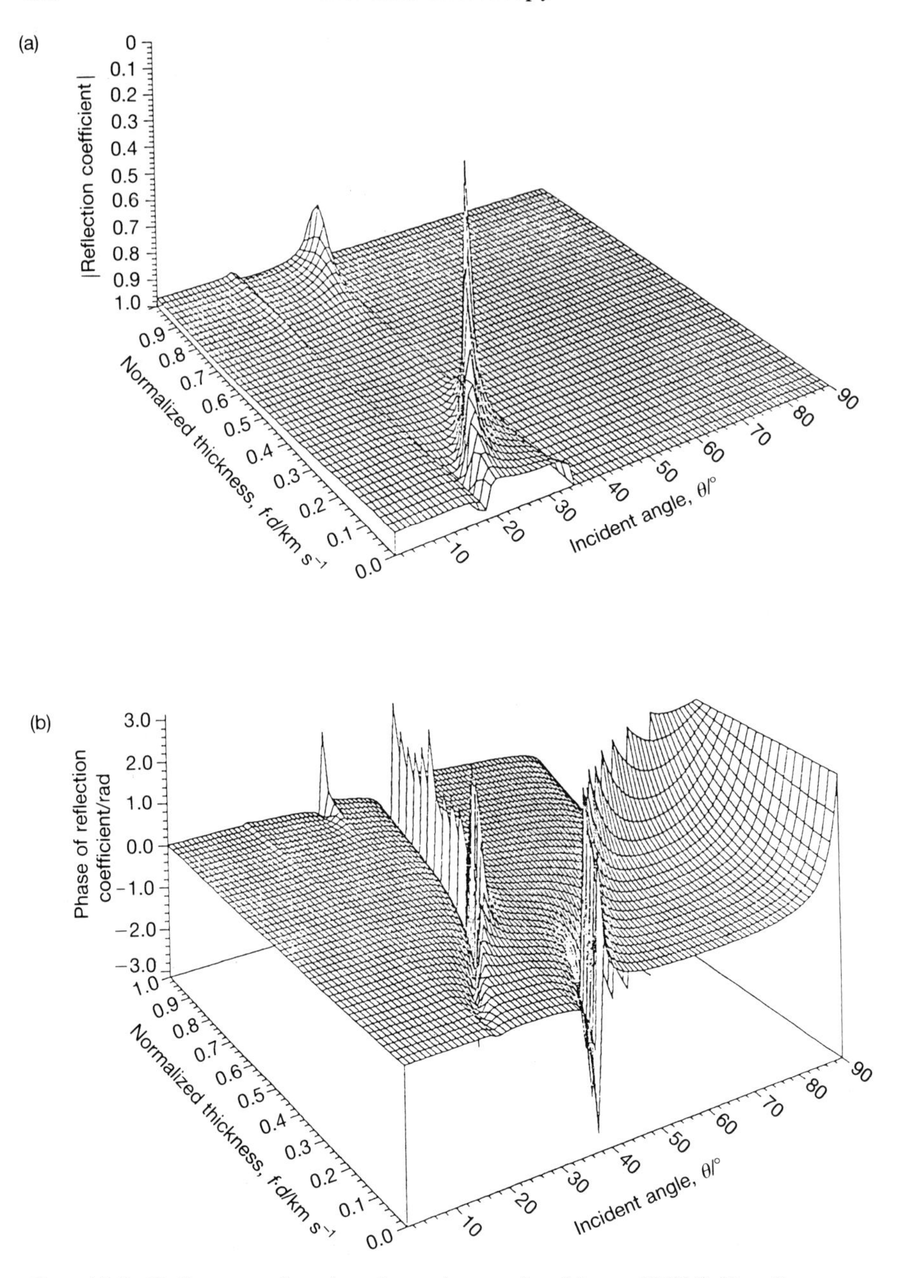

FIG. 10.9. Reflectance function for a layer of gold on 42%Ni–Fe alloy as a function of both the incident angle θ and the frequency–thickness product fd. (a) Magnitude, plotted as $1-|R(\theta)|$ to show the sharp minima in the magnitude; (b) phase (Tsukahara *et al.* 1989*a*).

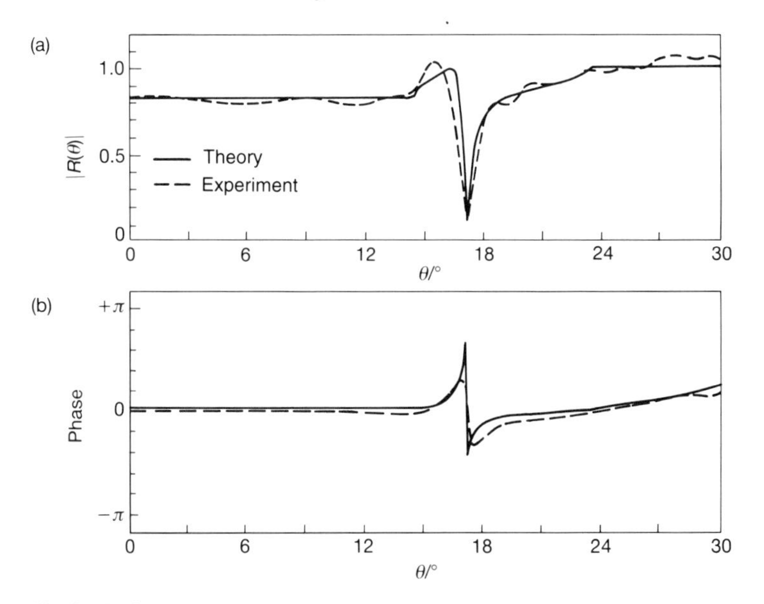

FIG. 10.10. Reflectance function for a 5 μm layer of gold on fused silica at 10.17 MHz ($fd = 50.85$ m s^{-1}), using the same system as in Fig. 8.2. (a) Magnitude; (b) phase. ---- Measured by inversion of $V(z)$; —— calculated with $v_0 = 1496$ m s^{-1}. For gold $v_l = 3240$ m s^{-1}, $v_s = 1200$ m s^{-1}, $\rho_1 = 19\,300$ kg m^{-3}; for fused silica $v_l = 5960$ m s^{-1}, $v_s = 3760$ m s^{-1}, $\rho_2 = 2200$ kg m^{-3}. (Liang *et al.* 1985*b*.)

curves for a 5 μm gold film on fused silica. The experimental magnitude and phase curves were deduced from a complex $V(z)$ measured with a spherical transducer at 10 MHz. Both the dip in the modulus and the associated phase change are present.

Measurements at higher frequencies have been made using the line-focus-beam technique (Kushibiki and Chubachi 1987). Figures 10.11(a) and (b) show $V(z)$ curves at 225 MHz for gold films of 100 and 300 nm, respectively, on fused silica, together with (Fig. 10.11(c)) theoretical dispersion curves for the velocity and attenuation of pseudo-Sezawa waves and Rayleigh waves. In the thinner coating the Rayleigh wave is well excited, and its velocity can be measured by analysing $V(z)$ in the range $-150\ \mu\text{m} \geq z \geq -25\ \mu\text{m}$. The pseudo-Sezawa wave is also excited, but not so strongly. In the thicker coating the pseudo-Sezawa wave is the strongest mode and its velocity can be measured by analysis of almost the whole range of z beyond $z < -75\ \mu$m. If the film is regarded as an elastic continuum with perfect surfaces then, from the dispersion curves in Fig. 10.11(c), a measurement sensitivity of about 1 m s^{-1} would correspond to a sensitivity to change in film thickness of 0.4 nm for the 100 nm film and 0.2 nm for the 300 nm film. Absolute

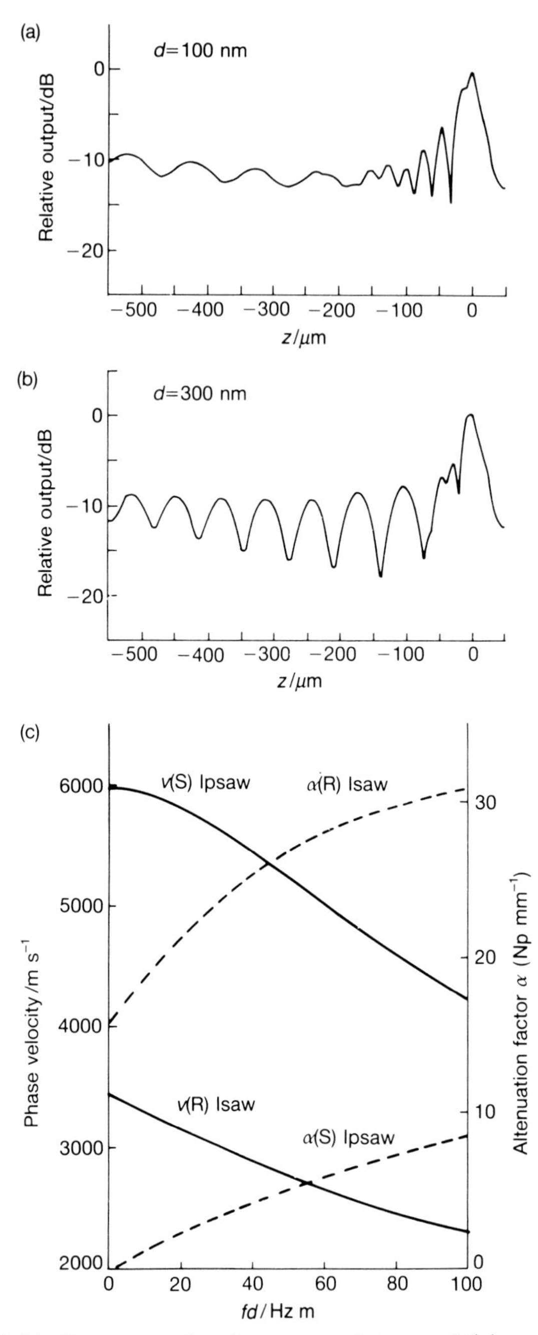

Fig. 10.11. Gold films on a fused quartz substrate. $V(z)$ measurements at 225 MHz on (a) 100 nm thick gold film; (b) 300 nm thick gold film. (c) Theoretical dispersion curves for leaky Rayleigh waves (R) and leaky pseudo-Sezawa waves (S): —— velocity; ---- attenuation (Kushibiki and Chubachi 1987).

interpretation should be carefuly checked against films of comparable thickness prepared in a similar way, because it can never be assumed that the properties of thin films are the same as those of bulk material.

There is a particularly abrupt change in the attenuation measured in an acoustic microscope at approximately the value of *fd* that corresponds to the transition from pseudo-Sezawa to Sezawa wave in the absence of a fluid, i.e. as the mode velocity crosses from above the shear velocity in the substrate to below it (Kushibiki *et al.* 1990). In simple terms, below that value of *fd* the attenuation is due to a combination of leaking into longitudinal waves in the fluid and shear waves in the substrate, whereas above that value of *fd,* only leaking into the fluid occurs. If the *fd* product for the given combination of materials is known, measurement of the frequency at which this abrupt drop in attenuation happens gives a particularly sensitive way of determining the thickness of the film. It might even be possible, by measuring Rayleigh wave dispersion as well, to determine completely the elastic constants, density, and thickness of an isotropic coating.

Special conical lenses have been developed for Lamb wave imaging of layered structures; because the waves are dispersive the frequency can be varied to tune the Lamb wave angle of the specimen to the angle for which the lens is designed (Atalar and Köymen 1989). Such lenses could similarly be used for Sezawa wave imaging and measurement, and also for other dispersive waves. It may be that for layered structures measurements of $V(f)$ at fixed defocus will become as prevalent as $V(z)$.

The viscoelastic properties of polymer films can be measured by their effect on the velocity and attenuation of Rayleigh waves. This has been demonstrated at 97 MHz with a 0.34 μm thick film of 28 per cent polystyrene–polybutadiene triblock copolymer in the temperature range 12–78°C (Martin and Frye 1990). The material and frequency were chosen to include the glass transition temperature of the copolymer within that range. A parametric plot of the change in normalized attenuation $\Delta\alpha/k$ against the fractional change in velocity $\Delta v/v_R$ for the measurements at different temperatures lay on a semicircle, indicating a simple Maxwellian solid behaviour. The radius of the semicircle was 1.4×10^{-3}, which is within the measurement capability of an acoustic microscope for both parameters. In an acoustic microscope it would be possible to vary frequency as well as temperature, and in this way it might be possible to study the properties of a variety of viscoelastic coating materials. Once again, the most sensitive way to look for such effects could be through Sezawa wave activity.

10.3 Near surface imaging

The effects of a thin surface layer on the acoustic image of the structure underneath it can be illustrated by pictures of strips of thin metal films on ceramics such as lead–lanthanum zirconate–titanate, PLZT (Hoppe and Bereiter-Hahn 1985; cf. Yin *et al.* 1982). If the specimens are unetched, then in optical images the grain structure is not visible, and all that can be seen are the metal electrodes. In one acoustic image at 2 GHz, the grain structure of the ceramic was revealed except where it was covered with gold. The grain structure of the gold was too small ($<0.01\ \mu$m) to be seen at this resolution ($\sim 0.7\ \mu$m), and it might be thought that, just as the gold electrode is opaque to light, so the gold also would prevent the grain structure of the ceramic from being seen by acoustic waves. But this depended on the particular defocus; at slightly greater defocus grain structure was revealed through the gold. The gold electrode on that specimen was $0.2\ \mu$m thick, which was considerably less than a wavelength in the solid. So there was no question of 'focusing underneath the gold'. Rather, the gold was acting as a perturbing layer on the ceramic, affecting the propagation of surface waves and hence changing $V(z)$. Defocusing may therefore be thought of as moving to a different part of $V(z)$ to optimize the differences between different grain orientations (cf. §9.4.1, especially Fig. 9.12). By reducing the frequency to 1 GHz the strength of the perturbation was roughly halved and, at a defocus $z = -5\ \mu$m (equivalent to $z = -2.5\ \mu$m at 2 GHz), the contrast of the ceramic under the gold showed greater variations between grains. Hard metal cermet wear-resistant coatings such as titanium nitride and titanium carbonitride can be studied by acoustic microscopy (Crostack *et al.* 1990; Matthaei *et al.* 1990; Vetters *et al.* 1989) though in one study intended to investigate the adhesion to the substrate, what was actually found was an extensive network of thermal stress relief cracks (Ilett *et al.* 1984, cf. §12.4).

Quality of adhesion at an interface is one of the properties that it would be most desirable to image by acoustic microscopy. There is some evidence that this may be possible (Bray *et al.* 1980; Addison *et al.* 1986). For example, samples of gold film were prepared of 500 nm thickness on a glass substrate, one half of which was first given a 20–30 nm flash of chromium. The chromium interlayer gives better adhesion to the glass. The Rayleigh velocity in the region with the chromium interface was slower by 2 per cent or more and the two regions gave different contrast in an acoustic image (linear elastic perturbation theory would predict an increase in velocity, due to the higher stiffness of the chromium, of a little less than 0.2 per cent). It is not altogether clear what aspect of adhesion the acoustic microscope should be sensitive to. Quality of adhesion is an elusive concept, but the strength of an adhesive bond in almost all cases involves the behaviour under deformation that is plastic or, in the case of polymer, viscoelastic. Using the quantities of §3.1.2 and a focal spot area of $1\ \mu\text{m}^2$ in eqn (6.72) and multiplying by a factor of 2 for the reflection from a high-impedance specimen, the normal component of surface stress is 35 MPa. This is relatively small compared with the adhesion between most metals, semiconductors, and ceramics. And the frequencies in the acoustic microscope are far higher than those of most viscoelastic processes that control polymer adhesion. So it would seem that the acoustic microscope is sensitive to only those aspects of adhesion that are governed by elastic properties. It is certainly sensitive to the true contact area. If the scale of the loss of contact is much less than the resolution, then the interface will appear as a compliant boundary (Thompson *et al.* 1983); this will affect $V(z)$ and therefore the contrast (Addison *et al.* 1986, 1987). On the other hand, if the scale of disbonding is greater than the resolution then the disbonded regions can be imaged directly.

Integrated circuits have been studied fairly extensively by acoustic microscopy (Miller 1983, 1985); in addition to disbonding the defects that have been reported also include resist/metal overhang, subsurface blisters and resist inclusions, hillocks shorting between metallization layers, spikes in epi-layers, thickness variations, and microcracks. Many of the most interesting high resolution pictures that have been taken of defects in integrated circuits cannot be published for proprietary reasons, which makes it impossible to discuss them. But it may be that the most important applications of acoustic microscopy to electronic devices are at lower frequencies, for inspection of packaging problems.

Semiconductor materials themselves have also been studied. One set of pictures, with associated $V(z)$ measurements, looked at the effects of ion implantation with As^+ and with Si^+ on the elastic properties of silicon, in the dose range 10^{14} to 10^{15} ion cm^{-2} at 180 keV (Burnett and Briggs 1986). These doses cause amorphization of the crystalline silicon, and the specimens were not annealed to remove such damage. The damage would be expected to extend to a depth of 0.3 μm or so, which is a small fraction of the Rayleigh wavelength $\lambda_R \approx 10\ \mu$m at the frequency of 550 MHz used in the acoustic microscope for these experiments. Half of each wafer was masked during the implantation. When the boundary between the masked and unmasked regions was examined using interference light microscopy, no step could be seen, so if there was any topography associated with volume change due to the ion implantation it was less than 15 nm. But when the same region was imaged in the acoustic microscope, it was possible to see where the boundary was, and it was also possible to measure $V(z)$ either side of the boundary for each specimen. Analysis of the $V(z)$ data indicated very approximately that the silicon suffered a reduction of 30 per cent in stiffness when it became amorphous. The analysis used was a useful, but somewhat crude one, especially because it treated the specimen as behaving isotropically under the spherical lens of the imaging acoustic microscope. A similar approximation has been used in measurements at 375 MHz of thin films of tungsten silicide co-sputtered on silicon wafers (Crean *et al.* 1987). The dependence of the elastic properties of silicon, and hence Rayleigh velocity, especially at high doping levels (Keyes 1982), makes it possible to study high doping concentrations by acoustic microscopy.

Although, occasionally, semiconductor materials are used in amorphous or polycrystalline forms, usually they are single crystals. In the discussion of images and imaging theory up to this point, everything has been isotropic. But in order to understand images of grain structures, and also quantitative measurements of surfaces of single crystals such as semiconductor wafers, it is necessary to consider the effect of anisotropy in the acoustic microscope.

10.4 Layers edge on

In many cases when the elastic properties of a material vary with depth, the most accurate way to study them is in cross-section. This has proved particularly powerful in the case of cladded glass fibres for optical waveguides. The refractive index of the fibres is made to increase away from the centre, so that by Snell's law light that starts off close to the axis cannot propagate with real wavenumber in the outermost part of the fibre. Measurements have been made of the Rayleigh velocity in cross-sections of fibres, and it is found that for a given dopant there is good correlation between changes in the optical refractive index and the Rayleigh velocity (Jen *et al.* 1989). The results of a series of calibration measurements on fused-silica fibre preforms with three different dopants are listed in Table 10.2. Apart from their importance for optical fibres,

Table 10.2
Optical fibre dopant measurements

Dopant	Range of fractional concentration, C	$\frac{1}{n}\frac{dn}{dC}$	$\frac{1}{v_R}\frac{dv_R}{dC}$
B_2O_3	0.05–0.10	−0.050	−1.54
GeO_2	0.0122–0.17	+0.063	−1.1
F	0.0048–0.0070	−0.010	−0.010

These values give an indication of the fractional variation of the optical refractive index and the Rayleigh velocity as a function of the dopant concentration. However, the relationship may not be linear over the range indicated (especially GeO_2) and the original data should be consulted for accurate purposes (Jen *et al.* 1989).

such measurements are directly applicable to fibres for use as acoustic waveguides (Jen 1989).

Fibres in composites often have properties that vary radially too. The remaining pictures in this chapter are of fibres in metal matrix composites (Lawrence 1990). The matrix is a Ti–6Al–4V (wt%) alloy. The fibres are Textron silicon carbide monofilaments. Their manufacture (Yajima *et al.* 1978) starts with a 33 μm diameter carbon fibre, on to which a 1 μm thick layer of pyrolitic graphite is coated by chemical vapour deposition to make the fibre more uniform. On to this core β-SiC is deposited to form radial columnar grains. After growth of about 22 μm the grain size changes abruptly from a diameter of 40–50 nm to double that, with a length of a micrometre or so. Finally, a 3 μm thick carbon-rich layer is deposited, tailored to give optimum compatibility with the matrix. These metal matrix composites are designed for high specific stiffness and strength. They are not particularly tough, and their maximum useful operating temperature is generally reckoned to be about 450°C. The pictures here are taken from a series of experiments to investigate the thermal degradation of this kind of composite.

The specimen shown in Fig. 10.12 had been kept at 450°C in air for 500 hours after manufacture had been completed. In the first acoustic image (Fig. 10.12(a)), there is a layer 1.4 μm thick surrounding the outer β-SiC; this may be the inner part of the carbon-rich layer. Beyond that there is another of about the same thickness, but slightly lighter in appearance, probably corresponding to the remaining part of the carbon-rich outer layer. There is a dark line around the fibre, which sometimes seems to run between the outer carbon-rich layer and the matrix, and sometimes between the two parts of the carbon rich layer itself. When the defocus is increased (Fig. 10.12(b)), the grain structure of the matrix shows up, and it becomes apparent that there is some correspondence between the grains in this picture and the crenelations of the carbon-rich layer in the previous one. It is possible that the carbon-rich layer has reacted chemically with the matrix alloy, and even that a particular phase is responsible—lower magnification acoustic micrographs confirmed that there are two phases present in the matrix.

The next specimen (Fig. 10.13) had been given heat treatment under the same conditions but for twice as long, 1000 hours. In low-magnification pictures it was found that about a third of the fibres had lost their cores. In the higher-magnification pictures in Fig. 10.13, although the core is present (with some scratches from the polishing), the interface between the core and the surrounding silicon carbide looks rather ragged. The difference between the inner and outer β-SiC can be seen, and the outer SiC has several radial cracks that are visible at both defocus settings. These cracks seemed to be tightly closed; it was impossible to see them with even the best light microscopy.

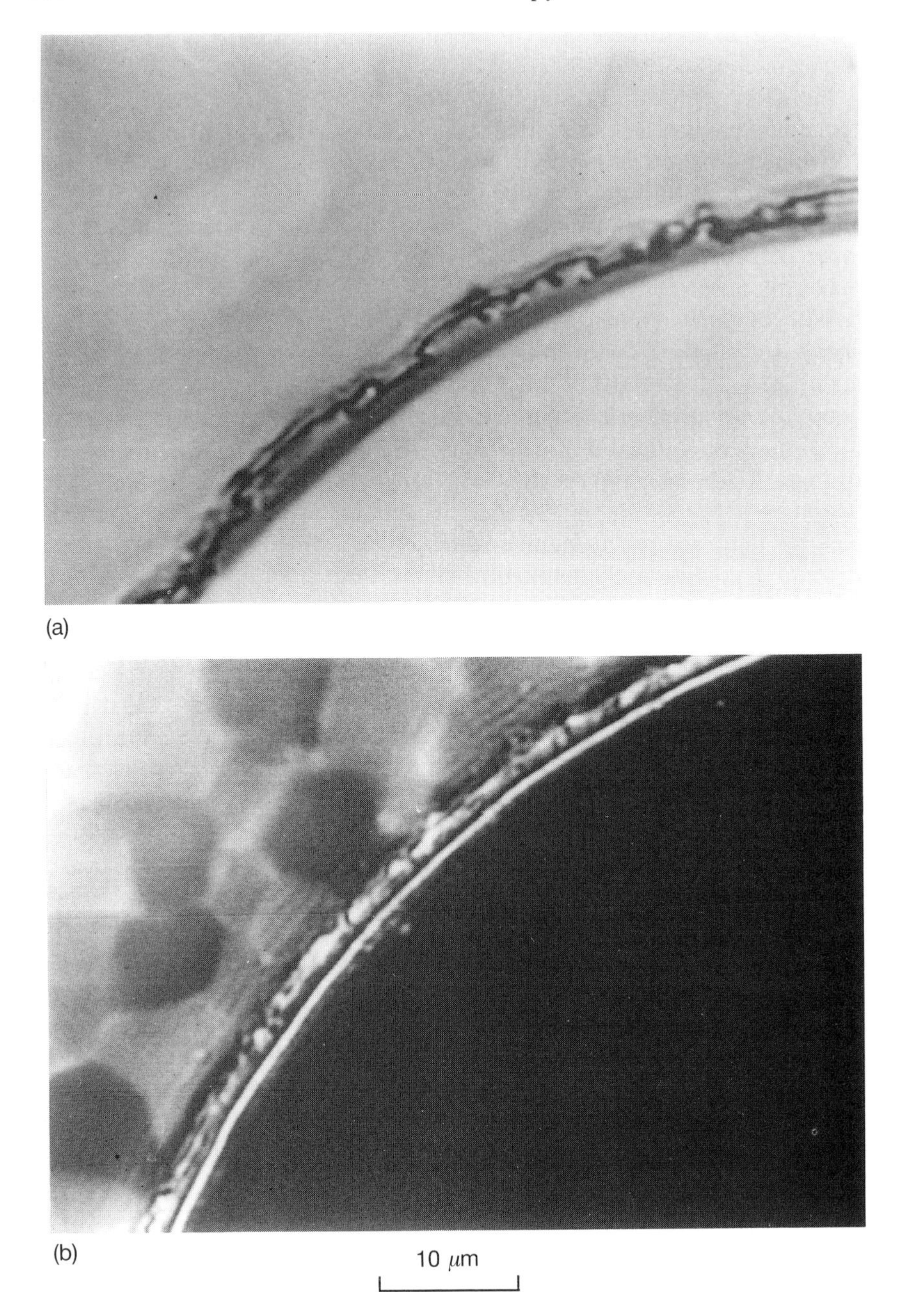

FIG. 10.12. Interface between matrix and fibre of a Ti–6Al–4V/SiC monofilament composite, after a post-fabrication heat treatment of 450°C for 500 h. (a) $z = 0$; (b) $z = -2.1\,\mu\text{m}$; ELSAM 1.9 GHz (Lawrence 1990).

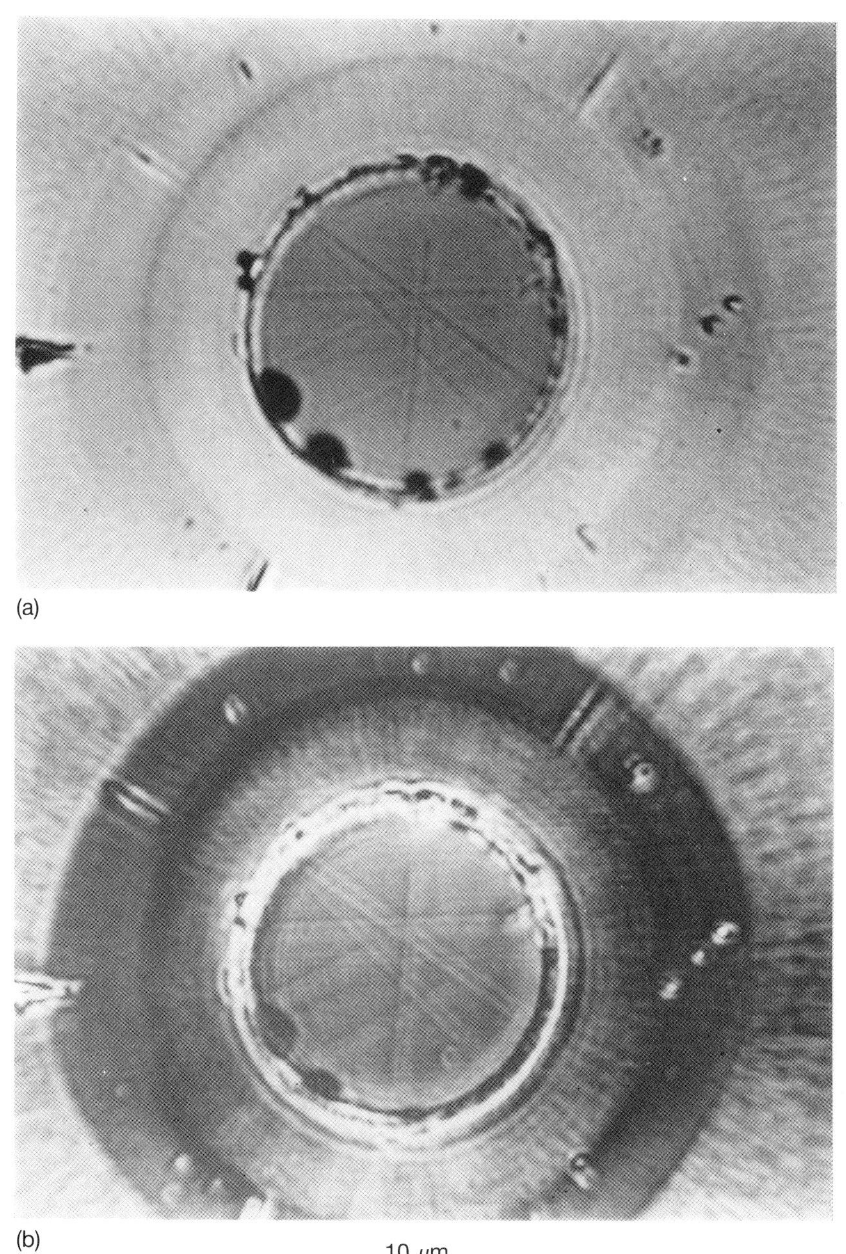

FIG. 10.13. Fibre in a Ti–6Al–4V/SiC monofilament composite, after a post-fabrication heat treatment of 450°C for 1000 h. (a) $z = 0$; (b) $z = -2.0\,\mu$m; ELSAM 1.9 GHz (Lawrence 1990).

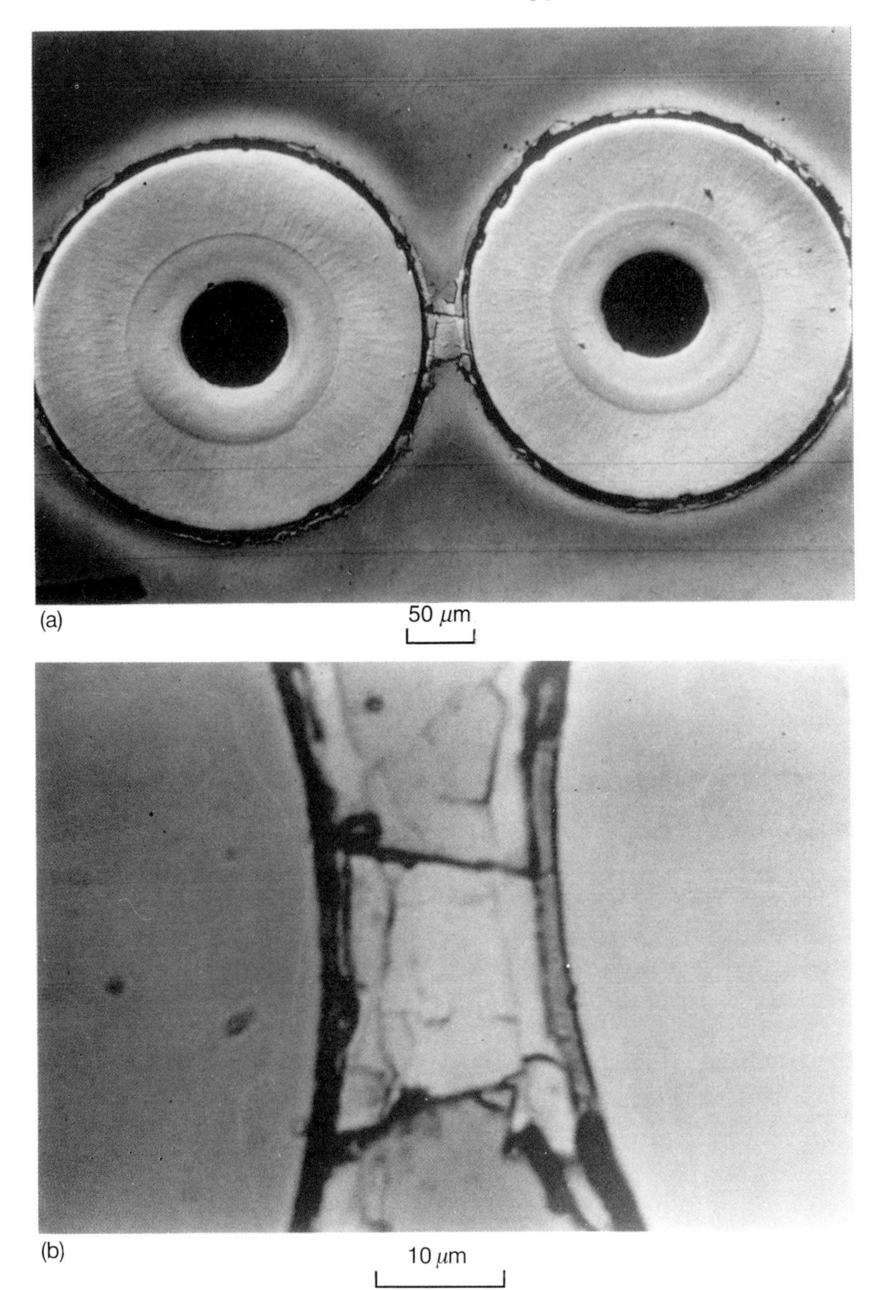

FIG. 10.14. Contact damage between two fibres in a Ti–6Al–4V/SiC monofilament composite, after a post-fabrication heat treatment of 600°C for 1000 h. (a) $z = 0$; (b) $z = -1.0\ \mu m$; ELSAM 1.9 GHz (Lawrence 1990).

One of the most important roles of the matrix in a composite is to keep the fibres apart: in the second specimen, degradation and damage seemed generally to be associated with contact between fibres. Even more severe case of neighbours touching each other were found in the final specimen (Fig. 10.14), which had been subjected to a sweltering 600°C for 1000 h. In a low-magnification image (Fig. 10.14(a)), the growth bands can again be seen, together with an annular region in the outer 6 μm or so of the fine columnar β-SiC region which was also seen in many of the other specimens: its nature remains to be explained. In other pictures of the monofilaments, radial microcracking was observed as in Fig. 10.13. In the contact region between the two fibres damage has occurred, and this is seen at higher magnification in Fig. 10.14(b). Once again, there are at least two layers at the circumference of the fibres. The inner layer is the residual carbon-rich coating, and is 1.5 μm thick. The outer coating is a TiO_2 reaction layer, 2–2.5 μm thick. The cracks pass through the TiO_2 reaction layer, but not through the residual carbon-rich coating.

The Rayleigh velocity of the fibres and the matrix were measured in this study, and had average values of 3074 m s^{-1} for the matrix (cf. the value for titanium in Table 6.3), 6610 m s^{-1} for the β-SiC in the fibres (somewhat higher than the range of values for most other solids), and 2766 m s^{-1} in the carbon core. Such measurements played an important part in the interpretation of results in the very extensive study from which these pictures were taken (Lawrence 1990). The anisotropy and the microcracking seen in pictures like these are of immense importance for the development of these kinds of materials. Their interpretation is the subject of Chapters 11 and 12.

11
Anisotropy

Then the little old Woman sate down in the chair of the Great, Huge Bear, and that was too hard for her. And then she sate down in the chair of the Middle Bear, and that was too soft for her. And then she sate down in the chair of the Little, Small, Wee Bear, and that was neither too hard, nor too soft, but just right. (*Story of the Three Bears,* Robert Southey 1837; 'Goldilocks' was introduced *c.* 1904)

When a polished polycrystalline specimen is examined in acoustic microscopy, the grain structure is usually revealed. This was illustrated by the appearance of the titanium alloy in Fig. 10.12. Since the surface had not been etched, the grain structure would not have been visible in an optical image. But since the acoustic microscope is sensitive to elastic properties, and since the orientation of the crystal elastic tensor differs from one grain to the next, different grains give different contrast. It turns out once again that the contrast is dominated by the propagation of surface waves.

11.1 Bulk anisotropy

All crystals are anisotropic; many other structures also have elastic anisotropy. The propagation of elastic waves in anisotropic media is described by the Christoffel equation. This still depends on Newton's law and Hooke's law, but it is expressed in tensor form so that elastic anisotropy may be included. The tensor description of elastic stiffness was summarized in §6.2, especially eqns (6.23)–(6.29). The Christoffel equation is

$$\rho\omega^2 u_i = k^2 \Gamma_{ij} u_j. \tag{11.1}$$

Γ_{ij} is the Christoffel matrix. The elements of Γ_{ij} for arbitrary (i.e. triclinic) symmetry are given in Table 11.1(a). The direction cosines l_i are defined by

$$\mathbf{k} \equiv (k_x, k_y, k_z) \equiv |k|\,(l_x, l_y, l_z). \tag{11.2}$$

In many cases considerable simplification is possible, because of the constraints imposed on the number of independent elastic constants. For cubic symmetry, for which the elastic stiffness tensor has only three

Table 11.1
Christoffel matrix elements†

(a) Triclinic symmetry

$$\Gamma_{11} = c_{11}l_x^2 + c_{66}l_y^2 + c_{55}l_z^2 + 2c_{56}l_yl_z + 2c_{15}l_zl_x + 2c_{16}l_xl_y$$
$$\Gamma_{22} = c_{66}l_x^2 + c_{22}l_y^2 + c_{44}l_z^2 + 2c_{24}l_yl_z + 2c_{46}l_zl_x + 2c_{26}l_xl_y$$
$$\Gamma_{33} = c_{55}l_x^2 + c_{44}l_y^2 + c_{33}l_z^2 + 2c_{34}l_yl_z + 2c_{35}l_zl_x + 2c_{45}l_xl_y$$
$$\Gamma_{12} = \Gamma_{21} = c_{16}l_x^2 + c_{26}l_y^2 + c_{45}l_z^2 + (c_{46} + c_{25})l_yl_z + (c_{14} + c_{56})l_zl_x + (c_{12} + c_{66})l_xl_y$$
$$\Gamma_{13} = \Gamma_{31} = c_{15}l_x^2 + c_{46}l_y^2 + c_{35}l_z^2 + (c_{45} + c_{36})l_yl_z + (c_{13} + c_{55})l_zl_x + (c_{14} + c_{56})l_xl_y$$
$$\Gamma_{23} = \Gamma_{32} = c_{56}l_x^2 + c_{24}l_y^2 + c_{34}l_z^2 + (c_{44} + c_{23})l_yl_z + (c_{36} + c_{45})l_zl_x + (c_{25} + c_{46})l_xl_y$$

(b) Cubic symmetry

$$\Gamma_{11} = c_{11}l_x^2 + c_{44}(l_y^2 + l_z^2) = (c_{11} - c_{44})l_x^2 + c_{44}$$
$$\Gamma_{22} = c_{11}l_y^2 + c_{44}(l_z^2 + l_x^2) = (c_{11} - c_{44})l_y^2 + c_{44}$$
$$\Gamma_{33} = c_{11}l_z^2 + c_{44}(l_x^2 + l_y^2) = (c_{11} - c_{44})l_z^2 + c_{44}$$
$$\Gamma_{12} = \Gamma_{21} = (c_{12} + c_{44})l_xl_y$$
$$\Gamma_{13} = \Gamma_{31} = (c_{12} + c_{44})l_zl_x$$
$$\Gamma_{23} = \Gamma_{32} = (c_{12} + c_{44})l_yl_z$$

(c) Hexagonal symmetry

$$\Gamma_{11} = c_{11}l_x^2 + c_{66}l_y^2 + c_{44}l_z^2$$
$$\Gamma_{22} = c_{66}l_x^2 + c_{11}l_y^2 + c_{44}l_z^2$$
$$\Gamma_{33} = c_{44}l_x^2 + c_{44}l_y^2 + c_{33}l_z^2$$
$$\Gamma_{12} = \Gamma_{21} = (c_{12} + c_{66})l_xl_y$$
$$\Gamma_{13} = \Gamma_{31} = (c_{13} + c_{44})l_zl_x$$
$$\Gamma_{23} = \Gamma_{32} = (c_{13} + c_{44})l_yl_z$$

† *Auld* 1973.

independent constants as given in (6.29), the elements of Γ_{ij} are given in Table 11.1(b), and for hexagonal symmetry the elements are given in Table 11.1(c). If $c_{12} = c_{11} - 2c_{44}$ were to be substituted in Table 11.1(b) the isotropic elements would be obtained, and the Christoffel equation would then give exactly the same solutions as eqns (6.40) and (6.41).

The solutions to the Christoffel equation can be found by rewriting it in the form

$$[k^2\Gamma_{ij} - \rho\omega^2\delta_{ij}][u_j] = 0. \tag{11.3}$$

This is solved by setting the characteristic determinant equal to zero,

$$|k^2\Gamma_{ij}(l_x, l_y, l_z) - \rho\omega^2\delta_{ij}| = 0. \tag{11.4}$$

Like the isotropic wave equation the Christoffel equation has three solutions, although in general there is no degeneracy except along symmetry directions. The motions of the particles are orthogonal for the three solutions, but not necessarily exactly parallel or perpendicular to the propagation direction, and so the waves are described as quasi-longitudinal or quasi-shear.

The solution to the Christoffel equation for a given material can be plotted in **k**-space as surfaces of $\mathbf{k}/\omega$. These are known as slowness surfaces, because they represent the reciprocal of phase velocity. Figure 11.1 is a representation of slowness surfaces of nickel in three-dimensional **k**-space. Because longitudinal waves have the greatest velocity and therefore the smallest slowness, the longitudinal surface is completely contained in the shear slowness surfaces, which are the ones seen in Fig. 11.1. Which polarization is seen in any given direction depends on which has the greater value of $|\mathbf{k}|/\omega$. A section through the slowness surfaces for nickel in an {001}-type plane is shown in Fig. 11.2. The polarizations are exactly longitudinal or shear only when the particle motion is along a symmetry direction (or certain other special directions, Auld 1973); otherwise they are quasi-longitudinal or quasi-shear. The slowness generally depends on the direction of **k**, although again for symmetry reasons an [001] polarized shear wave has constant slowness in an (001) section. The two shear waves are degenerate when **k** is an any $\langle 100\rangle$-type direction. Table 11.2 gives expressions in the form of combinations of elastic constants which enable slowness surfaces in symmetry planes of cubic and hexagonal materials to be calculated. Elastic stiffness constants of a number of anisotropic materials are given in Table 11.3. As with Table 6.3, the list is neither exhaustive nor definitive, but it does give an indication of the constants of the materials discussed in this chapter, together with a selection of other solids of general interest.

In an anisotropic material the direction of energy flow is not generally parallel to the wavevector. Just as in a dispersive medium the energy flow depends on the group velocity, $v_g = \partial\omega/\partial k$, so in an anisotropic medium the energy flow is related to the gradient of ω in **k**-space

$$v_g = \nabla_{\mathbf{k}}\omega \equiv \left(\frac{\partial\omega}{\partial k_x}, \frac{\partial\omega}{\partial k_y}, \frac{\partial\omega}{\partial k_z}\right). \tag{11.5}$$

Graphically, this means that the direction of energy flow, which may be represented by a ray, is normal to the slowness surface. It is apparent from Fig. 11.2 that the angle between the **k**-vector and the energy flow

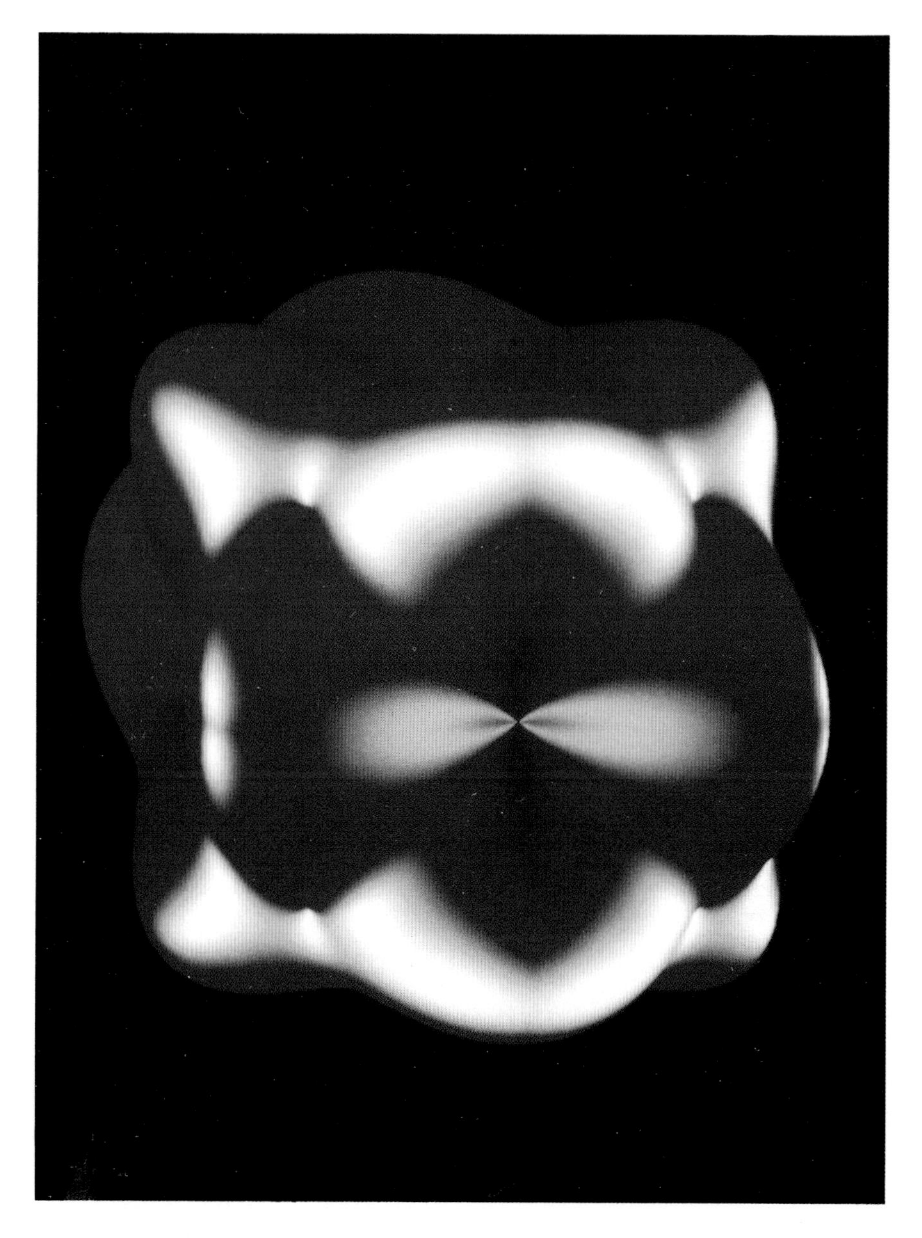

FIG. 11.1 Slowness surfaces, $\mathbf{k}/\omega$, for nickel. The representation is in $\mathbf{k}$-space, and only the outermost slowness surface in any direction is displayed (courtesy of David Barnett).

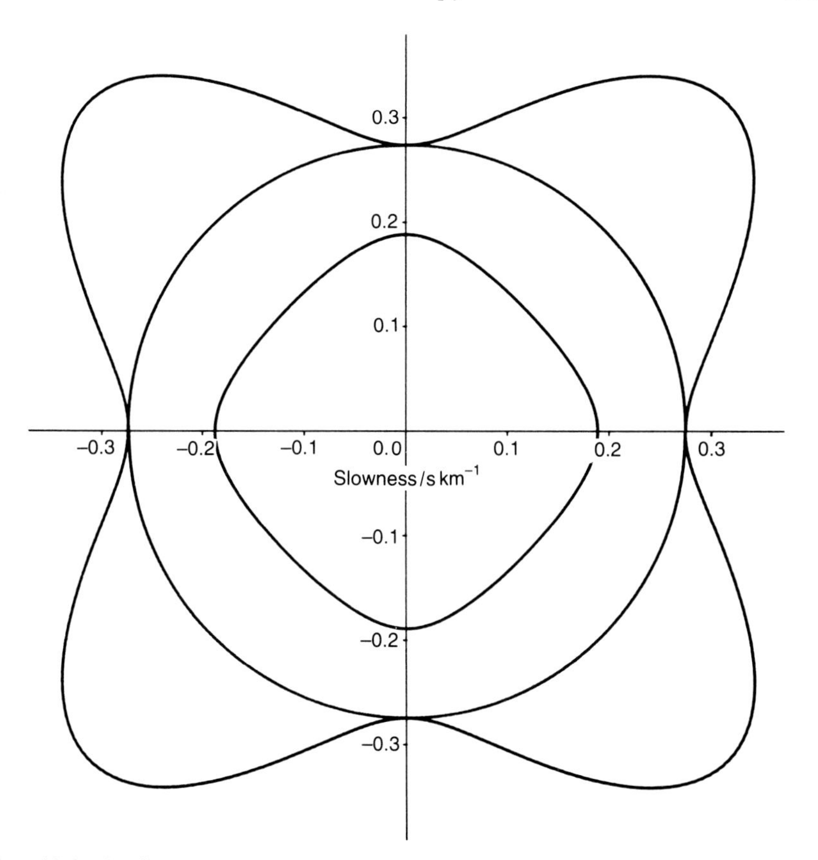

FIG. 11.2. Section through the nickel slowness surfaces of Fig. 11.1 in an {001} plane.

can be several degrees; in some tetragonal materials such as tellurium dioxide the angle can approach 45° for the quasi-longitudinal wave and nearly 90° for a quasi-shear wave (tellurium dioxide is an exceptional material with a pure shear-wave velocity faster than the longitudinal wave velocity in the same direction). This beam-steering effect, as it is sometimes called, means that if waves are excited in a crystal by a transducer or other source in a direction that is not a symmetry direction then, although the wavefronts will of necessity remain parallel to the source, the energy may propagate in a direction that is not normal to the source. This is why the sapphire of an acoustic lens must be very precisely aligned, in order that the waves from the transducer should fall evenly over the surface of the lens.

Snell's law of refraction may be solved graphically using slowness surfaces. It may be expressed as the requirement that the tangential

Table 11.2

Slowness surface sections

For each section, the quantity $\rho(\omega/k)^2$ is given for the pure shear wave, which is polarized perpendicular to the plane of the section, and for the quasi-shear and quasi-longitudinal waves, which each have particle motion in the plane of the section. The angle between the wavevector $\mathbf{k}$ and the lowest symmetry direction in the plane is denoted by ϕ.

Cubic, (001) section, $\mathbf{k}$ at an angle ϕ to [100]

pure shear	c_{44}
quasi-shear	$\frac{1}{2}\{c_{11}+c_{44}-\sqrt{(c_{11}-c_{44})^2\cos^2 2\phi+(c_{12}+c_{44})^2\sin^2 2\phi}\}$
quasi-longitudinal	$\frac{1}{2}\{c_{11}+c_{44}+\sqrt{(c_{11}-c_{44})^2\cos^2 2\phi+(c_{12}+c_{44})^2\sin^2 2\phi}\}$

Cubic, (011) section, $\mathbf{k}$ at an angle ϕ to $[0\bar{1}1]$

pure shear	$\frac{1}{2}(c_{11}-c_{12})\cos\phi+c_{44}\sin\phi$
quasi-shear	$\frac{1}{2}\{B-\sqrt{(B^2-C)}\}$
quasi-longitudinal	$\frac{1}{2}\{B+\sqrt{(B^2-C)}\}$
where	$B\equiv\frac{1}{2}(c_{11}+c_{12}+4c_{44})\cos^2\phi+(c_{11}+c_{44})\sin^2\phi$
	$C\equiv(c'c_{11}-c_{12}^2-2c_{12}c_{44})\sin^2 2\phi+4c_{44}(c'\cos^4\phi+c_{11}\sin^4\phi)$
and	$c'\equiv(c_{11}+c_{12})/2+c_{44}$

Hexagonal, in the basal plane, the velocities are all independent of propagation direction

shear, polarized perpendicular to the basal plane	c_{44}
shear, polarized in the basal plane	c_{66}
longitudinal	c_{11}

Hexagonal, in any plane normal to the basal plane, the velocity of each polarization depends only on the angle ϕ between the normal and $\mathbf{k}$

pure shear	$c_{66}\sin^2\phi+c_{44}\cos^2\phi$
quasi-shear	$\frac{1}{2}(B'-C')$
quasi-longitudinal	$\frac{1}{2}(B'+C')$
where	$B'\equiv c_{11}\sin^2\phi+c_{33}\cos^2\phi+c_{44}$
and	$C'\equiv\sqrt{[\{(c_{11}-c_{44})\sin^2\phi+(c_{44}-c_{33})\cos^2\phi\}^2+(c_{13}+c_{44})^2\sin^2 2\phi]}$

For hexagonal symmetry, $c_{66}=(c_{11}-c_{12})/2$: this may be used when calculating pure shear velocities using data in Table 11.3.

In symmetry directions, notably when $\phi=45°$ or $90°$, these expressions simplify considerably. Along all symmetry axes the shear modes are degenerate (i.e. the shear velocity is independent of the polarization) and all the modes are pure (Auld 1973).

Table 11.3
Anisotropic elastic constants

	ρ(kg m^{-3})	c_{11}(GPa)	c_{12}(GPa)	c_{13}(GPa)	c_{33}(GPa)	c_{44}(GPa)
Metals						
Aluminium	2 698	108	61.3			285
Copper	8 933	169	122			75.3
Gold	19 281	186	157			42.0
Iron crystal	7 873	237	141			116
Nickel	8 907	250	160			118.5
Silver	10 490	119	89.4			43.7
Stainless steel (316)	7 800	206	133			119
Titanium	4 500	162	92	69	181	46.7
Tungsten	19 200	502	199			152
Group IV semiconductors						
Germanium	5 323.4	124.0	41.3			68.3
Silicon	2 329.00	165.77	63.93			79.62
Group III–V semiconductors*						
Aluminium antimonide	4 260	87.69	43.41			40.76
Aluminium arsenide	3 679	120.2	57.0			58.9
Gallium antimonide	5 613.7	88.34	40.23			43.22
Gallium arsenide	5 317.6	118.8	53.8			59.5
Gallium phosphide	4 138	140.50	62.0			70.33
Indium antimonide	5 774.7	66.69	36.45			30.20
Indium arsenide	5 667	83.29	45.26			39.59
Indium phosphide	4 810	101.1	56.1			45.6
Group II–VI semiconductors*						
Telurium dioxide$^+$	5 990	55.7	51.2	21.8	105.8	26.5
Zinc oxide	5 675.26	209.7	121.1	105.1	210.9	42.47
Zinc sulphide	3 980	120.4	69.2	62.0	127.6	22.8
Insulators						
Barium fluoride	4 890	92.0	41.8			25.7
Barium titanate§	5 700	150	66	66	146	44
Beryllium oxide	3 009	460.6	126.5	88.48	491.6	147.7
Bismuth germanate*	7 095	115.8	27.0			43.6
Bismuth germanium oxide*	9 200	128.0	30.5			25.5
Diamond	3 515.25	1 076.4	125.2			577.4
Fluorapatite	3 214.7	143.4	44.5	57.5	180.5	51.5
Gadolinium gallium garnet		285	114			89.7
Hydroxyapatite	3 200.0	137.0	42.5	54.9	172.0	39.6
Lithium fluoride	2 601	111.2	42.0			62.8
Lithium niobate*	4 700	203	53	75	245	60
Lithium tantalate*	7 450	233	47	80	275	94
Magnesium oxide	3 650	286	87			148
PZT-2†§	7 600	135	113	22.2	67.9	68.1
PZT-5H†§	7 500	12.6	11.7	23.0	79.5	84.1
Silicon carbide (cubic)	3 166	360	151			198
(hexagonal)	3 211	500	92	56	564	168

Table 11.3—(*Continued*)

	$\rho(\mathrm{kg\,m^{-3}})$	c_{11}(GPa)	c_{12}(GPa)	c_{13}(GPa)	c_{33}(GPa)	c_{44}(GPa)
Sodium fluoride	2 558	97	24.4			28.1
Strontium titanate	5 110	318	102			124
Yttrium aluminium garnet	4 550	333.2	110.7			115.0
Yttrium gallium garnet	5 790	290.3	117.3			95.5
Yttrium iron garnet	5 170	268	110.6			76.6
Zirconia‡	6 075	406.5	103			55

Where only three elastic constants are given, the material has cubic symmetry. For hexagonal symmetry, $c_{66} = (c_{11} - c_{12})/2$. The number of digits does not necessarily indicate the accuracy of the values. Constants for ternary semiconductor compounds can often be interpolated linearly by atomic composition from the data here. All values are at room temperature and for semiconductors at low doping. Data from Auld (1973), Landolt–Börnstein (1979, 1984, 1987), and Ingel and Lewis (1988*b*; 1988*a* gives a general analysis of errors in the determination of anisotropic elastic constants from sound velocity measurements).
* Piezoelectric.
+ Tetragonal, $c_{66} = 65.9$ GPa.
§ Poled ceramic, *c*-axis along poling axis.
† Lead titanate–zirconate.
‡ 3.4 mole% Y_2O_3.

component of the **k**-vector be conserved across a refracting interface. This is illustrated for waves in water incident on an (001) GaAs surface in Fig. 11.3. The slowness surfaces for GaAs in a (010) plane are similar to those for Ni in Fig. 11.2, though the anisotropy is not so great; for water the slowness surface is simply a single sphere of radius greater than even the quasi-shear wave in GaAs. Angles of incidence of 10°, 20°, 30°, and 40° have been shown. At 10° there are six points of intersection with the GaAs slowness surfaces. Three of these are discarded, because the energy flow would be in the wrong direction. The [010] polarized pure shear wave could satisfy Snell's law and energy flow requirements, but the longitudinal waves in the water could not couple into the horizontal (i.e. parallel to the interface) motion of this polarization. Therefore, there are two refracted waves, one quasi-longitudinal and one quasi-shear. At 20° the angle of incidence is close to the longitudinal critical angle, at which a lateral longitudinal wave is excited with an angle of refraction of 90° (§7.2.3). A quasi-shear wave is also excited. Beyond the longitudinal angle no bulk longitudinal wave can be excited since the tangential component of the **k**-vector is constrained to lie outside the longitudinal slowness surface, though there may be an evanescent longitudinal wave with an imaginary component of the **k**-vector normal to the surface. At 30° the quasi-shear surface is cut at four places. Two non-degenerate solutions are allowed by energy flow considerations, and curiously enough one of those has the **k**-vector pointing back towards the fluid. This angle is above the angle normally associated with critical angle phenomena (i.e. where the angle of refraction is 90°), and this will

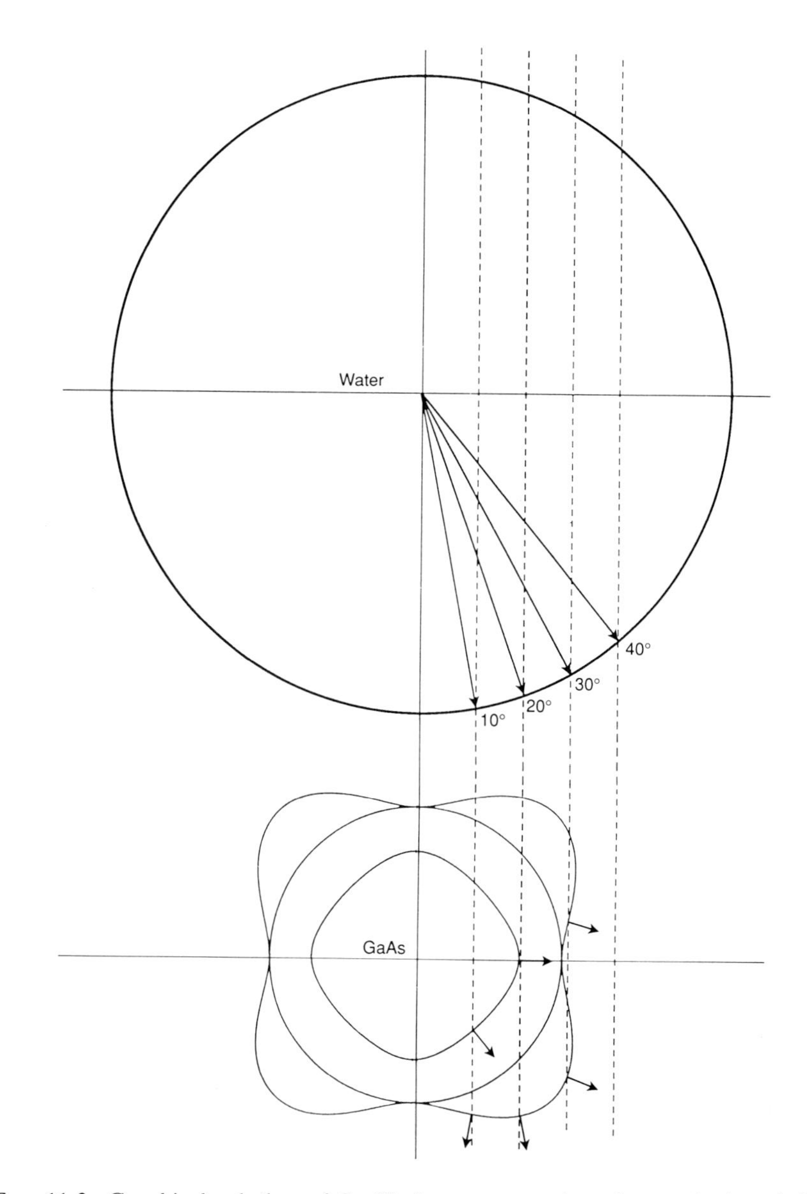

FIG. 11.3. Graphical solution of Snell's law, expressed as the continuity of the tangential component of **k** across an interface, which is equivalent to (6.75). On one side of the interface is water, with a single spherical slowness surface for an isotropic scalar longitudinal wave; on the other side is GaAs, with one longitudinal and two shear slowness surfaces (calculated from the data in Table 11.2). Solutions are illustrated for angles of incidence in the water of 10°, 20°, 30°, and 40°.

lead to unusual phenomena in the reflectance function in §11.3, especially in Fig. 11.6. At 40° only evanescent waves are excited. Around a certain angle beyond the shear critical angle Rayleigh waves may be excited, consisting of a mixture of evanescent longitudinal and shear waves. These and other surface wave modes play a vital role in the contrast of crystalline specimens in acoustic microscopy.

11.2 Waves in anisotropic surfaces

In the surface of a half space that is isotropic, the Rayleigh wave velocity is the same in all directions. If the surface is imagined to be in a horizontal plane, then the Rayleigh wave is composed of a shear wave component polarized in a vertical plane (SV) and a longitudinal wave component. Shear waves polarized horizontally (SH) can also exist, but they do not couple to the Rayleigh wave at all (nor, in the case of fluid loading, would they couple into waves in the fluid).

In the surface of an anisotropic solid the situation is more complicated. Pure Rayleigh waves can exist only along certain symmetry directions in which pure SV waves exist. Away from these directions, however, the two quasi-shear polarizations are not pure SV and SH; therefore, although the particle motions are orthogonal, at the surface they can be weakly coupled. If the SH mode has a higher velocity than the SV, then there can be no real solution to Snell's law for bulk waves, and the surface wave remains bound to the surface. But if the SH velocity is less than the SV velocity, then a surface wave can couple at the surface itself into waves of orthogonal polarization which can then propagate energy into the bulk. Because such coupling is weak the surface wave exists as a recognizable mode, but because its energy becomes gradually depleted it is not a pure surface wave, and so it is designated a pseudo-surface wave (Farnell 1970, 1978). (In some early literature these waves were called leaky surface waves, because energy leaks away from the surface into the solid. In the context of fluid-loaded surfaces, however, and specifically in acoustic microscopy, the term leaky is used for waves leaking energy into the fluid.)

The situation is illustrated for an $\{001\}$ surface of a cubic crystal in Fig. 11.4. From Table 11.2 the ratio of shear velocities in a $\langle 110 \rangle$ direction is $(v_{SV}/v_{SH})_{\langle 110 \rangle} = A_{an}^{1/2}$, where the anisotropy factor $A_{an} \equiv 2c_{44}/(c_{11} - c_{12})$. Figure 11.4(a) illustrates a case where $A < 1$ and so $v_{SH} > v_{SV}$; the particular values used are for cubic zirconia with a small addition of yttria. Although v_{SV} does not change with angle, the anisotropy in v_l leads to a small variation in the Rayleigh wave velocity with propagation direction. But there is only one surface wave, and it remains bound

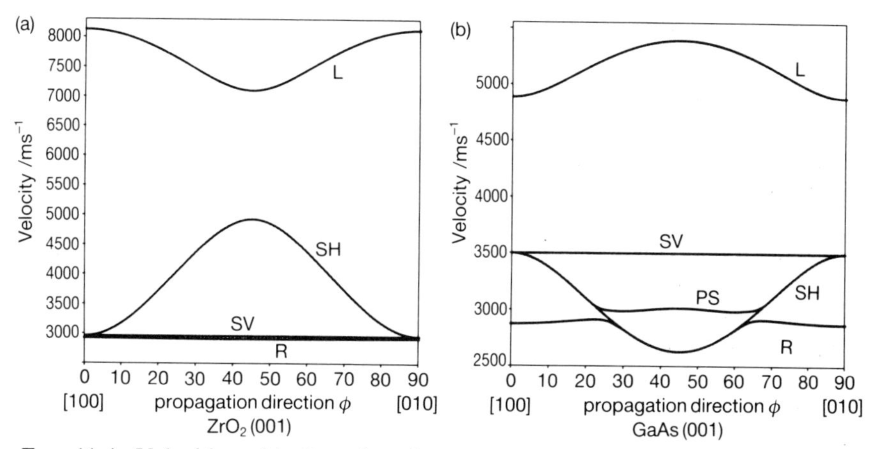

FIG. 11.4. Velocities of bulk and surface waves in an (001) plane; the angle of propagation ϕ in the plane is relative to a [100] direction. (a) Zirconia, anisotropy factor $A_{an} = 0.36$; (b) gallium arsenide, anisotropy factor $A_{an} = 1.83$; material constants taken from Table 11.3. Bulk polarizations: L, longitudinal; SV, shear vertical, polarized normal to the (001) plane; SH, shear horizontal, polarized in the (001) plane. Surface modes: R, Rayleigh, slower than any bulk wave in that propagation direction; PS, pseudo-surface wave, faster than one polarization of bulk shear wave propagating in that direction.

to the surface at all angles. Most metals and semiconductors have $A_{an} > 1$; their situation is illustrated in Fig. 11.4(b), with values corresponding to gallium arsenide. For propagation in the [100] direction, the Rayleigh wave is polarized vertically. As the angle from the [100] direction increases, the polarization remains quasi-vertical until the Rayleigh branch approaches the v_{SH} curve, when the quasi-shear horizontal mode takes over as the dominant component in the Rayleigh wave, whose polarization therefore twists over towards the horizontal. The Rayleigh velocity remains slightly lower than the bulk SH velocity but becomes indistinguishable from it as the [110] direction is approached, until in the [110] direction it is pure SH and the Rayleigh character has been lost. At directions of propagation other than [110] the Rayleigh wave remains bound to the surface with no leakage of energy into the bulk. Meanwhile, on the other side of the bulk SH curve, and with a velocity close to what the Rayleigh mode would have had if it had not been interrupted by the SH branch, another surface wave has developed. Its polarization is close to the vertical, and it has a character very similar to the Rayleigh wave in the case $A_{an} < 1$. The difference is, however, that Snell's law has a real solution for SH waves propagating away from the surface into the solid, so the energy is not bound to the surface and the wave is only a pseudo-Rayleigh wave, more commonly known as a pseudo-surface (PS) wave. In the [110] direction the coupling of the SH wave to all other modes vanishes, and so the wave becomes pure Rayleigh at the very point at which, for the same reason, the Rayleigh branch disappeared.

If, *per impossibile,* the coupling between modes at the surface could be switched off, there would be two independent waves that could propagate parallel to the surface. One would be Rayleigh in character, and the other quasi-shear horizontal. These two branches would cross each other at the point where they were degenerate. But because there is coupling, two new branches are formed. One starts off pure shear horizontal, but close to the point where degeneracy would have occurred, it twists its polarization to predominantly SV and follows the PS branch shown, finally ending up as a pure Rayleigh wave at [110]. The other branch is the Rayleigh branch, whose polarization starts off purely in the vertical direction and then twists to become quasi-horizontal, ending up as pure horizontal at [110]. The two branches are closest to one another where uncoupled degeneracy would have occurred. The profile of the particle displacements is displayed in Fig. 11.5, which is the anisotropic equivalent of Fig. 6.2(b). The characteristic depth of the excitation is comparable to the isotropic case, but the presence of transverse motion parallel to the surface is due to the anisotropy and corresponds to the twisting of the polarization.

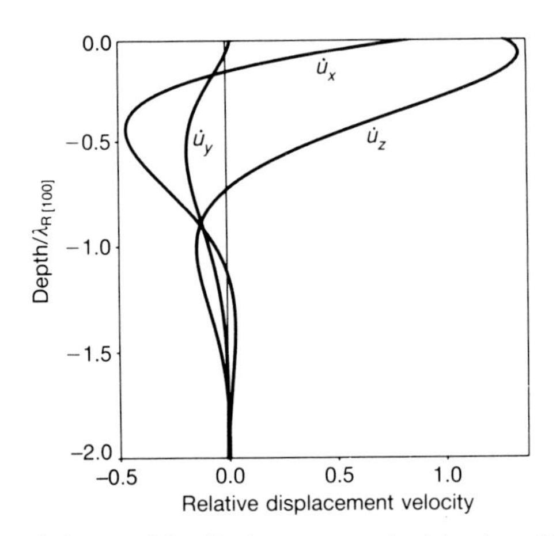

FIG. 11.5. Profile of the particle displacement velocities in a Rayleigh wave in a GaAs(001) surface, propagating at an angle $\phi = 15°$ to a [100] direction. This figure is analogous to Fig. 6.2(b) for a Rayleigh wave in an isotropic medium, but in this case there are components of displacement in both horizontal axes, because of the anisotropy, and so three curves are needed; they are normalized by setting the larger horizontal component to unity at the surface.

A code has been written to enable the velocities of surface waves in multilayered anisotropic materials, at any orientation and propagation and including piezoelectric effects, to be calculated on a personal computer (Adler *et al.* 1990). The principle of the calculation is a matrix approach, somewhat along the lines of §10.2 but, because of the additional variables and boundary conditions, and because the wave velocities themselves are being found, it amounts to solving a first-order eight-dimensional vector–matrix equation. A program of this type is invaluable both for calculating contrast from known or postulated structures and also for interpreting quantitative results from line-focus-beam $V(z)$ measurements.

Light fluid loading will perturb the surface wave modes, in general increasing their velocity slightly, but will not alter their basic character as far as propagation into the solid is concerned. In calculating the reflection coefficient and hence $V(z)$ for anisotropic surfaces, the surface waves are implicit in the equations and it is never necessary to add them in separately. But an understanding of them is invaluable in interpreting the contrast and analysing $V(z)$ curves. Waves that have a significant component of SV polarization will couple to waves in the fluid, but those that are SH in character will experience vanishingly weak coupling. Other orientations and symmetries will exhibit a pseudo-Rayleigh wave in different directions, but the principles remain the same. For example, a cubic $\{111\}$ surface exhibits both a Rayleigh wave and a pseudo-Rayleigh wave in all directions, except in exact $\langle 110 \rangle$ directions in which there is no coupling to SH waves.

11.3 Anisotropic reflectance functions

In order to calculate the reflectance function for waves from a fluid incident on an anisotropic surface, the Christoffel equation must be solved with the constraint imposed by Snell's law that the tangential component of the wave vector be the same as that in the fluid (Somekh *et al.* 1984).

The Christoffel equation may be written in matrix form

$$k^2 \begin{pmatrix} \Gamma_{xx} & \Gamma_{xy} & \Gamma_{xz} \\ \Gamma_{yx} & \Gamma_{yy} & \Gamma_{yz} \\ \Gamma_{zx} & \Gamma_{zy} & \Gamma_{zz} \end{pmatrix} \begin{pmatrix} u_x \\ u_y \\ u_z \end{pmatrix} = \rho\omega^2 \begin{pmatrix} u_x \\ u_y \\ u_z \end{pmatrix}. \qquad (11.6)$$

If the z direction is taken as the normal to the surface, and the x direction is taken as also lying in the plane of incidence, then in calculating the Christoffel matrix Γ for arbitrary crystallographic orientation the stiffness matrix must be transformed to these new coordinates. In the notation of (11.2), and for an incident wavevector (k_x', k_y', k_z') in the fluid, the requirements of Snell's law become

$$|k|\, l_x = k_x', \qquad l_y = 0. \tag{11.7}$$

The Christoffel equation yields a sextic in l_z. For each of these solutions the normalized particle velocity vector $\mathbf{v}$ can be found. Hence the traction is

$$-\tau_{ik} = Z_{ij}^K v_j. \tag{11.8}$$

Z_{ij}^K is the impedance matrix in the propagation direction K; it is simply the identity matrix multiplied by the product of the density ρ and the phase velocity ω/k. The energy flow vector is then

$$P_i = -\frac{v_j^* \tau_{ij}}{2}. \tag{11.9}$$

The energy flow vector is calculated for each of the six solutions, and the z component is tested. A positive real value indicates energy flowing from solid to liquid, and is disallowed. A positive imaginary value indicates a wave growing exponentially into the solid, and is also disallowed. The other two cases are permitted, yielding three allowable solutions, each real or imaginary, in all. These three solutions in the solid, together with the reflected wave in the fluid, give four unknown amplitudes to be calculated: A_1, A_2, A_3, and A_4. The boundary conditions at the surface are continuity of normal components of displacement and traction, and zero tangential components of traction.

$$\begin{aligned} A_1 v_{z1} + A_2 v_{z2} + A_3 v_{z3} - 1 - A_4 &= 0,\\ A_1 t_{x1} + A_2 t_{x2} + A_3 t_{x3} &= 0,\\ A_1 t_{y1} + A_2 t_{y2} + A_3 t_{y3} &= 0,\\ A_1 t_{z1} + A_2 t_{z2} + A_3 t_{z3} + (1 - A_4) t_{z4} &= 0. \end{aligned} \tag{11.10}$$

In isotropic materials, and along symmetry directions in anisotropic materials, all the traction components in the third equation vanish, as does the component of A associated with the SH mode. There are then three equations for three unknowns. For the general anisotropic case the four equations can be solved to give the amplitude A_4 of the reflected wave. The amplitude is a complex quantity, because there may be a change in phase upon reflection. For an incident wave at an angle θ to the normal and ϕ to some direction lying in the surface (usually the lowest index direction available) the reflectance function may be written

$$R_{hkl}(\theta, \phi) \equiv A_4 \tag{11.11}$$

where h, k, l are the Miller indices of the surface.

A series of reflectance functions for a GaAs(001) surface is shown in Fig. 11.6. In each case the modulus of the reflectance function is drawn as a solid curve referred to the left ordinate, and its phase is drawn as a broken curve referred to the right ordinate. The angle ϕ is referred to an [010] direction. Figure 11.6(a) is $R_{001}(\theta, 0°)$. Qualitatively it appears quite similar to an isotropic reflectance function, such as the one for fused silica in Fig. 6.3(b)i. Point A marks the longitudinal critical angle, where the reflection coefficient is unity. B marks the shear wave critical angle, beyond which no energy can propagate into the solid. C indicates the Rayleigh angle, with its familiar phase change. But even in this

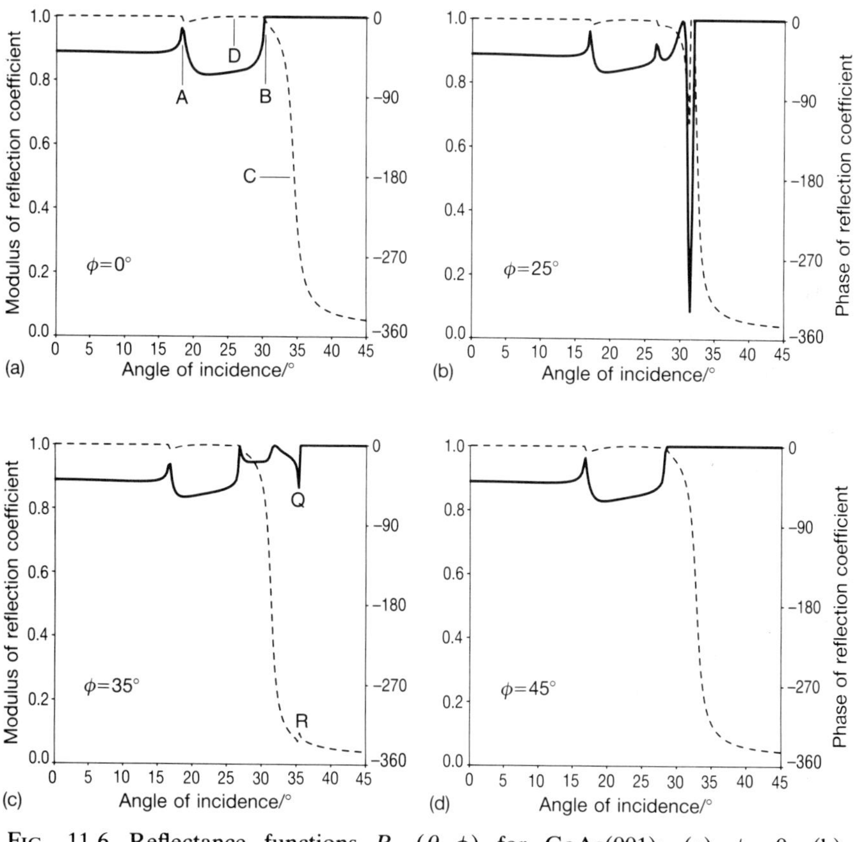

FIG. 11.6. Reflectance functions $R_{001}(\theta, \phi)$ for GaAs(001): (a) $\phi = 0$; (b) $\phi = 25°$; (c) $\phi = 35°$; (d) $\phi = 45°$.

symmetry direction, no isotropic material could ever generate this reflectance function (for example, the ratios of longitudinal, shear, and Rayleigh velocities would be quite impossible). D indicates the point at which the angle of refraction of the shear wave is 90° (the glitch associated with this is barely visible); in an isotropic material this would be the shear critical angle, but because of the re-entrant slowness surface the shear critical angle occurs beyond this point. For water/GaAs, $v_0/\sqrt{(c_{44}/\rho)} = 0.443$, corresponding to an angle $\theta_m = 26.3°$.

At $\phi = 25°$ (Fig. 11.6(b)) the effects of anisotropy are more obvious. The modulus approaches unity at the longitudinal critical angle, much as before. It has another peak at the quasi SV critical angle (about 27°). At $\theta = 31°$ the modulus again rises to unity and there is a phase change of a little over $-\pi/2$. This corresponds to the excitation of the pseudo-Rayleigh wave. There is a sharp dip in the magnitude to less than 0.1, because at this azimuthal angle the pseudo-Rayleigh wave is quite strongly coupled to the bulk quasi SH wave, which is slightly slower and can therefore leak energy into the bulk. The quasi SH critical angle is at $\theta \approx 32°$, and this is accompanied by the Rayleigh wave phase change of -2π. These phenomena in $R(\theta, \phi)$ can be correlated with the various modes by following down a vertical locus through $\phi = 25°$ in Fig. 11.4(b), and correlating the mode velocity v_m with angle via Snell's law for 90° refraction

$$\sin \theta_m = \frac{v_0}{v_m}. \tag{11.12}$$

At a larger azimuthal angle, $\phi = 35°$ (Fig. 11.6(c)), the pseudo-Rayleigh wave has become dominant. The Rayleigh wave has become almost SH in character and is only very weakly coupled to the fluid, so that both the quasi SH critical angle (Q) and the phase change at the Rayleigh angle (R) are almost insignificant. In the [110] direction, $\phi = 45°$ (Fig. 11.6(d)) the old Rayleigh wave vanishes due to merger into the SH mode, and the old pseudo-Rayleigh wave takes on pure Rayleigh character (not because it is slower than the SH wave, but because at exactly this angle there is no coupling to it). $R_{001}(\theta, 45°)$ again appears qualitatively similar to an isotropic reflectance function, but like the $R_{001}(\theta, 0°)$ curve it could not in fact have been generated by any combination of isotropic elastic constants.

11.4 Cylindrical lens anisotropic $V(z)$

The phenomena in the reflectance function associated with surface wave modes have a dominant effect on $V(z)$. Whenever a Rayleigh or pseudo-Rayleigh wave is excited, oscillations of corresponding period appear in $V(z)$. If both waves are excited, beats can occur in $V(z)$. These effects can be observed and measured as a function of angle in a uniform anisotropic surface using the line-focus-beam technique with a cylindrical lens (Kushibiki *et al.* 1983). Figure 11.7(a) is a measured $V(z)$ curve for a GaAs(001) surface, with $\phi = 26°$. At this angle both Rayleigh and pseudo-Rayleigh waves are strongly excited. Beating between the two modes is apparent, with cancellation occurring around $z \approx -110\ \mu\text{m}$, and partial cancellation beyond $-300\ \mu\text{m}$ (the two waves do not suffer the same rate of decay). At sufficiently large defocus oscillations due to longitudinal head waves also appear. The individual contributions can be analysed in the Fourier transform of $V(z)$ (Fig. 11.7(b)). The fastest mode (i.e. the peak at the lowest value in the transform) is the longitudinal headwave; the peaks at higher values are the pseudo-Rayleigh and Rayleigh waves, close together but nevertheless adequately separated.

In order to measure the velocity and attenuation of each wave using the method of §8.2.1, the peaks must be analysed one at a time, using an appropriate value for ξ'' in step 7 in each case. The results of a set of such measurements on a GaAs(001) surface, for ϕ from 0° to 90°, are presented in Fig. 11.8. For $\phi < 20°$ only the Rayleigh wave could be measured. In the range $20° \leq \phi \leq 35°$ both Rayleigh and pseudo-Rayleigh waves were detected. Below about 25° the attenuation of the pseudo-Rayleigh wave could not be measured, because the wave couples only weakly to fluid. Above 25° the Rayleigh wave coupling becomes weaker and, although the velocity can still be measured, the attenuation is so small that the measurements are again very uncertain. Part of the

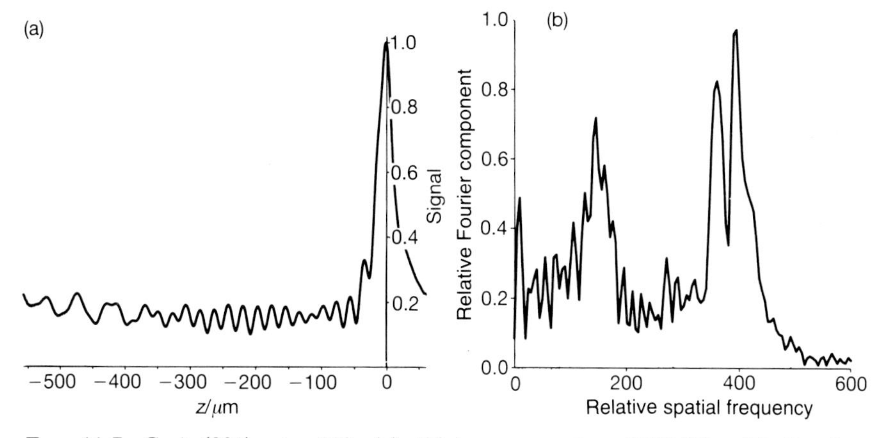

FIG. 11.7. GaAs(001), $\phi = 26°$. (a) $V(z)$, measured at 225 MHz; (b) Fourier transform of the $V(z)$ data, corresponding to step 6 in §8.2.1.

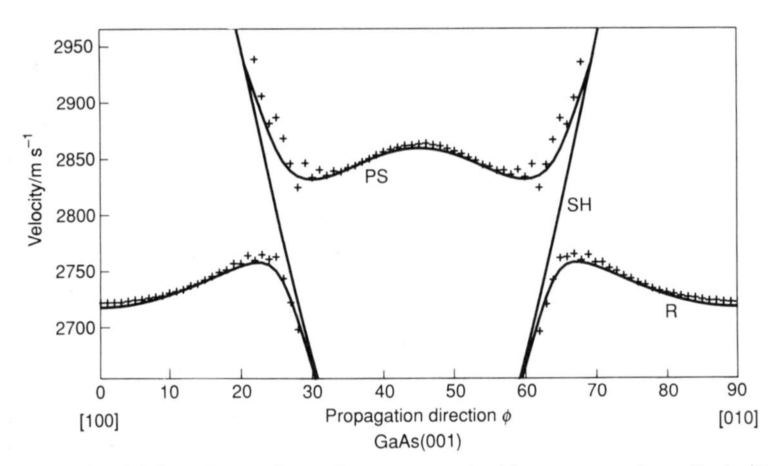

FIG. 11.8. Rayleigh and pseudo-surface wave velocities measured on GaAs(001), indicated by +. The solid lines are the calculated curves of Fig. 11.4(b) (without the longitudinal wave curve), plotted on the enlarged vertical scale used for the experimental measurements.

calculated curve of Fig. 11.4(b) has been included in Fig. 11.8, to show the relationship of the Rayleigh waves and pseudo-Rayleigh waves to the SH and SV waves, along with experimental results. An attenuation curve for the Rayleigh wave gives a direct indication of its coupling to the fluid, but the attenuation of the pseudo-Rayleigh wave includes the contributions both of coupling into the fluid and of coupling into bulk waves in the solid. GaAs{110} exhibits no pseudo-Rayleigh waves, but the coupling to Rayleigh waves is stronger and there is only one mode present in the analysis.

As well as semiconductor wafers, studies have been made of piezo-electric wafers by the line-focus-beam technique, in particular lithium niobate ($LiNbO_3$). Lithium niobate is an important optoelectronic material, because light beams in it can be switched by exciting surface acoustic waves in the surface so that the light suffers Bragg scattering. One way of guiding the light is to remove lithium atoms from the material in the vicinity of the surface, exchanging them for protons to maintain charge neutrality. The proton-exchanged material has a slower velocity of light (i.e. a higher optical refractive index), and so light can be confined to a channel, for much the same reason that a Rayleigh wave is confined to a surface, or indeed that any conventional waveguide works. But if the proton-exchanged material is to be used with surface acoustic wave switching, it is essential to know how the elastic properties are affected. Measurements were made of the three principal crystallographic surfaces, and the effects were quite large (Burnett *et al.* 1986). The behaviour was most straightforward in the Z-cut face, where the proton exchange caused a reduction in Rayleigh velocity of about $200\,\mathrm{m\,s^{-1}}$ in all propagation directions. The effects in the X-cut face were more complicated, but the maximum velocity change was about the same. In the technologically important Y-cut face, proton exchange had little effect on Rayleigh wave propagation within 45° of the Z-propagation direction. Beyond that there was a decrease in Rayleigh velocity relative to the virgin material: in the X-propagation direction the difference was $140\,\mathrm{m\,s^{-1}}$. The attenuation was also measured. The decrease in velocity within 45° of the X-propagation direction was accompanied by an increase in Rayleigh attenuation: in the X-propagation it was double its value in virgin $LiNbO_3$. Doping with MgO increases the optical power that $LiNbO_3$ can handle without damage, as well as increasing the refractive index, and this may make it a candidate for optical fibres. The correlation between the effects on the optical and acoustic velocities has been measured: 5 per cent doping causes an increase in the optical refractive index of 0.7 per cent and an increase in the Rayleigh wave velocity of 0.9 per cent (Jen 1989). Lithium niobate is also used for surface acoustic wave filters, both for military use and in domestic television sets. Measurements have been made area by area on whole wafers, and correlated with the performance of devices previously fabricated on the wafer and removed before the $V(z)$ measurements were made (Kushibiki *et al.* 1985, 1991*b*). The variation in Rayleigh velocity was about 0.6 per cent, and the correlation between regions on the wafer of high and low velocity exhibited by the devices and measured by $V(z)$ was excellent. Lithium tantalate and zinc oxide on glass have also been measured in this way (Kushibiki *et al.* 1986, 1991*a*).

Not all anisotropic materials are perfect single crystals, and not all of them give such tidy results as electronic and optoelectronic wafers. But measurements on these more difficult inhomogeneous anisotropic specimens may be significant nonetheless, especially if they can be related to similar measurements on homogeneous specimens of the constituents as they were with the dental enamel in §9.4.2. Angle-resolved Rayleigh wave measurements can also be used to characterize the anisotropy of heavily drawn metal–metal composites.

Dual phase alloys of Cu–20%Nb were originally developed to investigate their superconducting properties. It turned out that they also had exceptional mechanical strength, together with good high-temperature electrical conductivity. Cu–Nb composites are prepared by casting them so as to produce an array of Nb dendrites in a copper matrix. They are then extruded or rolled to break up the dendrites and align them into long ribbons of filaments or platelets with separation less than a micron. The local crystallographic orientations of the filaments and the matrix determine the directions in which dislocation motion can occur, and hence control the mechanical properties. The extent of the strengthening depends strongly on the draw ratio η, the logarithm to base e of the ratio of the initial to the final cross-sectional area. In a series of experiments on materials of this kind in a line-focus-beam instrument, specimens of Cu–20%Nb of varying η were measured, together with homogeneous samples of the parent metals Cu and Nb (Thompson *et al.* 1990a). The measurements proved extraordinarily difficult, and only the drawn samples of the composite alloys gave $V(z)$ curves that could be satisfactorily analysed to yield a Rayleigh velocity that could be trusted. Measurements were made at 5° intervals, and the results for the samples with $\eta = 3.6$ and $\eta = 5.4$ are given in Fig. 11.9. Theory suggests that the angular dependence of the Rayleigh velocity in single-phase polycrystalline metals with weak anisotropy should be a polynomial of the form $a + b\cos 2\phi + c\cos 4\phi$ (Sayers 1985). Best-fit curves of this form, with the zero of ϕ also chosen for best fit, have been drawn in

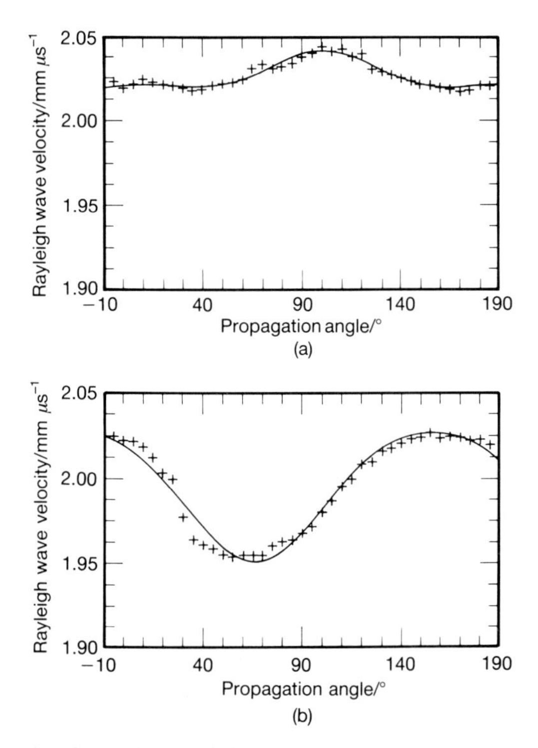

FIG. 11.9. Angular dependence of the Rayleigh velocity in Cu–20%Nb with draw ratio: (a) $\eta = 3.6$; (b) $\eta = 5.4$ (Thompson *et al.* 1990*a*).

Fig. 11.9. Of course, the specimens are not single phase and they may be more than weakly anisotropic; nevertheless, something can be made of the curves. They have been plotted with the same axes, and the anisotropy in the Rayleigh velocity of the less drawn specimen is only a quarter that of the other one. The Cu–20%Nb alloy gave considerably greater variation of v_R with ϕ than the Cu–20%Ta alloy. From the fitted curves, values of the parameters δW_{400} and W_{420} commonly used to describe texture in pole figures of X-ray or neutron diffraction measurements were calculated, and these made sense in terms of what would have been expected. It would be difficult to measure such a small specimen by either of those techniques, and it would be tedious to measure enough grains to get a sensible average by electron microscopy. The line-focus-beam technique can measure a comfortable size of specimen.

11.5 Spherical lens anisotropic $V(z)$

The LFB microscope gives good azimuthal resolution in ϕ, but is not suitable for imaging. The imaging microscope gives good spatial resolution, but the spherical lens averages over all ϕ. Therefore, the contrast for an anisotropic specimen becomes

$$V_{hkl}(z) = \frac{V_0}{2\pi} \int_0^{2\pi} \int_0^{\pi/2} R_{hkl}(\theta, \phi) P(\theta, \phi) \mathrm{e}^{-\mathrm{i}2kz\cos\theta} \sin\theta \cos\theta \, \mathrm{d}\theta \, \mathrm{d}\phi. \quad (11.13)$$

The integration could be performed by calculating $V(z)$ for each ϕ, corresponding to $V(z)$ curves such as the measured one in Fig. 11.7(a), and then integrating over ϕ. Alternatively, a complex mean reflectance function (denoted by prime) may be calculated as

$$R'_{hkl}(\theta) = \frac{1}{2\pi} \int_0^{2\pi} R_{hkl}(\theta, \phi) \, \mathrm{d}\phi. \quad (11.14)$$

By swapping the order of integration in (11.13), and assuming axial symmetry of the lens, the contrast may now be written

$$V_{hkl}(z) = V_0 \int_0^{\pi/2} R'_{hkl}(\theta) P(\theta) \mathrm{e}^{-\mathrm{i}2kz\cos\theta} \sin\theta \cos\theta \, \mathrm{d}\theta. \quad (11.15)$$

This is identical in form to the equation for $V(z)$ in an isotropic material. But of course $R'_{hkl}(\theta)$ is quite different from any isotropic reflectance function. For a cubic material an anisotropy factor can be defined as

$$A_{\mathrm{an}} \equiv \frac{2c_{44}}{c_{11} - c_{12}}. \tag{11.16}$$

Complex mean reflectance functions for aluminium, nickel, and copper are presented in Fig. 11.10. These represent a series of increasingly anisotropic materials, with anisotropy factors: $A_{\mathrm{Al}} = 1.22$, $A_{\mathrm{Ni}} = 2.63$, $A_{\mathrm{Cu}} = 3.20$. Iron has an anisotropy factor $A_{\mathrm{Fe}} = 2.32$, and ferrous alloys, most notably austenitic stainless steel, exhibit behaviour similar to nickel. For each metal, complex mean reflectance functions are shown for $\{100\}$, $\{110\}$, and $\{111\}$ surfaces. As the anisotropy increases, so does the difference between the curves at these three orientations. In aluminium (Fig. 11.10(a)), the modulus of $R'_{100}(\theta)$ looks rather like an isotropic material, except for a small dip just beyond the shear critical angle. This dip is much bigger in $R'_{110}(\theta)$, and it arises from the integration of $R_{hkl}(\theta, \phi)$ where there is a rapid phase change with θ that varies with ϕ (cf. §8.1.1). At all orientations of aluminium, the phase changes in the complex mean reflectance function are well marked but, because of the low anisotropy, the angle at which they occur does not change much from one orientation to another. In nickel, the differences between the angles at which the phase changes occur at the different orientations are greater than in aluminium, because of the greater anisotropy, but the phase changes themselves are small, again because of integration over rapidly varying phase. The difference between $R'_{110}(\theta)$ and $R'_{111}(\theta)$ is smaller than the difference between either of those two and $R'_{100}(\theta)$. In copper (Fig. 11.10(c)), the anisotropy is greater still; this is manifested in even greater differences between the three complex mean reflectance functions. But copper is a rather slow material; many of the significant features occur at rather large values of θ; indeed in both $R'_{110}(\theta)$ and $R'_{111}(\theta)$ the modulus is still considerably less than unity at $\theta = 60°$, which means that power is still propagating into the material beyond that angle. All these observations have consequences in $V(z)$.

Using the complex mean reflectance functions, $V(z)$ curves can be calculated. Figure 11.11 gives $V(z)$ curves for Al, Ni, and Cu, for the same orientations as Fig. 11.10. The curves for Al show strong oscillations associated with the clear phase changes in the complex mean reflectance functions. But, because of the weak anisotropy, the three curves do not differ much from one another. The curves for Ni show greater differences, with $V_{100}(z)$ differing more from the other two than they differ from each other. In copper the oscillations in $V(z)$ are very weak and, despite the strong anisotropy, there is little difference between the three curves. Or rather, it may be partly because the strong anisotropy causes the range of angles over which the phase change is averaged to be so great that extensive cancellation occurs and, except in the {100} orientation, the oscillations are so weak. In any case, the large angles characteristic of copper mean that the reflections of interest occur at angles where the lens response is relatively feeble.

Experimentally, although grains have been imaged in polished but not etched specimens of both aluminium (Weaver 1986; Fossheim *et al.* 1988) and copper (Bray 1981), in both cases the contrast was rather poor. But in nickel and iron and related alloys such as stainless steel, grain contrast can be superb (as it is indeed in many ceramics). It seems that, as in the *Story of the Three Bears,* aluminium is not anisotropic enough, copper is too anisotropic (though in copper there may also be other adverse factors such as strength of coupling and large Rayleigh angles), but nickel is just right!

Images of a polycrystalline specimen of nickel are presented in Fig. 11.12. At focus, $z = 0$ (Fig. 11.12(a)), there is little contrast, though grain boundaries can just be detected. At $z = -4\ \mu m$ (Fig. 11.12(b)), the contrast between grains is much better. At $z = -7\ \mu m$ (Fig. 11.12(c)), the contrast is even stronger, but contrast reversal has occurred. Of course, it is most unlikely that these grains had exact symmetrical crystallographic orientations, but in order to discuss the interpretation of the contrast and its relation to the $V(z)$ curves in Fig. 11.11(b), suppose that the grains α, β, and γ had orientations lying close to {111}, {100}, and {110}, respectively. At focus, the $V(z)$ curves from all three orientations coincide. No contrast would be expected here, and little is found in Fig. 11.12(a). At defocus $z = -4\ \mu m$ ($2\lambda_0$), the {100} curve shows a smaller signal than the other two, so that it would appear darker in an image. A {111} grain would be slightly brighter than a {110} grain, but the difference is much smaller. At $z = -7\ \mu m$ ($3.5\lambda_0$) defocus, complete reversal of the contrast would be expected. The $V(z)$ curve for {100} now indicates that this should be the brightest. These observations can be compared with what is seen in the pictures. Using a directional point-focus lens it is even possible to invert the process, and identify the approximate crystallographic orientation of grains from analysis of directional $V(z)$ measurements (Hildebrand and Lam 1983; Kushibiki *et al.* 1989; Ishikawa *et al.* 1990). Even with a conventional spherical lens it is possible, with accurate analysis, to identify specific velocity components in $V(z)$, because beam-steering effects can deflect all but the surface waves travelling in symmetry directions away from a return path to the transducer. For example, in measurements on Y-cut α-quartz at 225 MHz ($r_0 = 0.75$ mm), the Fourier transform of $V(z)$ data from an axially symmetric spherical point-focus lens showed three well-defined peaks corresponding to 3147, 3395, and 3878 m s^{-1} (Kushibiki *et al.* 1989).

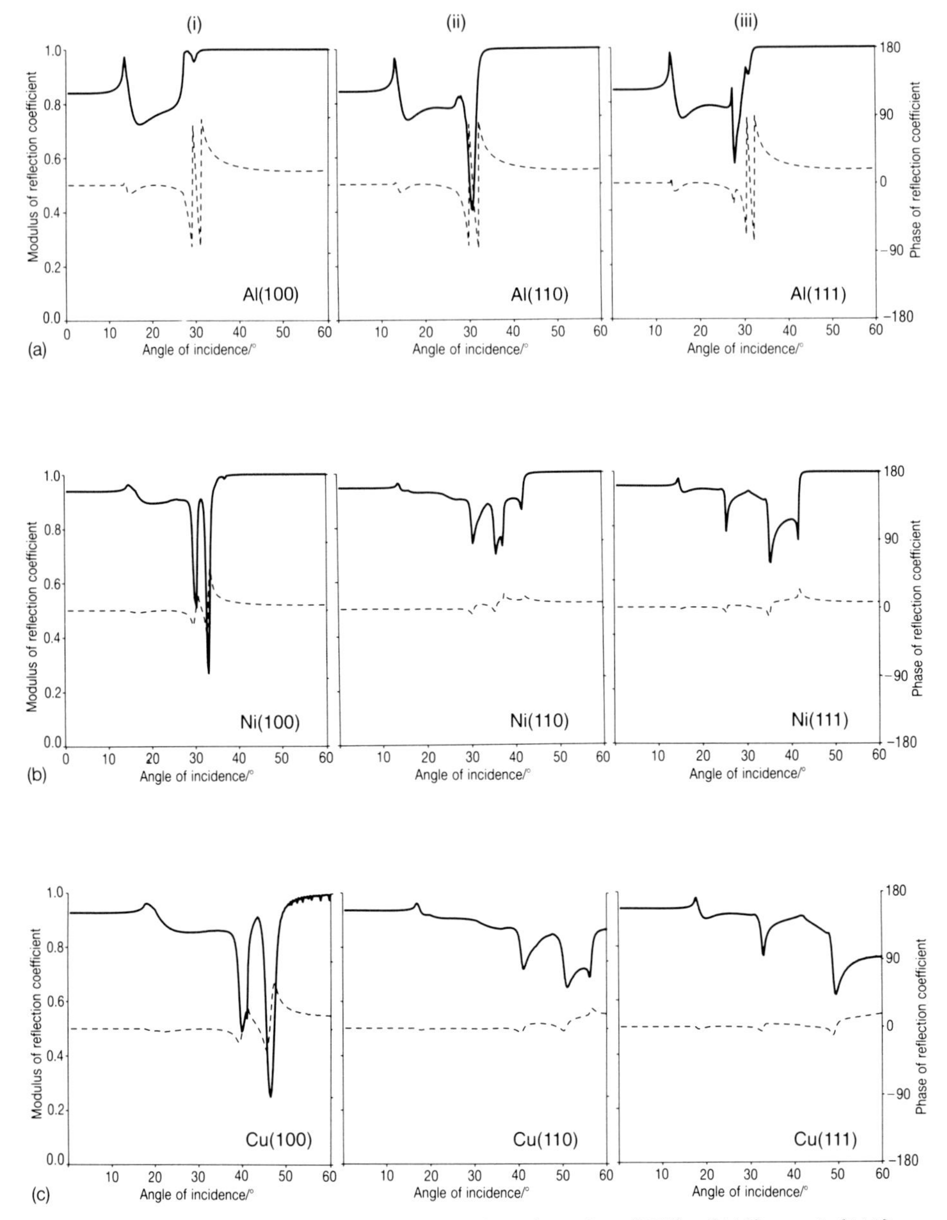

FIG. 11.10. Complex mean reflectance functions for {100}, {110}, and {111} surfaces: (a) Al; (b) Ni; (c) Cu. The horizontal axis is angle of incidence θ from 0 to 60°. —— Magnitude, related to the left axis, with a scale from 0 to 1; ---- phase, related to the right axis, with a scale from $-\pi$ to π. The origin of the glitches in the Cu(100) magnitude is not known. The difference between the nickel complex mean reflection function given by Somekh *et al.* (1984) and those given by Atalar (1989) and here was due to too large an increment in ϕ in the early curves.

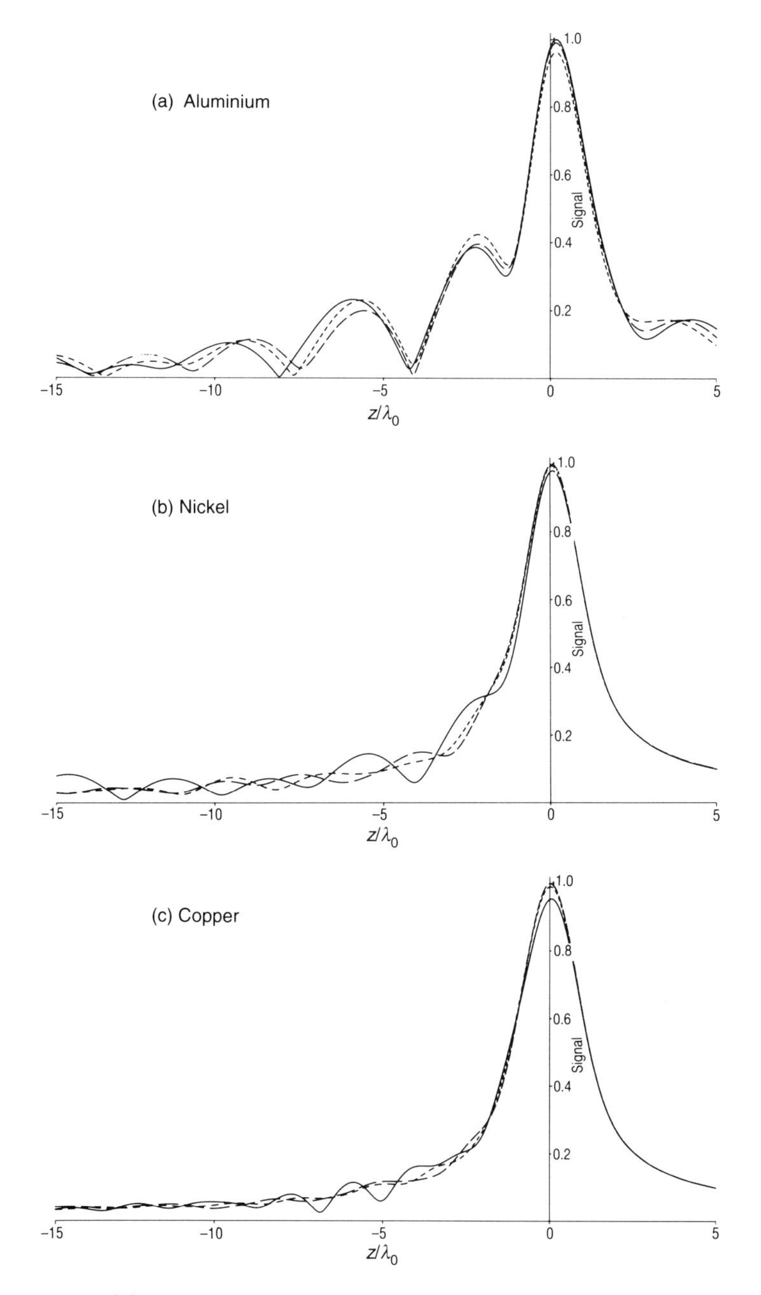

FIG. 11.11. $V(z)$ curves calculated from each of the complex mean reflectance functions in Fig. 11.10: (a) Al; (b) Ni; (c) Cu. Orientations —— (100); ---- (110); —— (111).

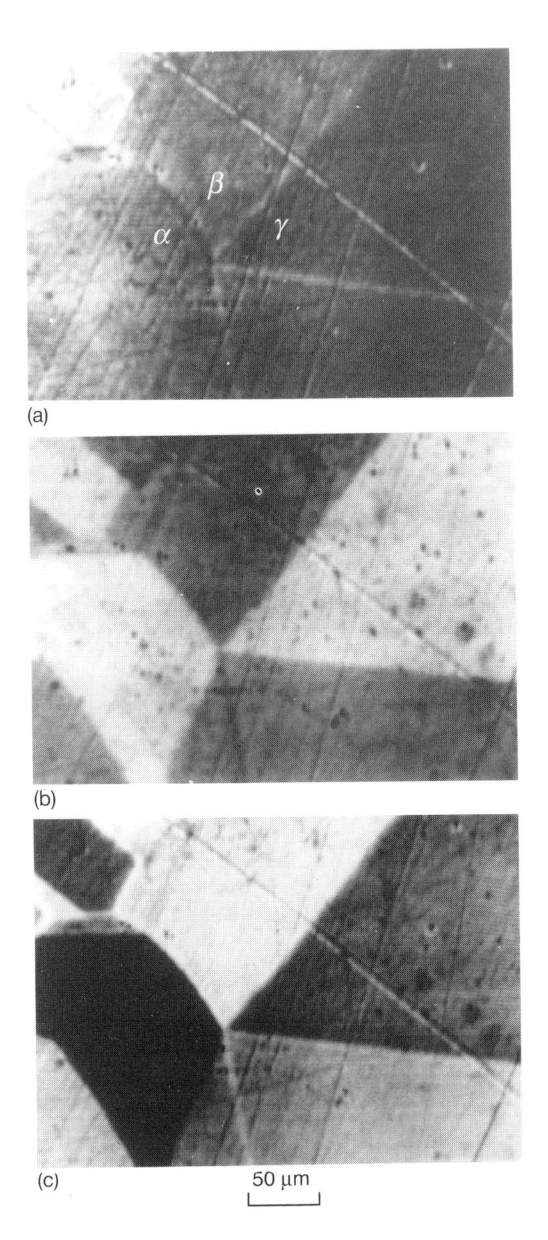

FIG. 11.12. Polycrystalline nickel: (a) $z = 0$; (b) $z = -4\,\mu$m; (c) $z = -7\,\mu$m; 0.73 GHz, $\lambda_0 = 2\,\mu$m (Somekh *et al.* 1984).

The calculation of $V(z)$ curves such as those in Fig. 11.11 for anisotropic surfaces is considerably more difficult than in the isotropic case, because of the need to sum the reflectance function over all angles in order to find the complex mean: for each value of θ and ϕ, Snell's law has to be solved numerically. But, once again, it is found that waves propagating in the surface play a key role. This is why the best contrast from grain structure, and indeed from most other variations in elastic properties, is found with negative defocus, $z < 0$ (in §12.5 an example will be given where it is useful to operate at positive defocus in order to eliminate grain contrast and reveal the other features that would otherwise be masked). It is theoretically possible to adjust the lens geometry to optimize the contrast from anisotropy in a given specimen (Atalar 1987, 1989). But in most practical microscopy it is a great deal easier to adjust the defocus to get the best contrast than to build a new lens.

11.6 Plastic deformation

The dislocations that are associated with plastic deformation of crystalline materials can attenuate sound (Granato and Lücke 1966). The dislocations act like violin strings in treacle; they are overdamped and, in trying to respond to the stresses associated with passing elastic waves in the solid, they dissipate energy from the waves through interactions with thermal phonons. It would be marvellous to be able to use this mechanism to image plastic deformation, for example around the tip of a crack. Unfortunately, it is not as easy as that. The mechanism is confined to materials in which the Peierls stress is small, and that largely means face-centred-cubic metals. Even then, the dislocation must still be able to move freely with the primary constraint coming from the thermal phonon interactions. Now much of the history of alloy development has gone into preventing dislocations from moving freely, so that the requirement for this contrast mechanism is almost the opposite of what is found in metals with good mechanical properties. The two chief candidate materials for model experiments are copper and aluminium. Many of the original dislocation damping experiments were performed on single crystals of 99.999 per cent pure copper. But, as Fig. 11.11 suggests, it is difficult to couple into surface waves in copper in an acoustic microscope. Aluminium is much more promising. Rayleigh waves couple strongly and the low anisotropy factor means that in a polycrystalline specimen the picture would be less cluttered with grain and grain boundary contrast. Changes in $V(z)$ associated with plastic deformation have indeed been seen in an aluminium specimen (Weaver *et al.* 1983). But, alas, even with 99.999 per cent pure aluminium the experiment is not easy. The activation energy for point defect creation and migration is relatively low in aluminium, and at the temperatures at which good acoustic microscopy measurements with water coupling can be made any dislocation damping following plastic deformation is quenched in a matter of minutes by point defects pinning the dislocations (Weaver and Briggs 1985). Experiments at 500 kHz on 1 mm thick plates of relatively pure copper and aluminium showed time-dependent effects in the velocity shift when the plates were subject to loading cycles that went beyond yield (Thompson *et al.* 1990*b*). These observations may shed further light on the nature of quenching of the interaction between elastic waves and disclocations.

There is one picture in which plastic deformation seems undoubtedly to have been captured (Ishikawa *et al.* 1989). Figure 11.13 shows a specimen of an Fe–3%Si alloy that has been deformed. Every precaution was taken to eliminate spurious sources of contrast; for example, the specimen was carefully repolished after deformation to avoid any topographical contrast due to bowing in or out of the surface around a stress concentrator as a result of deformation under plane stress. So the contrast seems indubitably to be due to interaction of Rayleigh waves with regions of plastic deformation. But the mechanism of the contrast in this picture has never been satisfactorily explained. The solubility of Si in Fe is low, as indeed it is in many metals, and so it is possible that the dislocations are not pinned too much by solution hardening. But the experience with pure metals raises at least a cautionary flag. Fe–Si alloys are widely used for transformer cores, and they exhibit twinning under

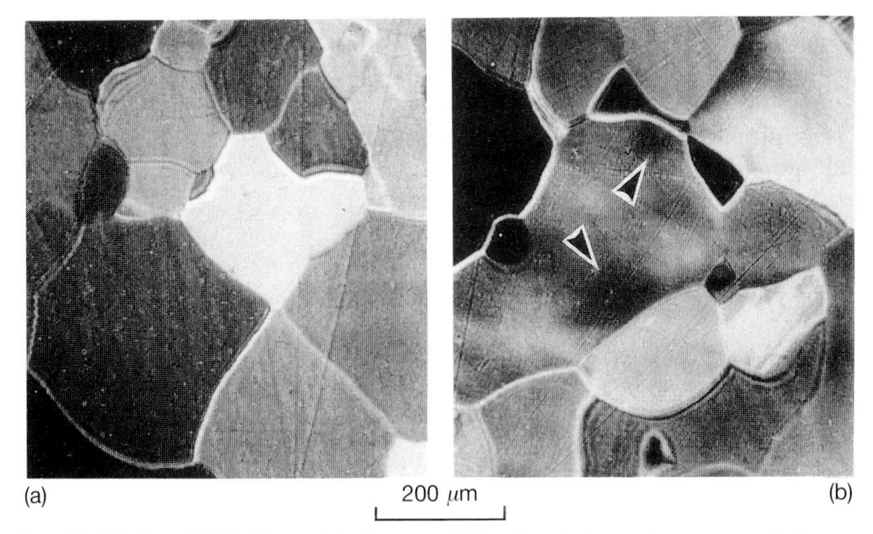

FIG. 11.13. Fe–3%Si alloy (a) before, and (b) after deformation, arrows indicate contrast attributed to deformation; 450 MHz, $z = -5\ \mu m$. (Courtesy of Isao Ishakawa.)

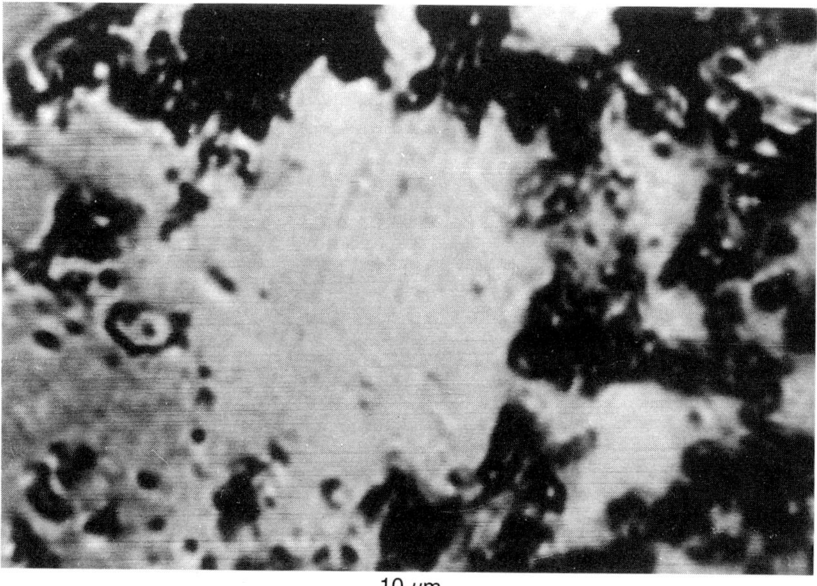

FIG. 11.14. High T_c superconductor ceramic, $YBa_2Cu_3O_{7-x}$, hot pressed; 1.7 GHz. (Bukhny *et al.* 1990)

plastic deformation. It may be that the contrast in Fig. 11.13 is due to twinning on a scale too small to be resolved, but which through the microscopic alteration of anisotropy causes significant perturbation and scattering of Rayleigh waves. It is also possible to observe texture anisotropy in deformed metal specimens using angle-resolved imaging lenses (Ishikawa *et al.* 1990).

11.7 Grain boundaries

It would be unthinkable to end this chapter without a picture of a high T_c superconductor, and so a picture of $YBa_2Cu_3O_{7-x}$ is given in Fig. 11.14. The difficulties in preparing such a specimen and protecting it from degradation under water are formidable. Twin structure was visible in some of the grains, and as the defocus was varied in the microscope so the boundaries between adjacent grains became more apparent. In some pictures of grain structure the grain boundaries seem to have a contrast of their own, quite distinct from the grains on either side. The theory presented for grain contrast so far cannot account for this; it would suggest that the transition from the contrast of one grain to the next should simply be gradual over a distance of about the resolution. In order to understand the grain boundary contrast it is necessary to have a theory of contrast from discontinuities in a surface.

12

Surface cracks and boundaries

MORE (*flinching*): Content? If they'd open a crack that wide (*between finger and thumb*) I'd be through it. (*A Man for All Seasons,* Robert Bolt 1960)

12.1 Initial observations

A plain bearing made of an experimental Al–20%Si alloy was subjected to a fatigue test and then examined for incipient fatigue cracks. The pictures in Fig. 12.1 are optical and acoustic images of a section through the bearing. Figure 12.1(a) is an optical image taken with a 10× objective. The bearing alloy is in the centre with a steel substrate on the left. No doubt with higher magnification and etching it might be possible to find the cracks. But with no further preparation, the acoustic image of the same area at the same magnification (Fig. 12.1(b)) is able to reveal the cracks with greatly enhanced contrast because of the Rayleigh wave scattering mechanism.

If a crack runs at an oblique angle θ_c to a surface, as illustrated in Fig. 12.2, then there can be interference between the reflection from the crack and the reflection from the surface (Ilett *et al.* 1984; Kojima 1987). As the lens is scanned from left to right, path A will remain constant, while path C will change by one wavelength when the lens moves a horizontal distance

$$\Delta x = \frac{\lambda}{2 \sin \theta_c}. \tag{12.1}$$

Fringes should therefore be expected with this spacing, analogous to wedge fringes seen in optics. An example is shown in Fig. 12.3, which contains images of part of a quartz grain in granite. The first acoustic image (Fig. 12.3(a)) shows such fringes beside two cracks; the right-hand crack is seen at higher magnification in Fig. 12.3(b) where quite a large number of fringes can be counted. Because quartz is transparent it was possible in this case to examine the crack optically (Fig. 12.3(c)) and to confirm the orientation of the crack.

A special case occurs when the crack is perpendicular to the surface. In this case $\sin \theta_c = 1$, and so the fringe spacing should be simply half a wavelength. It is; but whereas for smaller angles in eqn (12.1) the

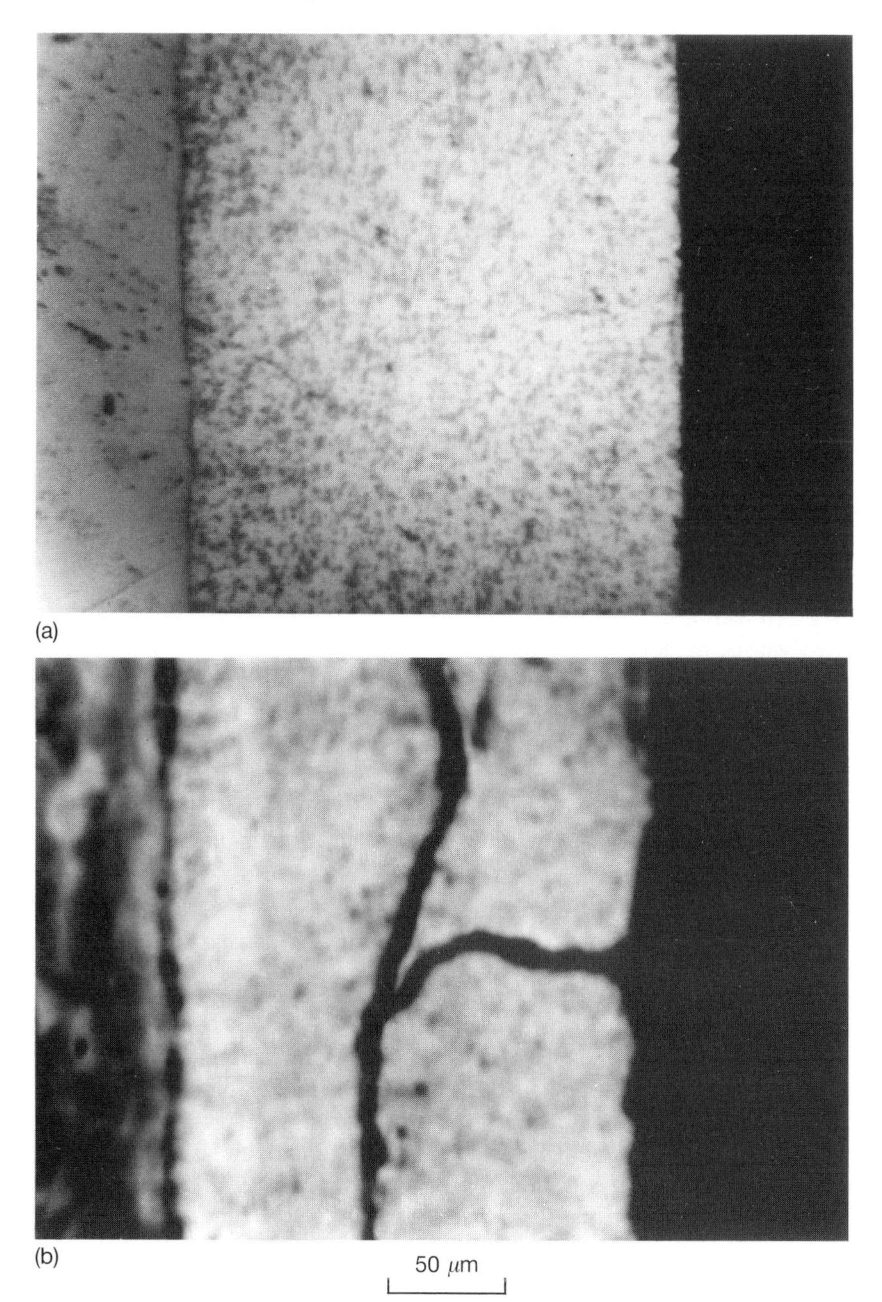

FIG. 12.1. Incipient fatigue cracks in an experimental Al–20%Si plain bearing alloy: (a) optical, 10× objective; (b) 0.73 GHz (Ilett *et al.* 1984).

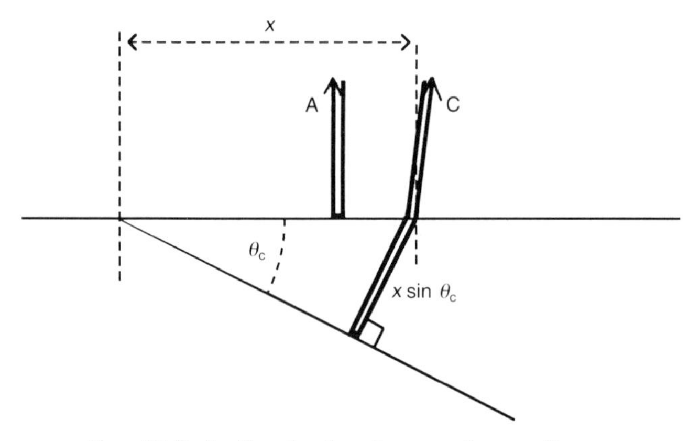

FIG. 12.2. Inclined subsurface crack: ray diagram.

relevant wavelength is the longitudinal wavelength in the solid, for angles close to 90° it is the Rayleigh wavelength (Yamanaka and Enomoto 1982). An example is shown in Fig. 12.4, which presents acoustic images at two values of defocus of cracks from an indent in GaAs. The cracks have taken up the symmetry of the $\{111\}$ surface rather than of the indentor. The fringes parallel to the cracks have a spacing of $2\,\mu$m, which at the frequency at which the picture was made (0.73 GHz) is half a Rayleigh wavelength. The presence of fringes spaced at half a Rayleigh wavelength like this is exceptionally strong confirmation that Rayleigh waves are playing a dominant role in the contrast mechanism.

12.2 Contrast theory of surface cracks

The presence of a crack or other discontinuity presents a serious difficulty for the standard $V(z)$ theory, because the reflectance function is defined for infinite plane waves that are reflected into infinite plane waves, and this requires a reflecting surface that is uniform. If a surface contains a crack then this requirement is violated, and an incident plane wave may be scattered into a whole family of waves (Tew *et al.* 1988). This scattering can be described in **k**-space by a scattering function $S(k_x, k_x')$, where the prime refers to incident waves and the unprime to scattered waves. The x-direction is taken as tangential to the surface, and at this stage the theory is confined to two dimensions in the plane normal to both the surface and the crack. The response of the microscope can then be written in terms of the scattering function by integrating over the

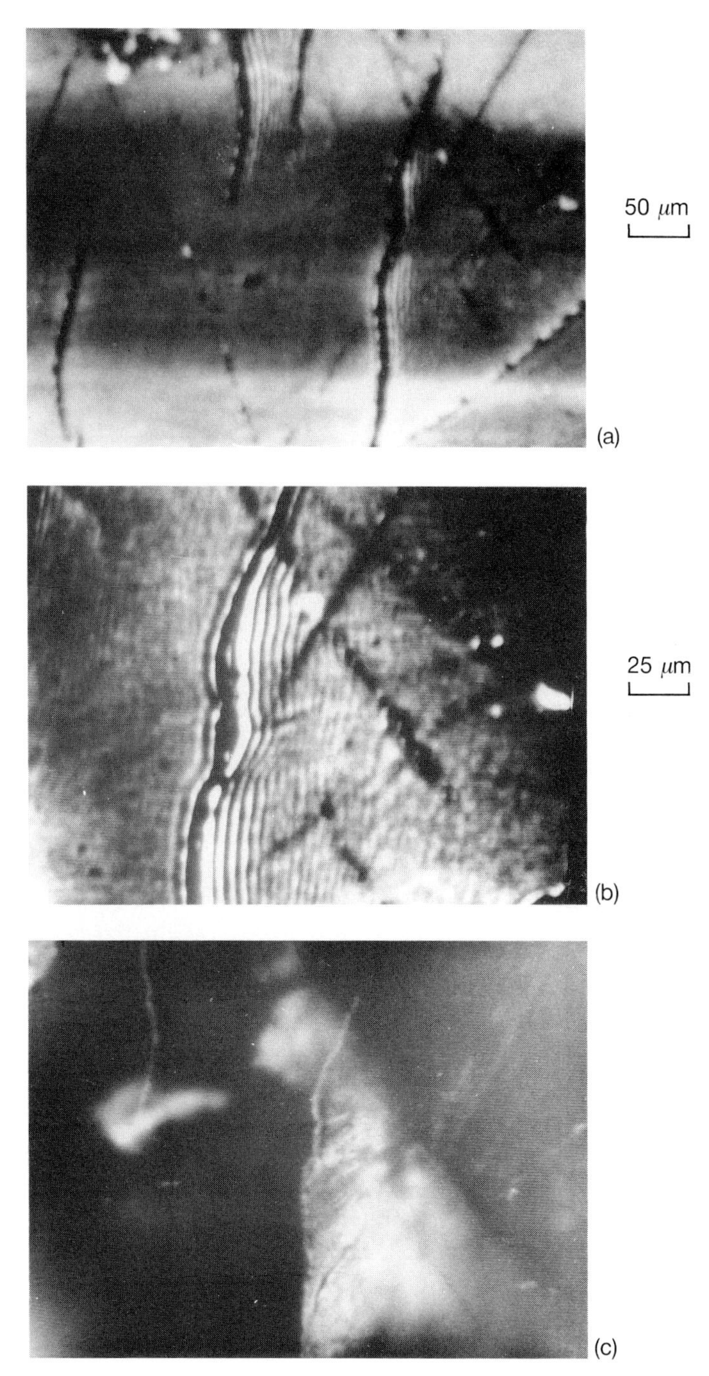

FIG. 12.3. A quartz grain in granite: (a) and (b) acoustic, 0.73 GHz; (c) optical, dark field (Ilett *et al.* 1984).

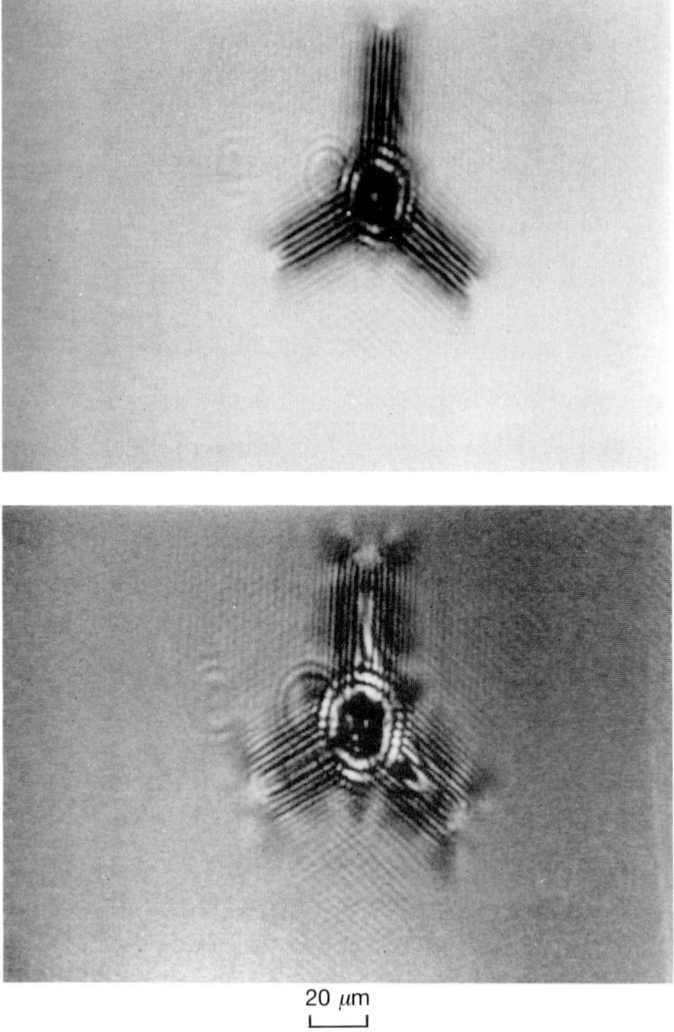

FIG. 12.4. Radial cracks from an indent in GaAs(111); 0.73 GHz.

incident and reflected waves separately (Somekh *et al.* 1985),

$$V(x, z) = \int_{-k}^{k} \int_{-k}^{k} \exp[\mathrm{i}(k_z' - k_z)z] L_1(k_x') L_2(k_x) S(k_x, k_x') \times \exp[\mathrm{i}(k_x' - k_x)x] \, \mathrm{d}k_x \, \mathrm{d}k_x', \tag{12.2}$$

where $k_z' = \sqrt{[(\omega/v_0)^2 - k_x'^2]}$, $k_z = \sqrt{[(\omega/v_0)^2 - k_x^2]}$ are taken as negative and positive, respectively, $L_1(k_x')$, $L_2(k_x)$ are the lens functions for outgoing and incoming waves, and the crack is taken to be at the origin. Thus the first term of the integrand allows for phase shifts due to defocus, and the last term allows for phase shifts due to lateral displacement of the lens from the position overhead of the crack. It can be seen that in the absence of a crack, when $S(k_x, k_x')$ vanishes except at $k_x = k_x'$, the expression reduces to the simple formula for $V(z)$. The problem now is to find $S(k_x, k_x')$.

The experimental observation that Rayleigh wave scattering plays a dominant role in the contrast suggests that the reflection coefficient should be separated into the geometrical and Rayleigh parts in the way described in §7.2.1;

$$R(k_x) \approx R_0(k_x) + \frac{k_p^2 - k_0^2}{k_x^2 - k_p^2}, \tag{12.3}$$

where the approximation is made that over the range of values of k_x for which the second term on the right contributes significantly, the value of $R_0(k_x)$ is unity. If the crack is very fine, much less than a wavelength wide, then by conventional resolution criteria the geometrically reflected component will not be affected by the crack very much. On the other hand the Rayleigh reflected component may be strongly affected, because the pole in the reflectance function corresponds to the excitation of Rayleigh waves in the surface of the specimen, and these can strike the crack broadside. The method of calculation is first to find the effect of the crack on the Rayleigh wave field in real space, and then to transform this to k space to give the scattering function $S(k_x, k_x')$. As in (7.33), if the incident pressure field at the surface of the specimen is $p_{\mathrm{inc}}(x)$, and the reflected pressure field at the surface due to Rayleigh wave interactions is $p_{\mathrm{R}}(x)$, then in the absence of a crack these are related by

$$p_{\mathrm{R}}(x) = -2\alpha_{\mathrm{R}} \int_{-\infty}^{\infty} p_{\mathrm{inc}}(x') \exp[\mathrm{i}k_p\,|x - x'|]\,\mathrm{d}x'. \tag{12.4}$$

The exponential term can be thought of as a Green function, with the time dependence always implicit. Thus an excitation at x' causes a response at x whose phase is delayed by the distance between them multiplied by the real part of k_p (this corresponds approximately to $2\pi/\lambda_{\mathrm{R}}$), and whose amplitude is decreased exponentially by the distance between them multiplied by the imaginary part of k_p (this corresponds to the decay associated with the propagation of the leaky Rayleigh wave). The magnitude $|x - x'|$ is used because only Rayleigh waves propagating towards the point of observation contribute, so the distance between the point of excitation and the point of observation is taken as positive whatever the direction of propagation. The term $-2\alpha_R$ outside the integration is a measure of the strength of the coupling between waves at the Rayleigh angle in the fluid and Rayleigh waves in the surface; for lossless materials it is just twice the rate of attenuation due to leaking into the fluid; the minus sign is associated with the phase change of π in the reflection of plane waves at the Rayleigh angle.

When a crack is present eqn (12.4) must be extended to include a total of four terms. Let the crack be situated at the origin, and let it be characterized by coefficients of reflection R_R and transmission T_R for Rayleigh waves. Consider first the response at a point $x < 0$ to the left of the crack. The contributions are:

(1) excitation from everywhere to the left of x, propagating to the right;
(2) excitation from points between x and the crack, propagating to the left;
(3) excitation from everywhere to the left of the crack, propagating to the right, reflected by the crack and propagating back to x;
(4) excitation from everywhere to the right of the crack, propagating to the left, transmitted by the crack, and propagating on to x.

These four terms give an expression for the complete response at x;

$$p_R(x) = -2\alpha_R \bigg[\int_{-\infty}^{x} p_{inc}(x') \exp[ik_p(x - x')]\, dx' + \int_{x}^{0} p_{inc}(x') \exp[-ik_p(x - x')]\, dx' + \exp[-ik_p x] \bigg\{ R_R \int_{-\infty}^{0} p_{inc}(x') \exp[ik_p(x - x')]\, dx' + T_R \int_{0}^{\infty} p_{inc}(x') \exp[-ik_p(x - x')]\, dx' \bigg\} \bigg]. \tag{12.5}$$

A similar expression can be written for the response to the right of the crack

$$p_{\mathrm{R}}(x) = -2\alpha_{\mathrm{R}}\bigg[\int_{x}^{\infty} p_{\mathrm{inc}}(x')\exp[-\mathrm{i}k_{\mathrm{p}}(x-x')]\,\mathrm{d}x'$$

$$+\int_{0}^{x} p_{\mathrm{inc}}(x')\exp[\mathrm{i}k_{\mathrm{p}}(x-x')]\,\mathrm{d}x'$$

$$+\exp[\mathrm{i}k_{\mathrm{p}}x]\bigg\{R_{\mathrm{R}}\int_{0}^{\infty} p_{\mathrm{inc}}(x')\exp[-\mathrm{i}k_{\mathrm{p}}(x-x')]\,\mathrm{d}x'$$

$$+T_{\mathrm{R}}\int_{-\infty}^{0} p_{\mathrm{inc}}(x')\exp[\mathrm{i}k_{\mathrm{p}}(x-x')]\,\mathrm{d}x'\bigg\}\bigg]. \tag{12.6}$$

These two equations give the reflected field at the surface due to the Rayleigh wave interaction for any incident field.

The incident and reflected fields in the spatial frequency domains, $P_{\mathrm{inc}}(k_x)$ and $P_{\mathrm{R}}(k_x)$, are related to the fields $p_{\mathrm{inc}}(x)$ and $p_{\mathrm{R}}(x)$ by simple Fourier transforms

$$p_{\mathrm{inc}}(x') = \frac{1}{2\pi}\int_{-\infty}^{\infty} P_{\mathrm{inc}}(k_x')\exp(\mathrm{i}k_x'x')\,\mathrm{d}k_x' \tag{12.7}$$

and

$$P_{\mathrm{R}}(k_x) = \int_{-\infty}^{\infty} p_{\mathrm{R}}(x)\exp(-\mathrm{i}k_x x)\,\mathrm{d}x. \tag{12.8}$$

Equations (12.5) and (12.6) can be regarded as constituting a Green function, relating the Rayleigh reflected field to the incident field

$$p_{\mathrm{R}}(x) = \int_{-\infty}^{\infty} G(x, x')p_{\mathrm{inc}}(x')\,\mathrm{d}x'. \tag{12.9}$$

Then the incident and Rayleigh reflected fields in the spatial frequency domain are related by

$$P_{\mathrm{R}}(k_x) = \frac{1}{2\pi}\int_{-\infty}^{\infty}\int_{-\infty}^{\infty}\int_{-\infty}^{\infty} G(x, x')\exp(\mathrm{i}k_x'x')\exp(-\mathrm{i}k_x x)P_{\mathrm{inc}}(k_x')\,\mathrm{d}x\,\mathrm{d}x'\,\mathrm{d}k_x'. \tag{12.10}$$

Performing the inner integrals first this can be written

$$P_{\mathrm{R}}(k_x) = \int_{-\infty}^{\infty} S(k_x, k_x')P_{\mathrm{inc}}(k_x')\,\mathrm{d}k_x', \tag{12.11}$$

with

$$S(k_x, k_x') = \frac{1}{2\pi} \int_{-\infty}^{\infty} \int_{-\infty}^{\infty} G(x, x') \exp(\mathrm{i}k_x'x') \exp(-\mathrm{i}k_x x) \, \mathrm{d}x \, \mathrm{d}x'. \quad (12.12)$$

Thus the required scattering function can be obtained from eqns (12.5) and (12.6) by a double Fourier transform. This is straightforward but somewhat lengthy to perform; if the geometrical term is added back as unaffected by the crack, then the result is

$$S(k_x, k_x') = \left[R_0(k_x) + \frac{\mathrm{i}4\alpha_{\mathrm{R}} k_{\mathrm{p}}}{k_x^2 - k_{\mathrm{p}}^2} \right] \delta(k_x - k_x')$$
$$+ \frac{2\alpha_{\mathrm{R}}}{\pi} \left[\frac{(T_{\mathrm{R}} - R_{\mathrm{R}} - 1)k_x k_x' + (T_{\mathrm{R}} + R_{\mathrm{R}} - 1)k_{\mathrm{p}}^2}{(k_x^2 - k_{\mathrm{p}}^2)(k_x'^2 - k_{\mathrm{p}}^2)} \right]. \quad (12.13)$$

The two terms in the first square bracket on the right-hand side describe the reflection in the absence of a crack. The first is the geometrically reflected component. The second, allowing for the approximation of eqn (7.28),

$$\mathrm{i}(k_{\mathrm{p}}^2 - k_0^2)/(2k_{\mathrm{p}}) \approx -2\alpha_{\mathrm{R}},$$

is the Rayleigh reflected component. The term in the second square bracket of (12.13) represents the scattering due to the crack. A crack that has $T_{\mathrm{R}} = 1$ and $R_{\mathrm{R}} = 0$ has no effect, and in that case this term vanishes. But any other combination of T_{R} and R_{R} allows this term to contribute, and to scatter waves into different values of k_x. Because this is a Rayleigh wave effect, this term contributes most when both the incident wave and the scattered wave are close to the Rayleigh angle (where α_{R} is the measure of what is meant by close).

The approximations contained in eqn (12.13) are explicit in its derivation. First, it assumes that Rayleigh wave scattering mechanisms dominate the contrast. Second, it assumes that the fluid loading is light, i.e. $\alpha_{\mathrm{R}} \ll k_{\mathrm{R}}$. Third, the effect of the crack on the geometrical reflection is neglected. There may well be other terms. For example, a cylindrical wave radiating into the fluid is expected (Tew *et al.* 1988; Tew 1990). Implicit in $R_0(k_x)$ are the branch points located at $\pm\omega/v_{\mathrm{l}}$ and $\pm\omega/v_{\mathrm{s}}$ in the complex k_x plane, corresponding to critical excitation of lateral longitudinal and shear waves. Lateral waves can be scattered by the crack in a way analogous to Rayleigh waves, and this may be important in low-impedance materials (especially if they are too slow for Rayleigh waves to be excited in the acoustic microscope) or at very large values of defocus. The theory of lateral wave crack contrast can be tackled along the lines introduced in §7.2.3. In anisotropic materials pseudo-surface waves can be excited, and it might well be possible to incorporate these in the model. But for the large number of cases where the dominance of the Rayleigh wave contrast is confirmed by the experimental evidence of Rayleigh wave fringes parallel to a crack, or simply Rayleigh periodicity in $V(z)$ of the material, the theory presented here seems to be useful.

The values of R_R and T_R to be used in the expression for $S(k_x, k_x')$ are external to the theory presented here. This has the great advantage that the large library of solutions for different configurations of cracks (e.g. Achenbach 1987), each of which may be a major computational exercise in itself, can be drawn on for the particular crack configuration of interest. These solutions (and comparable experimental measurements, Dong and Adler 1984) are generally for Rayleigh waves from infinity incident on cracks in free surfaces, in the absence of any fluid loading. In the use of these for contrast from cracks in the acoustic microscope it should be appreciated that there will be a perturbation due to the fluid loading. Also, because of the exponential attenuation of leaky Rayleigh waves they cannot be generated at infinity; indeed, in the microscope they may be generated very close to the crack. As with the ray calculation of $V(z)$ on a uniform surface, therefore, this calculation of crack contrast may be expected to work best when the defocus is not too small.

Starting from a crack that is small compared with a wavelength, the effect on $V(z)$ when the lens is directly over the crack initially increases with crack depth. The chief information about the crack depth lies in the phase; there is a weak resonance when the crack depth is comparable with the Rayleigh wavelength, and beyond that the contrast tends to become independent of the crack depth. In a perfect two-dimensional situation a Rayleigh wave propagating along a free crack face would be unattenuated, and so resonance effects could still exist even for very deep cracks; but it would be unusual to encounter such ideal conditions experimentally, and in any case this would not apply to the three-dimensional situation with an imaging lens. If it is assumed that for very large depths of crack the scattering of the incident Rayleigh wave at the corner where the crack meets the surface is dominant, and that transmission across the crack mouth and diffraction of a scattered Rayleigh wave at the crack tip may both be neglected, then approximate values for the crack parameters are (Achenbach *et al.* 1980)

$$\begin{aligned} T_R &= 0, \\ R_R &= 0.4 \exp(\mathrm{i}0.6). \end{aligned} \tag{12.14}$$

$V(z)$ curves calculated for these parameters have been compared with measurements with a cylindrical lens at 215 MHz of a crack in a glass specimen (Rowe *et al.* 1986). Directly over the crack only the sum of T_R

and R_R is relevant because of the symmetry. In both theory and experiment there was a reduction in the amplitude of the oscillations compared with $V(z)$ over a defect-free region, and also a phase shift in the oscillations. When the lens was displaced to one side of the crack by an amount $(n/2+1/8)\lambda_R$, where n is an integer that was 1 in the comparison, in both theory and experiment the oscillations had their smallest amplitude, because of the phase cancellation between the Rayleigh waves reflected from the two sides of the crack. When the lens was moved rather farther away from the crack ($2.25\lambda_R$), both theory and experiment showed departures from the defect-free $V(z)$ a little before the defocus at which the area contained by the Rayleigh rays would be expected to include the crack on simple geometrical arguments. Comparisons were also made of theoretical and experimental line scans at constant height, or $V(x)$ curves. Once again the most important features of the experiments, such as contrast reversal and fringes associated with the crack, corresponded nicely to what was calculated.

There are considerable difficulties in comparing theory and experiment even in such model experiments. The theoretical calculations are subject to the approximations inherent in the method, and also to uncertainties in the pupil function used to characterize the lens and in the two parameters used to characterize the crack. The experiments are subject to the difficulties of making a crack that is straight and flat to a fraction of the acoustic wavelength used, over the length measured by the line-focus-beam lens, and to the sensitivity of the results in some cases to small changes in x or z. Nevertheless, when all these considerations are taken into account it does seem that the theory gives a substantially accurate account of the experimental phenomena.

12.3 Extension to three dimensions

The theory developed so far has been for two dimensions only. The form of the three-dimensional scattering function can be obtained by substituting $k_p^2 - k_y^2$ for k_p^2 in eqn (12.13); because then this new quantity will continue to give the correct Rayleigh pole behaviour. The Cartesian coordinate system is extended to three dimensions, with the crack lying in the plane $x=0$, and the component of wavevector parallel to the crack is k_y. Since the crack is continuous along its length, the component k_y parallel to the crack must be conserved in any reflection from the crack, and therefore throughout the problem. Equation (12.4) applies for a two-dimensional incident field having no phase variation along y. The generalization of (12.4) to the three-dimensional case when there is a phase variation $\exp(ik_y y)$ along a line of constant x in the surface at $z=0$

is obtained by letting $\sqrt{(k_p^2 - k_y^2)}$ replace k_p in the exponent and multiplying by an extra factor $k_p/\sqrt{(k_p^2 - k_y^2)}$. The derivation of this begins by considering a distribution of pressure $p_{inc}(x', y')$ over the entire x–y plane due to the incident wave. The net pressure field immediately above the surface due to re-radiation of the leaky wave is, by analogy with (12.4),

$$p_R(x, y) = C \int_{-\infty}^{\infty} \int_{-\infty}^{\infty} \frac{\exp[ik_p |\mathbf{r} - \mathbf{r}'|]}{\sqrt{|\mathbf{r} - \mathbf{r}'|}} p_{inc}(x', y') \, dx' \, dy', \quad (12.15)$$

where $\mathbf{r}' \equiv (x', y')$ is a point of excitation of the Rayleigh wave, and $\mathbf{r} \equiv (x, y)$ is a point of re-radiation. The factor $1/\sqrt{|\mathbf{r} - \mathbf{r}'|}$ is due to cylindrical spreading of the Rayleigh wave in the surface; in the two-dimensional problem such spreading did not occur. C is a constant to be determined by manipulating (12.15) so that it can be directly compared with (12.4), via (12.16–27) as follows.

Since k_y is conserved the Rayleigh reflected field can be separated into the product of a term in x and a simple exponential term in y, i.e. $p_R(x, y) = p_R(x) \exp(ik_y y)$. Rewriting (12.15) in this form gives

$$p_R(x, y) = \exp(ik_y y) C \int_{-\infty}^{\infty} p_{inc}(x') \left\{ \int_{-\infty}^{\infty} \frac{\exp[i\phi]}{\sqrt{|\mathbf{r} - \mathbf{r}'|}} dy' \right\} dx', \quad (12.16)$$

where

$$\phi = k_p\{(x - x')^2 + (y - y')^2\}^{1/2} - k_y(y - y'). \quad (12.17)$$

The integration over y' can be evaluated by the method of stationary points. The stationary point is the solution of $d\phi/dy' = 0$, and is found to be

$$y - y' = \frac{k_y |x - x'|}{\sqrt{(k_p^2 - k_y^2)}}. \quad (12.18)$$

Near the stationary point,

$$\phi \approx |x - x'| \sqrt{(k_p^2 - k_y^2)} + \left\{ \frac{(k_p^2 - k_y^2)^{3/2}}{2k_p^2 |x - x'|} \right\} (y - y')^2 \quad (12.19)$$

and

$$\frac{1}{\sqrt{|\mathbf{r} - \mathbf{r}'|}} \approx \frac{(k_p^2 - k_y^2)^{1/4}}{\sqrt{(k_p |x - x'|)}}, \quad (12.20)$$

so that

$$\int_{-\infty}^{\infty} \frac{\exp[\mathrm{i}\phi]}{\sqrt{|\mathbf{r}-\mathbf{r}'|}}\,\mathrm{d}y' \approx \frac{\exp[\mathrm{i}\sqrt{(k_\mathrm{p}^2-k_y^2)}\,|x-x'|]}{\sqrt{(k_\mathrm{p}\,|x-x'|)}}(k_\mathrm{p}^2-k_y^2)^{1/4}$$
$$\times \int_{-\infty}^{\infty} \exp\left\{\frac{\mathrm{i}(k_\mathrm{p}^2-k_y^2)^{3/2}}{2k_\mathrm{p}^2\,|x-x'|}\right\}(y-y')^2\,\mathrm{d}y'. \quad (12.21)$$

If the change of variable is made,

$$u \equiv \left\{\frac{\mathrm{i}(k_\mathrm{p}^2-k_y^2)^{3/2}}{2k_\mathrm{p}^2\,|x-x'|}\right\}^{1/2}(y-y'), \quad (12.22)$$

then

$$\int_{-\infty}^{\infty} \frac{\exp[\mathrm{i}\phi]}{\sqrt{|\mathbf{r}-\mathbf{r}'|}}\,\mathrm{d}y' \approx -\left\{\frac{2k_\mathrm{p}}{\mathrm{i}(k_\mathrm{p}^2-k_y^2)}\right\}^{1/2} \exp[\mathrm{i}(k_\mathrm{p}^2-k_y^2)^{1/2}\,|x-x'|]$$
$$\times \int_{-\infty}^{\infty} \exp(-u^2)\,\mathrm{d}u. \quad (12.23)$$

Substituting this expression into (12.16),

$$p_\mathrm{R}(x, y) = -\exp(\mathrm{i}k_y y)C\left\{\frac{2k_\mathrm{p}}{\mathrm{i}(k_\mathrm{p}^2-k_y^2)}\right\}^{1/2}$$
$$\times \int_{-\infty}^{\infty} p_\mathrm{inc}(x')\exp[\mathrm{i}(k_\mathrm{p}^2-k_y^2)^{1/2}\,|x-x'|]\,\mathrm{d}x' \int_{-\infty}^{\infty} \exp(-u^2)\,\mathrm{d}u. \quad (12.24)$$

When $k_y = 0$, this must reduce to (12.4). Thus,

$$-2\alpha_\mathrm{R} = -C\sqrt{\left(\frac{2}{\mathrm{i}k_\mathrm{p}}\right)} \int_{-\infty}^{\infty} \exp(-u^2)\,\mathrm{d}u. \quad (12.25)$$

The value of the integral is $\sqrt{\pi}$ (although that later cancels out) so

$$C = \alpha_\mathrm{R}\sqrt{\left(\frac{\mathrm{i}2k_\mathrm{p}}{\pi}\right)}. \quad (12.26)$$

Substituting back into (12.24)

$$p_\mathrm{R}(x, y) = \exp(\mathrm{i}k_y y)\frac{k_\mathrm{p}}{\sqrt{(k_\mathrm{p}^2-k_y^2)}}$$
$$\times\left\{-2\alpha_\mathrm{R}\int_{-\infty}^{\infty} p_\mathrm{inc}(x')\exp[\mathrm{i}(k_\mathrm{p}^2-k_y^2)^{1/2}\,|x-x'|]\,\mathrm{d}x'\right\}. \quad (12.27)$$

This is the generalized three-dimensional scattering relationship for the response just above the surface at x to an oscillatory pressure just above the surface at x', due to Rayleigh wave excitation, in the case where the y component of the wavevector is constant. The three-dimensional scattering function can now be calculated.

In (12.13) the R_0 term enters separately from the terms derived from $p_R(x)$, and it is unaffected by the phase factor $\exp[ik_y y]$. Thus the generalization of (12.13) is found by replacing k_p by $\surd(k_p^2 - k_y^2)$ and multiplying all terms except R_0 by $k_p/\surd(k_p^2 - k_y^2)$. The result is

$$S(k_x, k_x', k_y) = R(\theta)\,\delta(k_x - k_x')$$
$$+\frac{2\alpha_R}{\pi}\left\{\frac{(T_R - 1 - R_R)k_x k_x' + (T_R - 1 + R_R)(k_p^2 - k_y^2)}{(k_x^2 + k_y^2 - k_p^2)(k_x'^2 + k_y^2 - k_p^2)}\right\}\frac{k_p}{\surd(k_p^2 - k_y^2)}. \qquad (12.28)$$

In this expression T_R and R_R refer to Rayleigh waves propagating obliquely to the crack, i.e. with phase variation $\exp[ik_y y]$ along y. The first term on the right is simply the reflectance function for a defect-free surface (with $\theta = \cos^{-1}\{k_z/k\}$). The second term describes departures from this due to the presence of the crack. It is explicit in this formalism that $S(k_x, k_x', k_y)$ is a function not only of k_x and k_x', but also of k_y. This dependence occurs through the dependence of T_R and R_R on the angle of incidence ϕ of the Rayleigh wave (Angel and Achenbach 1984). Those results are for excitation of the Rayleigh waves at infinity, in which case the angles of transmission and reflection in the surface by a uniform crack must each be equal to the angle of incidence. In the acoustic microscope excitation is not from infinity, and therefore k_x and k_x' are not necessarily the same; indeed this was the basis of the derivation of $S(k_x, k_x')$. But the available calculations of $T_R(\phi)$ and $R_R(\phi)$ are undefined when $k_x \neq k_x'$. To overcome this difficulty the further approximation can made that, provided that $T_R(\phi)$ and $R_R(\phi)$ do not vary too rapidly with the angle of incidence of the Rayleigh wave, they may be written as functions of k_y only, by putting $\phi = \sin^{-1}(k_y/k)$. In most cases this will be a good approximation since, although formally k_x and k_x' are each summed independently between $\pm k$, in fact there is significant contribution to the sum only when they are both close to $\surd(k_p^2 - k_y^2)$; this observation may also be exploited when computing the triple integral in (12.29) below. 'Close' in this context means differing by not more than α_R, and for most materials in the acoustic microscope in which Rayleigh waves play a significant part in the contrast $\alpha_R \ll |k_p|$; indeed this was necessary for the derivation of eqn (12.4). Therefore, to a good approximation the dependence of $S(k_r, k_r', k_y)$ on k_y may be expressed by giving k_y dependence to T_R and R_R in eqn (12.28).

In order to calculate $V(x, z)$ for a spherical lens in the presence of a crack a summation must be made over k_y. For each value of k_y a double summation is first made over k_x and k'_x. A wave is considered to be incident with components of wavevector k'_x and k_y. It is then transmitted by the crack with components of wavevector k_x and k_y, and reflected with components $-k_x$ and k_y; transmitted and reflected waves may be summed in the same integration. If the axis of the lens is displaced a distance x from the crack, the resulting phase change is $k'_x x$ for the incident wave and $-k_x x$ for the scattered wave. Then, by extension of eqn (12.2),

$$V(x, z) = \int_{-k}^{k} \int_{-k}^{k} \int_{-k}^{k} \exp[\mathrm{i}(k'_z - k_z)z] L_1(k'_x, k_y) L_2(k_x, k_y) \times S(k_x, k'_x, k_y) \exp[\mathrm{i}(k'_x - k_x)x] \, \mathrm{d}k_x \, \mathrm{d}k'_x \, \mathrm{d}k_y. \quad (12.29)$$

This gives the contrast in an imaging acoustic microscope due to a crack in a specimen, for any defocus and lens position. Once again a great advantage of expressing the result in this form, with T_R and R_R as external parameters in the Green function, is that, when different geometries are encountered, such as varying depth or partial crack closure (for example at the mouth of the crack), solutions independently obtained for those situations may be incorporated directly. As always, T_R and R_R are complex quantities. An over-zealous editor removed the phase information when the definitive angle-dependent parameters for a simple surface crack were published (Angel and Achenbach 1984). Experimental measurements are available for this configuration, and they are in good agreement with the calculations.

In the special case where the lens is directly over a scattering feature, certain further simplifying assumptions can be made to obtain a very elementary approximate solution (Ilett *et al.* 1984). When the lens is directly above a crack, then a Rayleigh wave reflected by the crack propagates back to the lens by a path exactly symmetrical to the one that it would have followed if the crack had not been there. In that case the simple formula for $V(z)$ can be used with the reflection coefficient split up into the geometrical and Rayleigh parts, and the Rayleigh part multiplied by a scattering factor S. There is no difference between the path lengths travelled by transmitted and reflected waves, so S is simply the complex-valued sum of the transmission and reflection coefficients T_R and R_R of the crack for Rayleigh waves. Thus in two dimensions a local effective scattering reflectance function may be defined as

$$R_\mathrm{S}(k_x) \equiv R_0(k_x)\{1 + S(k_\mathrm{p}^2 - k_0^2)/(k_x^2 - k_\mathrm{p}^2)\}. \quad (12.30)$$

This may be inserted into the $V(z)$ formula for a cylindrical lens to give

$$V_c(z) = \int_{-k}^{k} P(k_x)R_s(k_x)\exp(i2k_z z)\,dk_x. \tag{12.31}$$

In terms of the four contributions to eqns (12.5–6) in §12.2, 1 and 2 have been multiplied by the scattering factor, S, in (12.30–31), as well as 3 and 4, whereas in fact only 3 and 4 actually encounter the crack. But within the ray model these two terms should contribute very little when the lens is directly above the crack, because they correspond to rays leaking into the fluid at the Rayleigh angle which miss the transducer at sufficient defocus. Also, the possibility of rays incident at one angle scattering into rays reflected at another angle is neglected, but again within the ray model only rays close to the Rayleigh angle will interact strongly with the crack. On the other hand in this calculation it is not necessary to approximate that $R_0(k_x) = 1$ near the Rayleigh pole.

This simple expression for the contrast over a crack may be extended to three dimensions by a procedure similar to the one used for anisotropy. If T_R and R_R are known as functions of incident azimuthal angle ϕ, then R_{eff} may be written as a function of ϕ and the component of $\mathbf{k}$ parallel to the surface $k_{\|}$:

$$R_s(k_{\|}, \phi) = R_0(k_{\|})\{1 + S(\phi)(k_p^2 - k_0^2)/(k_x^2 - k_p^2)\}. \tag{12.32}$$

Changing the variables, this may be used in the expression for $V(z)$ for a spherical lens with axial symmetry, and integrated over ϕ

$$V_c(z) = \int_{-\pi}^{\pi}\int_{0}^{\pi/2} P(\theta)R_s(\theta, \phi)\exp(i2kz\cos\theta)\cos\theta\sin\theta\,d\theta\,d\phi. \tag{12.33}$$

If the order of integration is changed, a complex mean scattering reflectance function may be defined analogous to the complex mean reflectance function:

$$R_{cms}(\theta) \equiv R_0(\theta)\{1 + S_{cm}(k_p^2 - k_0^2)/(k_x^2 - k_p^2)\} \tag{12.34}$$

where

$$S_{cm} = \int_{-\pi}^{\pi} S(\phi)\,d\phi. \tag{12.35}$$

Then the contrast over the crack may be calculated as a single integral,

$$V_c(z) = \int_{0}^{\pi/2} P(\theta)R_{cms}(\theta)\exp(i2kz\cos\theta)\cos\theta\sin\theta\,d\theta. \tag{12.36}$$

This simple expression may sometimes be useful for a quick calculation of the contrast to be expected directly over a crack. An alternative approach to this special case considers the reflected field in real space at the specimen surface, and multiplies all the reflected field that has crossed the lens axis by S_{cm} on the assumption that it has propagated there by a Rayleigh wave mechanism (Cox and Addison 1984). Equation (12.36) can be used to calculate the change in the Rayleigh wave oscillations in $V(z)$, but, of course, for the apparent width of the crack in images and the Rayleigh wave interference fringes on either side of the crack the full theory for $V(x, z)$ must be used. Boundary element computations may allow new geometries to be calculated (Achenbach *et al.* 1991). It turns out that the two-dimensional theory gives remarkably accurate predictions of phenomena found in images.

The use of a line scan facility on an imaging microscope enables the applicability of the two-dimensional theory to be tested directly. In Fig. 12.5, images of glass at three different values of defocus are presented. Superimposed on each picture are line scans measured along the line indicated in the middle. Beside each picture is a line scan corresponding to that value of defocus calculated using the two-dimensional theory. The agreement is not too bad, and suggests that the two-dimensional theory does indeed give a useful account of what is going on.

Grinding damage in ceramics often leads to cracks with a slightly different geometry. Even in the most brittle materials, the multiple abrasion by fine particles creates an array of vertical surface cracks of surprisingly uniform depth, and also causes plastic flow to occur so that near the surface the faces of the cracks are in compression. These cracks with closed mouths are elastically equivalent to cracks normal to the surface but starting some way below it. Rayleigh wave scattering coefficients for such geometries are available, and their frequency dependence has features that are determined both by the total depth of the crack, and by the extent of the crack closure at the mouth (Achenbach and Brind 1981). This effect has been used to investigate the nature of machining damage in brittle materials by measuring the reflection of Rayleigh waves; the guinea-pig material was hot-pressed silicon nitride, chosen because it is free from stress corrosion cracking at room temperature and also because it is such an important strutural ceramic (Tien *et al.* 1982; Marshall *et al.* 1983). Such measurements have been extended to imaging grinding damage, again in hot-pressed silicon nitride, by acoustic microscopy (Clarke *et al.* 1985). There were two particularly interesting aspects of the experiments. The first arose from the frequency dependence of the scattering from subsurface cracks. Pictures were taken over a range of frequencies from 25 MHz to 60 MHz, and amazingly enough the distribution of crack geometries was sufficiently small to enable systematic differences to be seen: in one

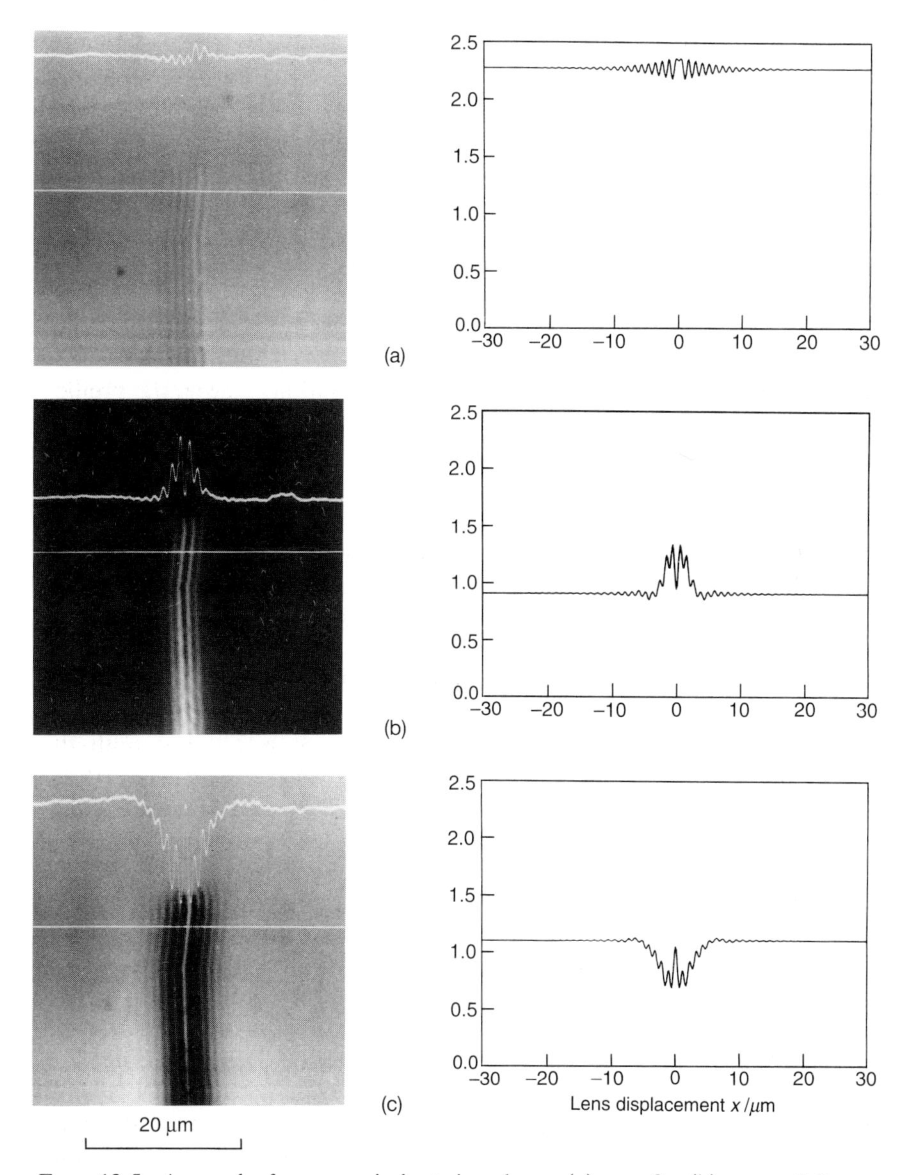

FIG. 12.5. A crack from an indent in glass: (a) $z = 0$; (b) $z = -3.8\,\mu\text{m}$; (c) $z = -5.2\,\mu\text{m}$; ELSAM, 1.5 GHz. The experimental line-scans superimposed on the images can be compared with the plots calculated using two-dimensional theory (eqns (12.2), (12.13), and (12.14)) with elastic constants from Table 6.3 and values of defocus: (a) $z = 0$; (b) $z = -4.2\,\mu\text{m}$; (c) $z = -6.8\,\mu\text{m}$. The values of z in the calculations were chosen for best fit; the reason for the discrepancy is not known, though no doubt there are the usual uncertainties associated with thermal drift, the measurement of z, and the frequency and pupil function used (Briggs *et al.* 1990).

specimen the picture at 35 MHz was almost blank, corresponding to a crack depth of 35 μm with the top 7 μm or so of the crack faces in contact. Second, it was found that by blanking off 180° of the lens, and also the central portion of the aperture, the microscope could be operated in a dark field mode, so that only backward scattered Rayleigh waves contributed any signal. Scans made with this lens gave a significantly greater level of contrast and were therefore somewhat easier to interpret. Such a situation would naturally lend itself to a Rayleigh wave axicon lens system (Atalar and Köymen 1987).

12.4 How fine a crack can you see?

The reflection coefficients of (12.14) are for a crack with nothing in it, so that the two surfaces of the crack could move freely with respect to one another. In that case the calculations unequivocally give strong contrast from cracks regardless of their thickness (or thinness!). In the microscope there is water in contact with the specimen, and this may penetrate the crack. If the crack is very thin, does the presence of the water eliminate the scattering?

Suppose that the effect of the fluid is to transmit normal components of displacement and traction across the crack, but not tangential components. With the crack in the half plane $x = 0$, $z \leq 0$, and with the notation

$$u(x, z) = \text{displacement in } x\text{-direction},$$
$$w(x, z) = \text{displacement in } z\text{-direction},$$
$$\tau_{xz}(x, z) = \text{shear stress in } xz \text{ plane},$$

the boundary conditions of the problem are that at $x = 0$, $u(z)$ is continuous and $\tau_{xz}(z) = 0$. Consider a wave incident from the left. In the absence of the crack this would give rise to a displacement $u^{\text{in}}(0, z)$ and a shear stress $\tau_{xz}^{\text{in}}(0, z)$. To satisfy the boundary conditions we must add a scattered field travelling to the left defined at $x = 0$ by

$$\begin{aligned} u^{\text{sc}}(z) &= 0, \\ \tau_{xz}^{\text{sc}}(0, z) &= -\tau_{xz}^{\text{in}}(0, z). \end{aligned} \tag{12.37}$$

A superposition of this scattered field and the incident field would satisfy the boundary conditions. An approximation to the scattered field can be generated by a distribution of body forces Q_z acting in the z-direction in the half-plane $x = 0$, $z \leq 0$, i.e. where the crack would be. Because of symmetry this distribution of body forces gives $u = 0$ at $x = 0$, and also $\tau_z(z) = Q_z(z)/2$ just to the left of the plane of the crack. Hence for $x \leq 0$

the scattered field is found by selecting

$$Q_z(z) = -2\tau^{\text{in}}_{xz}(0, z). \tag{12.38}$$

Now the field due to a delta function source $Q_z = \delta(z - z_0)\,\delta(x)$ is known as Lamb's problem, and at some distance from $x = 0$ the solution is known (Achenbach 1973). If the surface displacement of the surface wave generated by the concentrated load is $w_c(x, \zeta)$ and the depth of the crack is d, then for the distributed body forces, by superposition,

$$w^{\text{sc}}(x) = -2\int_0^d w_c(x, \zeta)\tau^{\text{in}}_{xz}(0, \zeta)\,\mathrm{d}\zeta. \tag{12.39}$$

In this expression, $\tau^{\text{in}}_{xz}(0, \zeta)$ decays exponentially with ζ (eqn (6.62)). For $x < 0$, $z = 0$, the net field is the sum of the incident and the reflected wave,

$$u(x, 0) = w^{\text{in}}(x, 0) + w^{\text{sc}}(x). \tag{12.40}$$

The transmitted field to the right of the crack can be calculated similarly, with a change of sign in eqn (12.38) and in the x-dependence of $w_c(x, \zeta)$ and $w^{\text{sc}}(x)$. In this way approximate values of T_R and R_R can be found for the two-dimensional crack contrast theory, and possibly for the three-dimensional theory as well. The calculated field is reasonably good near the free surface but not near the crack tip, so the approximations are better the deeper the crack is compared with the wavelength.

A particular conclusion from this theoretical analysis is that, if a crack has faces that are separated by a thin layer of fluid, so that normal components of traction and displacement are transmitted across the crack but the faces are free with regard to shear components of traction and displacement, then there will be a scattered wave however thin the fluid layer is. This is perhaps not surprising. A Rayleigh wave can exist only because solids can support both longitudinal and shear waves, and the greater part of the displacement in a Rayleigh wave is shear in character (§6.3). Of course, liquids can support shear stress over a short distance. In a liquid of viscosity η, and density ρ_0, at a frequency ω the amplitude of a shear wave decays by a factor e over a distance

$$d_e = \sqrt{\left(\frac{2\eta}{\rho_0\omega}\right)}. \tag{12.41}$$

At 1 GHz in water at 20°C this gives a value of 18 nm, falling to half that value at 2 GHz and 60°C. In many situations asperity contact would be more important at that separation.

If, therefore, a discontinuity is introduced across which negligible shear stress is transmitted, then it is to be expected that there will be considerable Rayleigh wave scattering. The importance of this is that it suggests that almost no surface crack is too narrow to give contrast in the acoustic microscope. This suggestion seems to be confirmed by the experimental evidence.

Figure 12.6 shows pictures of cracks around indents in a partially stabilized zirconia (PSZ) ceramic. Figure 12.6(a) is an s.e.m. image, and Fig. 12.6(b) is a light microscope image. At these magnifications the lengths of the cracks can be seen. The cracks running from top to bottom look shorter than those running from left to right, so there may be some anisotropy in the fracture toughness. The acoustic images in Fig. 12.6(c) and (d) were taken at 400 MHz; in zirconia this gives an average Rayleigh wavelength $\lambda_R = 8\ \mu m$. The cracks running between the two indents in Fig. 12.6(c) show up quite well, and the contrast is adequate to reveal the overlap that can also be seen in the s.e.m. and light microscope pictures. When the defocus is increased, Fig. 12.6(d), a strong $\lambda_R/2$ fringe pattern develops. The fringes stop at a well defined distance along each crack, so that their length can be measured unambiguously. Further observation at higher magnification in s.e.m. and at higher frequency and magnification in an ELSAM confirmed that no part of a crack that was seen in the s.e.m. failed to give contrast in the acoustic microscope, and that the length of a crack indicated by the acoustic microscope was the same as was seen in the s.e.m. (Briggs *et al.* 1990). Similar conclusions were reached for thermal stress relief cracks in titanium nitride coatings on hard metals, cracks from indents in silicon carbide, hot-pressed silicon nitride, reaction-bonded silicon nitride, and silicon wafers, as well as the gallium arsenide specimen in Fig. 12.4 and the glass specimen in Fig. 12.11 below.

In each of these examples the fringes stop at a very well defined length along the crack. If the contrast somehow depended on the width of the crack, then it might be expected that the fringes would gradually decrease in strength as the tip of the crack was approached. But this is not what is seen, and pictures like the ones in Figs 12.4–6 tend to confirm that the strength of the contrast depends only on whether a crack is present, and not on its width. It would, of course, be expected that crack depth and crack closure would affect the contrast, but in all the examples cited the Rayleigh fringes parallel to the crack seem to continue right up to the tip, and then stop abruptly. Where this does not happen, the inference must be that crack closure has occurred. For example, in studies of stable crack growth in alumina ceramics by acoustic microscopy at 1 GHz (Quinten and Arnold 1989; Quinten *et al.* 1991), it was found that often microcracks could not be imaged at all even when it was known from acoustic emission measurements that microcracking had occurred.

Measurements were also made of acoustic velocity in the process zone in the vicinity of the crack tip (both by $V(z)$ and by a bulk wave technique), and the drop in velocity was too large to be accounted for by strain and the third-order elastic constants. It was therefore concluded that extensive crack closure must be taking place, and that this could be used in accounting for the observed R-curve behaviour. The effect of crack closure on the interaction of a Rayleigh wave with a crack can be modelled using techniques developed for non-destructive testing (Thompson *et al.* 1983, 1984). Around cracks and indents in partially stabilized zirconia (PSZ) ceramics, a superabundance of fringes is sometimes seen (Fagan *et al.* 1991). Figure 12.7(a) shows the area around an indent in zirconia containing 2 (weight per cent) MgO. Some of the fringes are readily identified, notably water-ripple fringes marking contours of spacing $\lambda_0/2$ in height, and Yamanaka fringes with spacing λ_R, but some of the other fringes are not so easy to account for (Fagan *et al.* 1990). At higher magnification (Fig. 12.7(b)), transformation of the zirconia from the tetragonal to the monoclinic phase is visible, and no doubt the fringes contain information about this. Indeed, it is the combination of the compressive stresses and the microcracking caused by this transformation, and its associated volume increase, that gives the PSZ ceramics their enhanced toughness. A remarkable picture of this in another zirconia ceramic with 4 (weight per cent) CaO, similarly indented to induce the transformation, is given in Fig. 12.7(c). Wear and delamination mechanisms in the surfaces of a number of engineering ceramics have also been studied by acoustic microscopy (Yamanaka *et al.* 1985).

In some acoustic images of cracks, such as Fig. 12.1(b), interference fringes are not found. This must be because for some reason there is little reflection of the Rayleigh waves by the crack. It is possible that the geometry at the mouth of the cracks was affected in some way when they were polished in preparation for microscopy, and that this has reduced the Rayleigh wave reflection. In more brittle materials fringes are almost always seen. Even when fringes are not present, good contrast can still be obtained. In a study of thermal stress relief cracks in a titanium nitride coating on a hard metal cutting tool, the cracks, which are in compression because the metal contracts more than the ceramic on cooling, could be seen without difficulty in an acoustic microscope with a 2 μm wavelength (0.73 GHz), even though in the s.e.m. they looked less than a tenth of that width (Ilett *et al.* 1984). This kind of observation confirms that the cracks are visible, not because of any scattering of geometrically reflected waves, but because of their effect on the Rayleigh waves that are incident broadside.

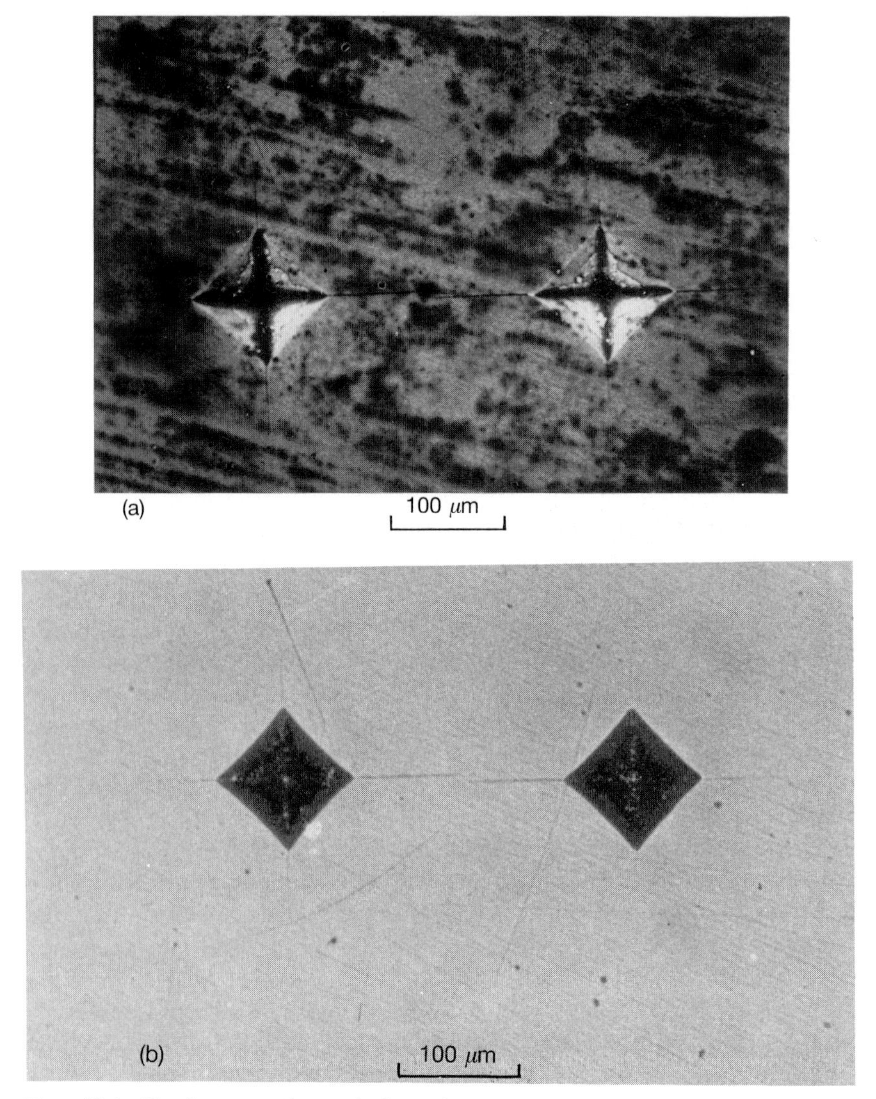

FIG. 12.6. Cracks around two indents in a partially stabilized zirconia (PSZ) ceramic: (a) s.e.m.; (b) light microscope; (c) 400 MHz, $z = -8$ μm; (d) 400 MHz, $z = -26$ μm (Briggs *et al.* 1990).

What about defects that are smaller than a wavelength in all dimensions? This can be considered by an extension of Lord Rayleigh's original explanation of why the sky is blue. If the scattered amplitude A_b is proportional to the volume a_b^3 of the scatterer, and inversely proportional to the distance r of the observer, by simple dimensional analysis the

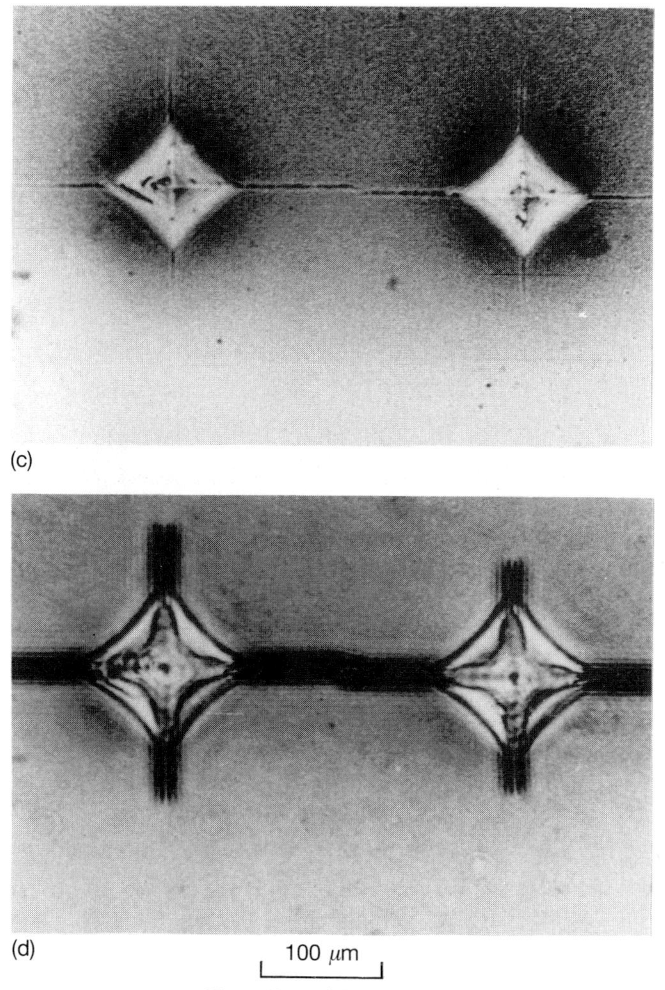

FIG. 12.6. (*Continued*)

dependence on the wavelength λ must be

$$A_b = \frac{A_0}{r}\frac{a_b^3}{\lambda^2}. \tag{12.42}$$

When this is expressed in squared form to give the scattered intensity, it gives the familiar a_b^6/λ^4 dependence (Morse and Ingard 1987). A similar dimensional argument can be given for the scattering of Rayleigh waves from inhomogeneities in a surface. In this case the scattered amplitude varies as $1/\sqrt{r}$. If an inhomogeneity extended down from the surface uniformly to a depth much greater than a wavelength, then the scattered

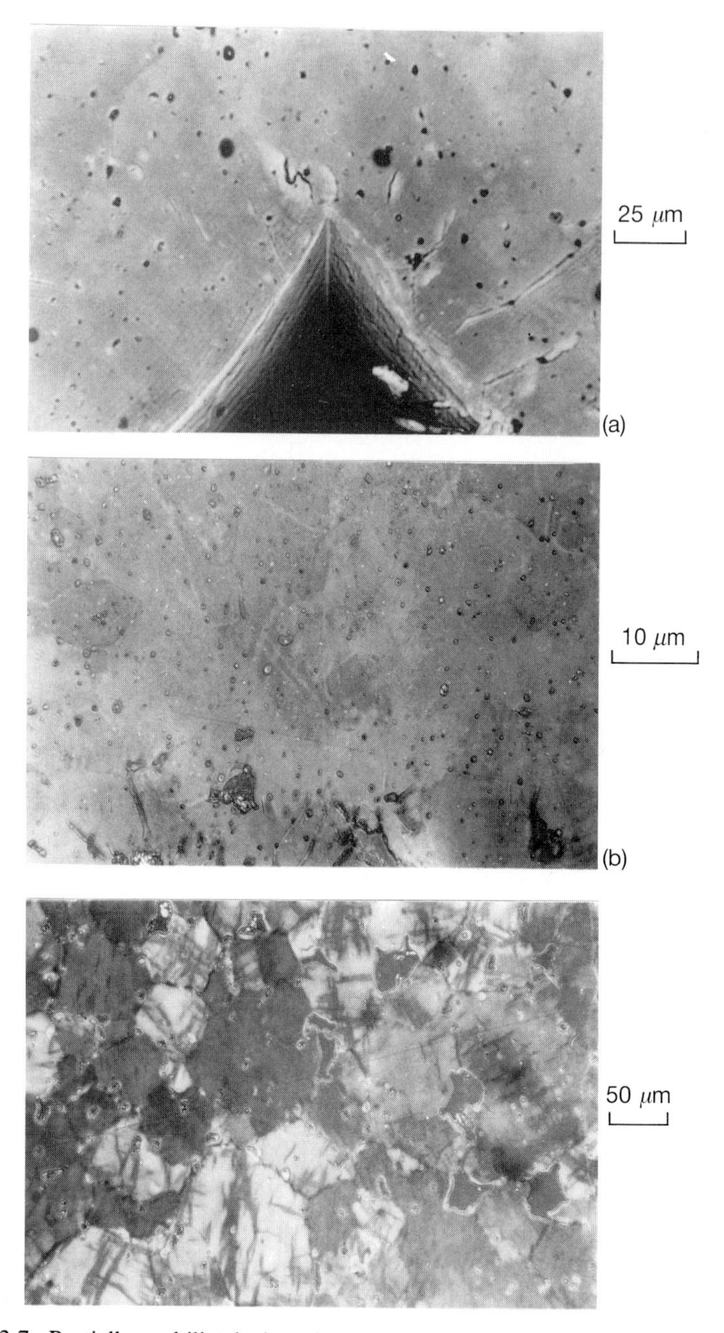

FIG. 12.7. Partially stabilized zirconia ceramics, with cracking and monoclinic to tetragonal transformation induced by a hardness indent: (a) sub-eutectoid aged ZrO_2–2wt%MgO, $z = -1\,\mu m$, 1.3 GHz; (b) $z = -1.2\,\mu m$, 1.9 GHz; (c) ZrO_2–4wt%CaO, $z = -4.6\,\mu m$, 1.6 GHz: ELSAM (Fagan 1990).

amplitude would vary as the cross-sectional area, giving a wavelength dependence

$$A_b = \frac{A_0}{r^{1/2}} \frac{a_b^2}{\lambda^{3/2}}. \tag{12.43}$$

This would give an a_b^4/λ_R^3 dependence, and would correspond to a small but deep drilled hole. If, on the other hand, the defect were small in all three dimensions, and were at a depth much less than the wavelength, then the scattered amplitude would be

$$A_b = \frac{A_0}{r^{1/2}} \frac{a_b^3}{\lambda^{5/2}}. \tag{12.44}$$

This would give a scattered intensity proportional to a_b^6/λ_R^5.

12.5 Contrast at boundaries

The scattering function of eqn (12.13) can be extended to the more general case of different materials on either side of the boundary; indeed it was originally derived in that form (Somekh *et al.* 1985). The two sides are denoted by subscripts 1 and 2, having Rayleigh wavenumbers k_{p1} and k_{p2} with imaginary components α_1 and α_2. Transmission and reflection coefficients T_{R1}, T_{R2}, and R_{R1}, R_{R2} are defined for waves incident from sides 1 and 2, respectively. Then the scattering Green function becomes

$$\begin{aligned}
S(k_x, k_x') = {} & R_0(k_x)\delta(k_x - k_x') + \mathrm{i}2\,\delta(k_x - k_x')\left[\frac{\alpha_1 k_{p1}}{k_x^2 - k_{p1}^2} + \frac{\alpha_2 k_{p2}}{k_x'^2 - k_{p2}^2}\right] \\
& + \frac{1}{2\pi}\left\{2\surd(\alpha_1\alpha_2)\left[\frac{T_{R2}}{(k_x + k_{p1})(k_x' + k_{p2})} + \frac{T_{R1}}{(k_x - k_{p2})(k_x' - k_{p1})}\right]\right. \\
& - \frac{2\alpha_1}{k_x + k_{p1}}\left[\frac{1}{k_x' + k_{p1}} + \frac{R_{R1}}{k_x' - k_{p1}}\right] \\
& - \frac{2\alpha_2}{k_x - k_{p2}}\left[\frac{1}{k_x' - k_{p2}} + \frac{R_{R2}}{k_x' + k_{p2}}\right] \\
& \left. + \frac{4}{k_x' - k_x}\left[\frac{\alpha_1 k_{p1}}{k_x^2 - k_{p1}^2} - \frac{\alpha_2 k_{p2}}{k_x'^2 - k_{p2}^2}\right]\right\}.
\end{aligned} \tag{12.45}$$

The Green function can be used in eqn (12.2) to calculate the contrast in the vicinity of a boundary between two different materials. Like the two-dimensional theory for cracks, this may be used to help to understand the contrast that is seen at boundaries. Once again the values of the reflection and transmission coefficients are external to the theory, and while this has the advantage that solutions that have been calculated elsewhere can be used, unfortunately values are not readily available in general. However, a Rayleigh impedance can be defined as the Rayleigh velocity multiplied by the density and, in cases where the difference in Rayleigh impedance across the boundary is not too great, approximate values for the Rayleigh wave transmission and reflection coefficients can be taken as

$$T_{R1} = \frac{2Z_2}{Z_2 + Z_1}; \qquad T_{R2} = \frac{2Z_1}{Z_2 + Z_1},$$
$$R_{R1} = -R_{R2} = \frac{Z_2 - Z_1}{Z_2 + Z_1}. \tag{12.46}$$

If calculations are performed for a boundary between hypothetical materials that have different Rayleigh velocities, and whose impedances can be varied, then the difference in the real parts of k_p on either side of the crack can give strong contrast at the boundary even in the absence of scattering, i.e. even when $T_1 = T_2 = 1$, $R_1 = R_2 = 0$. Where the lens straddles the boundary, the path taken by the Rayleigh ray corresponds neither to one material by itself nor the other. It has a Rayleigh angle for one material on one side of the boundary and to the other on the other side. Thus a $V(z)$ scan over the boundary would have a periodicity intermediate between the periodicity of each material by itself. In $V(x)$ this manifests itself as oscillations of period not π/k_R, but $2\pi/(k_{R1} - k_{R2})$. This period is present only for a range of x such that a Rayleigh wave that is excited in one material is detected after it has crossed the boundary and re-radiated from the other. Self-evidently this effect cannot occur at positive defocus.

The model can also be used to give an approximate indication of the contrast to be expected at grain boundaries. In this case the density of the material is the same on both sides of the boundary. A full theory would have to take account of all the phenomena described for anisotropic materials in Chapter 11; in particular, the non-degeneracy of the transverse waves means that the reflection of surface waves would have to be treated as a fully anisotropic problem: calculations for reflection and transmission of bulk waves in anisotropic media (Auld 1973) would give approximate values in the absence of a full theory for surface waves. As with a boundary between two different materials there is contrast not only between one grain and another (as described in Chapter 11), but also between grains and grain boundaries. In some cases the crystal anisotropy is so great that fringes are seen parallel to the grain boundaries, a notable example being zirconia which has a crystal anisotropy greater that almost any other common ceramic (Ingel and Lewis 1988*b*). But generally the scattering of surface waves at grain boundaries is considerably less than at cracks.

This gives a useful way to distinguish between boundaries and cracks. Figure 12.8(a) shows an image of a stainless steel specimen that had been subject to fatigue. It is difficult to find the fatigue cracks, because they are camouflaged against the grain boundaries. But the contrast mechanisms have different dominant contributions. The chief mechanism for contrast from the cracks is the scattering of surface waves by the crack, whereas the chief contribution to contrast from the grain boundaries is

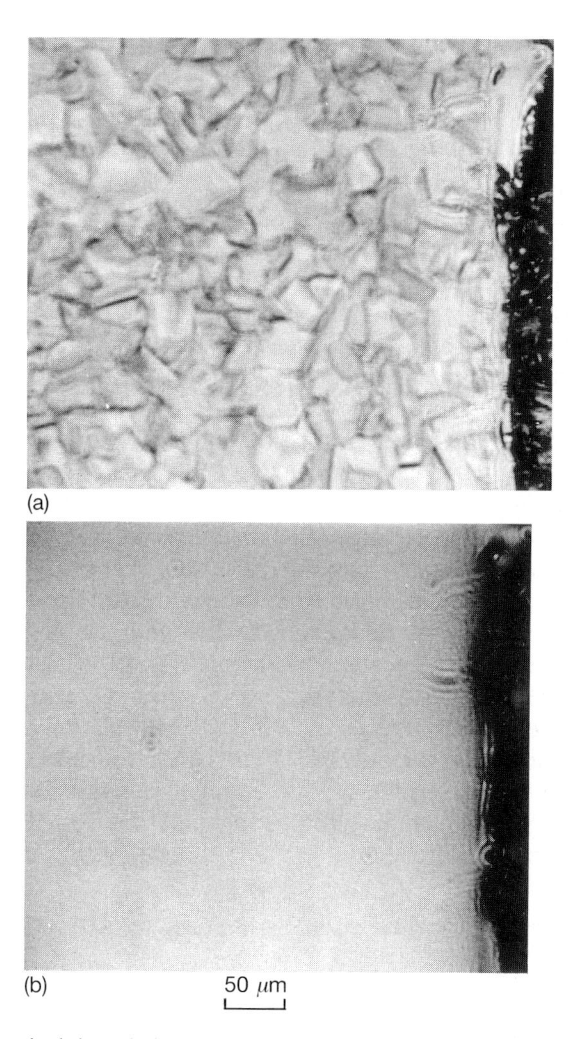

FIG. 12.8. Short incipient fatigue cracks in stainless steel: (a) negative defocus, $z = -20\,\mu$m; (b) positive defocus, $z = +32\,\mu$m; 0.37 GHz (Rowe *et al.* 1986).

the change of surface wave propagation parameters from one grain to the next. Contrast that depends on the change in Rayleigh angle across a boundary will be present only at negative defocus. On the other hand contrast that depends on reflection of surface waves from a crack will be present at zero and positive defocus (with a dead zone either side of the crack of approximately $2z \tan \theta_R$). Thus by using positive defocus it should be possible to distinguish small fatigue cracks from grain boundaries. This is illustrated in Fig. 12.8(b), where the cracks do indeed appear with a small dead zone and then Rayleigh fringes either side. Thus, by studying how the contrast varies with defocus it is possible to distinguish between different kinds of elastic discontinuity that give contrast in the acoustic microscope. It may even be possible to identify persistent slip bands and relate them to crack nucleation and growth at the earliest stages of fatigue (Jenkins 1990; Zhai *et al.* 1990). The ideal would be to be able to measure the detailed geometrical and interfacial parameters of cracks and boundaries.

Figure 12.9(a) is a picture of a silicon carbide fibre (Nicalon) in a calcium–aluminosilicate glass matrix. This picture is taken from a detailed study of a large number of ceramic matrix, glass–ceramic matrix, and metal matrix composites (Lawrence 1990). The interface between the fibre and the matrix is not where it appears at first sight to be. Starting from the middle of the fibre, there are a few Rayleigh fringes with a rather dark mean background; then the mean level becomes brighter and there are four or five fringes before they fade further out into the matrix. But the interface is not where the average contrast changes from dark to bright; it is one or two fringes further out than that. A clue to this is where the fringes from a crack in the matrix stop. The best way to understand it is to calculate the contrast. Figure 12.9(b) shows the results of such a calculation, plotted as a line scan with the vertical axis at the interface. The material constants of the matrix and the fibre were used to calculate the Rayleigh parameters from (12.46); $V(x)$ was then calculated for the frequency and defocus used in Fig. 12.9(a), and plotted with the same horizontal scale for direct comparison. In the calculated curve, not only do the mean levels vary in about the right way, but the big oscillations in $V(x)$ continue to the right of the boundary just as the fringes do in the picture.

12.6 Time-resolved measurements and crack tip diffraction

If any reader still needs convincing of the dominant role played by Rayleigh waves in the imaging of cracks, final evidence is given by time-resolved measurements of short pulses as a lens is scanned past a crack. Examples of such measurements are shown in Fig. 12.10,

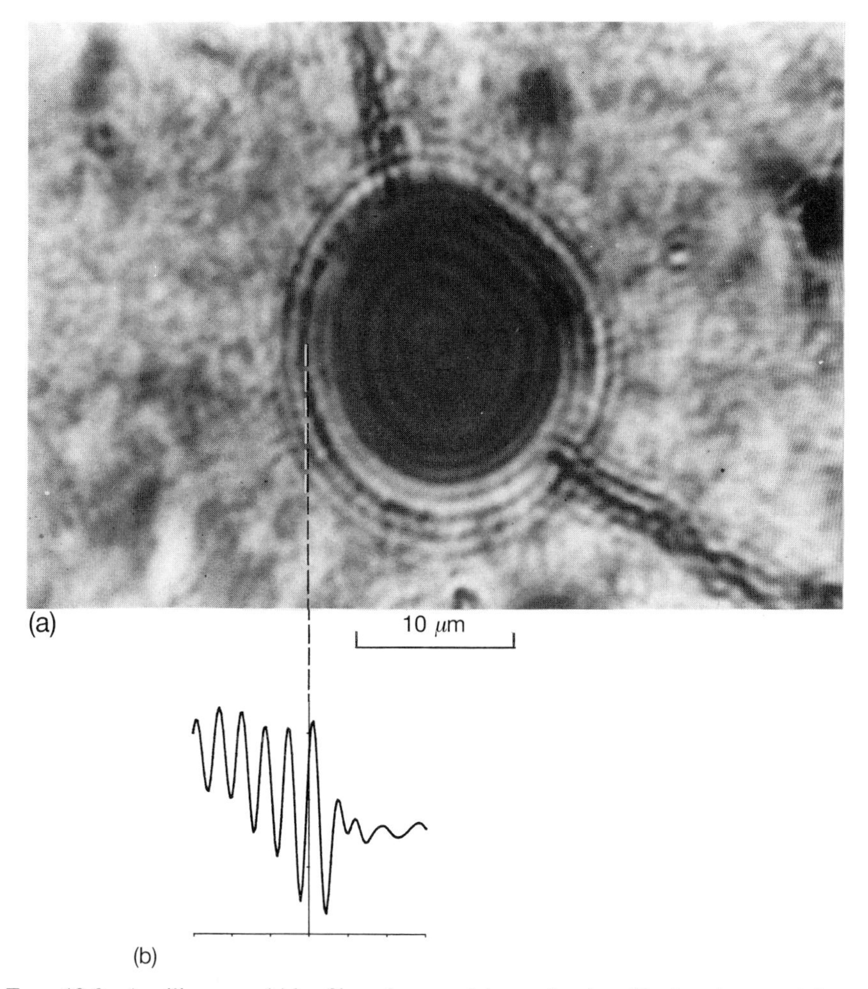

FIG. 12.9. A silicon carbide fibre in a calcium–aluminosilicate glass matrix: (a) ELSAM, 1.9 GHz, $z = -3.2\,\mu$m; (b) calculated contrast from eqns (12.2), (12.45), and (12.46), horizontal scale the same as (a), with tick spacing 2.5 μm; this time the experimental parameters gave the best fit without any adjustment! (Lawrence 1990).

presented as $S(t, y)$ displays, as described in §9.2.3. Each horizontal line represents an oscilloscope trace, but with the signal being indicated by the brightness instead of by vertical deflection. The vertical axis corresponds to displacement of the lens parallel to a specimen surface. There is a crack in the surface perpendicular to this scan direction. At the

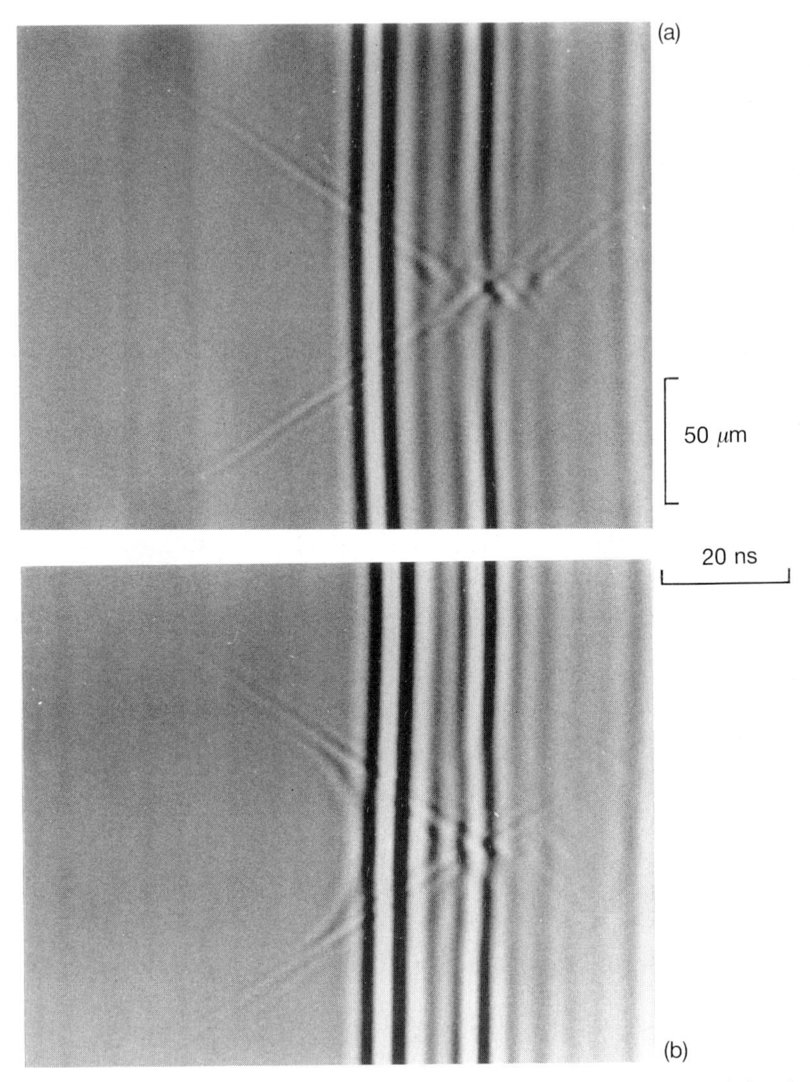

FIG. 12.10. Time-resolved $S(t, y)$ along a line perpendicular to a crack in glass, scanning across the crack: (a) some distance from the end of the crack; (b) 75 μm from the end of the crack. As in Fig. 9.3(b), the horizontal axis is time t; the vertical axis is y, and the value of $S(t, y)$ is indicated by the intensity, with mid-grey as zero and dark and light as negative and positive values of S. In both figures, the first echo (seen as the first stripy vertical bar) is the geometric reflection from the surface of the specimen, and the second echo (seen as the second stripy vertical bar) is the Rayleigh reflection (§7.2). The patterns forming a × are the reflections from the near and the far sides of the crack, which cross over when the lens is directly above the crack. In (b), where the scan passes quite near to the tip of the crack, the hyperbolic pattern is due to the crack-tip-diffracted wave. (Weaver *et al.* 1989.)

top of Fig. 12.10(a) there are only two contributions to the reflected signal. The first is the geometrically reflected ray, and the second is the Rayleigh reflected ray. As the lens is scanned towards the crack, and the crack begins to cut the circle of Rayleigh wave excitation, a new echo appears that can arrive before any of the others. This is a Rayleigh wave that is reflected by the crack almost as soon as it is excited in the surface. Although only a small proportion of the circular Rayleigh wavefront in the surface may be reflected back to the transducer in this way, the signal can still be quite strong because of the reduced pathlength in which attenuation due to leaking occurs. Next a later echo begins to appear that is due to a Rayleigh ray that has crossed the axis of the lens and been reflected by the crack on the opposite side of the axis from that on which it was excited, and subsequently propagates back to the lens. At first this reflection is weak, because it is subject both to the algebraic decay of a spreading wavefront and to exponential attenuation due to the fluid, but as the lens comes closer to the crack the echo becomes stronger. Finally, as the lens approaches a position overhead of the crack, so the three Rayleigh reflections converge, crossing over when the lens is directly over the crack. By interpolating the amplitudes of the Rayleigh reflections, and their interference, at the cross-over point, it may be possible to deduce the Rayleigh reflection and transmission coefficients of the crack (Weaver *et al.* 1989).

It is sometimes possible to observe Rayleigh interference from diffraction around the tip of a crack. This can be particularly clearly demonstrated at the ends of radial cracks from indents in glass, and an example is shown in Fig. 12.11. This provides exceptionally strong evidence that the tip of the crack is a well defined point as far as the acoustic image is concerned, and that in the acoustic microscope the crack is indeed imaged right up to its end, no matter how narrow it becomes. This can be exploited for the measurement of the length of cracks from indents in order to determine the fracture toughness of low-ductility materials such as ceramics, as in the pictures of indentation cracks in partially stabilized zirconia in Fig. 12.6 where, although crack tip diffraction is not visible, the end of each set of fringes is well defined to within the resolution of the microscope. It is even possible to relate anisotropy in the lengths of the cracks to anisotropy in the Rayleigh velocity measured by a line-focus-beam microscope (Yamanaka *et al.* 1984). Where crack tip fringes are present, as in Fig. 12.11, it may be possible to use advanced image analysis techniques to locate the centre of the circle of fringes, and thus determine the length of a crack to rather better than the resolution of the microscope (Fatkin *et al.* 1989). Similarly, where fringes are seen on either side of a boundary it may be possible to use two-dimensional Fourier analysis to find the spatial phase of the fringes and hence deduce additional information about the character and geometry of the discontinuity (Block *et al.* 1989, 1990).

The diffraction or scattering of Rayleigh waves by very small features as well as by cracks has proved fruitful in the study of the fracture of low-ductility materials. Figure 12.12 shows pictures of two specimens of a cermet composed of WC–6%Co. The specimens had brittle pre-cracks introduced at room temperature, and then were subjected to a load and heated to 850°C, which is in the vicinity of their ductile–brittle transition temperature. The specimen in Fig. 12.12(a) was held at that temperature for less than a minute and then allowed to cool. No ductile crack growth has occurred and there are nice straight fringes along the

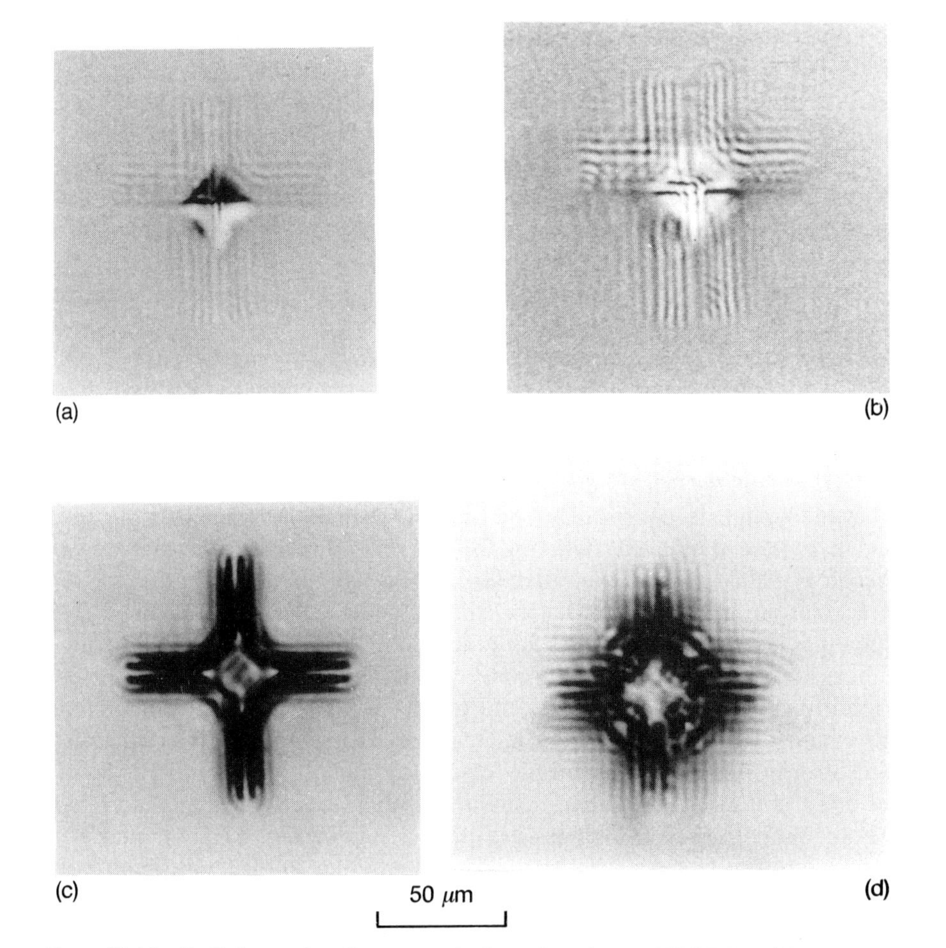

FIG. 12.11. Radial cracks from an indent in glass; 400 MHz: (a) $z = 0$; (b) $z = +6\,\mu m$; (c) $z = -20\,\mu m$; (d) $z = -46\,\mu m$ (Briggs *et al.* 1990).

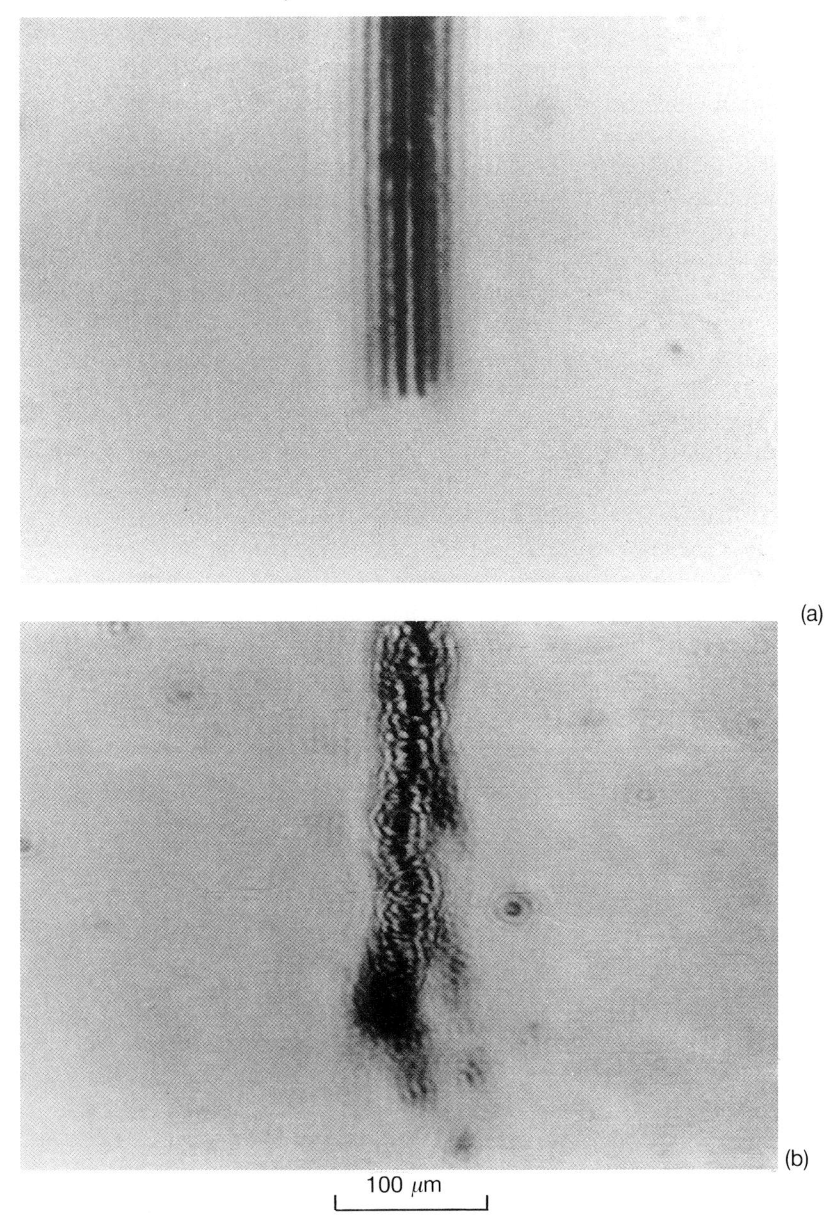

FIG. 12.12. Cracks in specimens of WC–6%Co hardmetal held under load at 850°C: (a) for less than 1 min, $z = -66\,\mu m$; 250 MHz; (b) for 12 min, $z = -34\,\mu m$; 370 MHz. In (a) only the straight brittle pre-crack introduced at room temperature is present; in (b) creep crack growth has occurred by secondary microcracking and void nucleation and coalescence (Briggs *et al.* 1989; Fagan 1990).

original brittle crack. The specimen in Fig. 12.12(b) was held under load at 850°C for 12 min. In this case the signs of ductile creep crack growth are apparent; there is evidence of secondary microcracking and void nucleation and coalescence. There are strong Rayleigh fringes from those features, including some nearly circular fringe patterns around certain voids. Similar contrast has been exploited for the study of WC–11%Co (Schmid *et al.* 1987; Schmid 1987).

Is it possible to combine crack tip diffraction and time-resolved microscopy for the measurement of cracks? This could be of great significance both for measuring very small fatigue cracks and for quantitative evaluation of small flaws in low-ductility advance engineering materials. The answer may well be 'yes'. Figure 12.10(b) shows a $S(t, y)$ image measured in the same way as Fig. 12.10(a), except that in this case the lens crossed the crack close to its tip at the surface. In this case a reflection is seen that appears with a hyperbolic shape in the display of $S(t, y)$. This is the crack tip diffracted signal. In this case it is the diffracted Rayleigh wave signal. If techniques can be developed in the acoustic microscope for the detection of bulk waves diffracted from a crack tip inside a specimen, as has been done so successfully in non-destructive testing at lower frequencies (Silk 1987), then it may become possible to measure the depth into the surface of very small cracks.

13

So what happens when you defocus?

On one occasion his prayer was full of the wonders of the microscope. (*George MacDonald*, William Raiper 1987)

Once a frequency has been selected, and an area of the specimen chosen, the most important parameter under the control of the microscopist is the defocus. Changing the defocus does not simply change the contrast; it actually selects what information is contained in the image in a way that no amount of subsequent image processing could ever do. So the central question when operating an acoustic microscope and interpreting the images is, 'What is changing as I defocus?' The answer must be considered not in terms of geometrical optics, but in terms of $V(z)$.

In high-velocity materials such as metals, semiconductors, and ceramics, the dominant effect of moving the lens towards the surface is not to focus below the surface, but rather to alter the interference conditions between the rays that are reflected geometrically from the surface and those that have interacted with Rayleigh waves excited in the surface. This is amply confirmed by analysis of $V(z)$ over a large range of defocus, but it applies equally at small values of defocus. As a rule of thumb, good contrast is often found by starting at the focal position and moving the lens towards the specimen until the signal has fallen to a half or a third of its level at focus. This occurs typically at a dimensionless defocus in the range $10 < kz < 20$, i.e. a defocus of two or three water wavelengths. On a geometrical model this would give a paraxial longitudinal focus a sixth of a wavelength below the surface, which is scarcely in the realm of ray optics. The contrast must therefore be understood in terms of how Rayleigh waves propagate; indeed for many applications an acoustic microscope can be thought of as a Rayleigh wave microscope. Changes in stiffness or density, changes in thickness or uniformity of surface layers, changes in crystallographic orientation, scattering from cracks, pores, inclusions, or other surface defects, all give contrast because they affect the propagation of Rayleigh and other surface and pseudo-surface waves. When fringes of half Rayleigh wavelength appear around a defect or an edge in an image, that gives the strongest possible confirmation of the origin of the contrast.

In materials of lower velocity the situation becomes simpler. If the shear critical angle is outside the lens angle, then there will be no Rayleigh phenomena. Such materials would include many polymers and composites. There may still be longitudinal head waves, and these can reveal some of the features associated with Rayleigh wave imaging, such as surface cracks. But two further factors enhance subsurface imaging. First, because the velocity is lower, the focal depth below the surface is greater for a given amount of defocus and aberrations are reduced. Second, more power can be transmitted into bulk waves, both because this can occur over a wider angle of incidence and also because the impedance is lower, especially if the density is also low. Subsurface imaging therefore becomes much more important. Finally, in materials, such as biological soft tissue, that are so slow that even the longitudinal critical angle either lies outside the lens or does not exist, the situation is the most straightforward of all. Scalar waves dominate, and simple diffraction optics may be used. If the specimen is layered or on a substrate, then the image may exhibit interference between reflections from different layers. Of course, the reflection from any layer is still in the confocal optical configuration, with its associated $V(z)$ curve.

So there is no escaping from a good understanding of the $V(z)$ phenomenon. All microscopy is best done when the microscopist has a clear idea of what he expects and how he would recognize it, and how he would know if he saw something different from what he expected. Acoustic microscopy is no exception and, specifically, the microscopist should have ever in his mind a clear idea of the $V(z)$ behaviour of the expected structure. Only in that way can he choose the optimum settings for examining a specimen. And only in that way can he subsequently look at a micrograph and ask, 'What does this tell me?'

References

References with three or more authors and the same first-named author are listed chronologically after one- and two-author references. Pages on which a reference is cited are indicated by numbers in square brackets at the end of the reference.

Achenbach, J. D. (1973). *Wave propagation in elastic solids.* North Holland, Amsterdam. [279]

Achenbach, J. D. (1987). Flaw characterisation by ultrasonic scattering methods. In *Solid mechanics research for quantitative non-destructive evaluation* (ed. J. D. Achenbach and Y. Rajapakse), pp. 67–81. Nijhoff, Dordrecht. [269]

Achenbach, J. D. and Brind, R. J. (1981). Scattering of surface waves by a sub-surface crack. *J. Sound and Vibration* **76,** 43–56. [276]

Achenbach, J. D., Gautesen, A. K., and Mendelsohn, D. A. (1980). Ray analysis of surface-wave interaction with an edge crack. *IEEE Trans.* **SU-27,** 124–9. [269]

Achenbach, J. D., Ahn, V. S., and Harris, J. G. (1991). Wave analysis of the acoustic material signature for the line focus microscope. *IEEE Trans.* **UFFC 38,** 380–7. [276]

Addison, R. C., Somekh, M. G., and Briggs, G. A. D. (1986). Techniques for the characterisation of film adhesion. *IEEE 1986 Ultrasonics Symposium,* pp. 775–82. IEEE, New York. [224, 225]

Addison, R. C., Somekh, M. G., Rowe, J. M., and Briggs, G. A. D. (1987). Characterisation of thin-film adhesion with the scanning acoustic microscope. In *Pattern recognition and acoustical imaging* (ed. L. A. Ferrari), pp. 275–84. SPIE 768. [225]

Addison, R. C., Kendig, M. W., and Jeanjaquet, S. J. (1989). In-situ measurement of cathodic disbonding of polybutadiene coating on steel. In *Acoustical imaging,* Vol. 17 (ed. H. Shimizu, N. Chubachi, and J. Kushibiki), pp. 143–52. Plenum Press, New York. [212]

Adler, E. L., Slaboszewicz, J. K., Farnell, G. W., and Jen, C. K. (1990). PC software for SAW propagation in anisotropic multilayers. *IEEE Trans.* **UFFC 37,** 215–29. [243]

Angel, Y. C. and Achenbach, J. D. (1984). Reflection and transmission of obliquely incident Rayleigh waves by a surface breaking crack. *J. Acoust. Soc. Am.* **75,** 313–39. [273, 274]

Angmar-Månsson, B. and ten Bosch, J. J. (1987). Optical methods for the detection and quantification of caries. *Adv. Dent. Res.* **1,** 14–20. [188]

Atalar, A. (1978). An angular spectrum approach to contrast in reflection acoustic microscopy. *J. Appl. Phys.* **49,** 5130–9. [112]

Atalar, A. (1979). A physical model for acoustic signatures. *J. Appl. Phys.* **50,** 8237–9. [112, 113]

Atalar, A. (1985). Penetration depth of the scanning acoustic microscope. *IEEE Trans.* **SU-32,** 164–7. [215]

Atalar, A. (1987). Increasing the sensitivity of the scanning acoustic microscope to anisotropy. *IEEE 1987 Ultrasonics Symposium*, pp. 791–4. IEEE, New York. [58, 257]

Atalar, A. (1989). Improvement of anisotropy sensitivity in the scanning acoustic microscope. *IEEE Trans.* **UFFC 36,** 264–73. [58, 254, 257]

Atalar, A. and Hoppe, M. (1986). High-performance acoustic microscope. *Rev. Sci. Instrum.* **57,** 2568–76. [23, 31, 50, 67, 69]

Atalar, A. and Köymen, H. (1987). Use of a conical axicon as a surface acoustic wave focusing device. *IEEE Trans.* **UFFC 34,** 53–63. [58, 278]

Atalar, A. and Köymen, H. (1989). A high efficiency Lamb wave lens for subsurface imaging. *IEEE 1989 Ultrasonics Symposium*, pp. 813–16. IEEE, New York. [58, 59, 223]

Attal, J. (1980). Acoustic microscopy with liquid metals. In *Scanned image microscopy* (ed. E. A. Ash), pp. 97–118. Academic Press, London. [33]

Attal, J. (1983). La microscopie acoustique. *Recherche* **14,** 664–7. [11, 33]

Attal, J. and Quate, C. F. (1976). Investigation of some low ultrasonic absorption liquids. *J. Acoust. Soc. Am.* **59,** 69–73. [31]

Attal, J., Saied, A., Saurel, J. M., and Ly, C. C. (1989). Acoustic microscopy: deep focusing inside materials at gigahertz frequencies. In *Acoustical imaging*, Vol. 17 (ed. H. Shimizu, N. Chubachi, and J. Kushibiki), pp. 121–30. Plenum Press, New York. [33]

Auld, B. A. (1973). *Acoustic fields and waves in solids*, John Wiley, New York. [59, 82, 84, 90, 95, 97, 203, 216, 217, 233, 234, 236, 238, 286]

Auld, B. A. (1985). Rayleigh wave propagation. In *Rayleigh-wave theory and applications* (ed. E. A. Ash and E. G. S. Paige), pp. 12–28. Springer-Verlag, Berlin. [87, 203]

Bates, R. H. T. and McDonnell, M. J. (1986). *Image restoration and reconstruction.* Clarendon Press, Oxford. [132]

Bennett, S. D. (1982). Approximate materials characterization by coherent acoustic microscopy. *IEEE Trans.* **SU-29,** 316–20. [Reprinted in Lee, H. and Wade, G. (1986). *Modern acoustical imaging*, pp. 389–93 IEEE, New York.] [114]

Bennett, S. D. and Ash, E. A. (1981). Differential imaging with the acoustic microscope. *IEEE Trans.* **SU-28,** 59–64. [Reprinted in Lee, H. and Wade, G. (1986), *Modern acoustical imaging*, pp. 203–8. IEEE, New York.] [181]

Bereiter-Hahn, J. (1987). Scanning acoustic microscopy visualizes cytomechanical responses to cytochalasin D. *J. Microsc.* **146,** 29–39. [166–8]

Bereiter-Hahn, J. and Buhles, N. (1987). In *Imaging and visual documentation in medicine* (ed. K. Wamsteker *et al.*), pp. 537–43. Elsevier, Amsterdam. [165]

Beresford, J. N., Gallagher, J. A., Gowen, M., McGuire, M. K. B., Poser, J., and Russel, R. G. G. (1983*a*) Human bone cells in culture. A novel system for the investigation of bone cell metabolism. *Clin. Sci.* **64,** 38–9. [9]

Beresford, J. N., MacDonald, B. R., Gowen, M., Couch, M., Gallagher, J. A., Sharpe, P. T., and Poser, J. (1983*b*). Further characterisation of a system for the culture of human bone cells. *Calcif. Tiss. Int.* **35,** 637 A. [9]

Bertoni, H. L. (1984). Ray-optical evaluation of $V(z)$ in the reflection acoustic microscope. *IEEE Trans.* **SU-31,** 105–16. [115, 118]

Bertoni, H. L. and Somekh, M. G. (1985). Ray-optical analysis of spherically focussing transducers for acoustic microscopy. *IEEE 1985 Ultrasonics Symposium*, pp. 715–19. IEEE, New York. [131]

Bertoni, H. L. and Tamir, T. (1973). Unified theory of Rayleigh angle phenomena for acoustic beams at liquid–solid interfaces. *Appl. Phys.* **2,** 157–72. [143]

Bjørnø, L. and Lewin, P. A. (1986). Measurement of nonlinear acoustic parameters in tissue. In *Tissue characterization with ultrasound* (ed. J. F. Greenleaf) Vol. 1, pp. 141–63. CRC Press. [43, 45, 186]

Bleaney, B. I. and Bleaney, B. (1983). *Electricity and magnetism.* Oxford University Press, Oxford. [59, 94]

Block, H., Heygster, G., and Boseck, S. (1989). Determintion of the OTF of a reflection scanning acoustic microscope (*SAM*) by a hair crack in glass at different ultrasonic frequencies. *Optik* **82,** 147–54. [29, 292]

Block, H., Heygster, G., and Boseck, S. (1990). Investigations on the defocusing effects in a SAM at straight amplitude and phase edges. In *Advanced materials and processes* (ed. H. E. Exner and V. Schumacher), pp. 1431–6. DGM Informationsgesellschaft mbH, Oberursel. [292]

Born, M. and Wolf, E. (1980). *Principles of optics: electromagnetic theory of propagation, interference and diffraction of light.* Pergamon, Oxford. [43]

Boyde, A. (1987). Colour-coded stereo images from the tandem scanning reflected light microscope (TSRLM). *J. Microsc.* **146,** 137–42. [182, 194, 196, 204]

Bracewell, R. N. (1978). *The Fourier transform and its applications.* McGraw-Hill, New York. [64, 65, 113]

Bray, R. C. (1981). Acoustic and photoacoustic microscopy. *Ph.D. thesis,* Ginzton Lab. Report 3243, Stanford University. [30, 253]

Bray, R. C., Quate, C. F., Calhoun, J., and Koch, R. (1980). Film adhesion studies with the acoustic microscope. *Thin Solid Films* **74,** 295–302. [224]

Breazeale, M. A. and Torbett, M. A. (1976). Backward displacement of waves reflected from an interface having superimposed periodicity. *Appl. Phys. Lett.* **29,** 456–8. [120]

Brekhovskikh, L. M. (1980). *Waves in layered media* (2nd edn). Academic Press, New York. [95, 213, 214]

Brekhovskikh, L. M. and Godin, O. A. (1990). *Acoustics of layered media I: plane and quasi-plane waves.* Springer-Verlag, Berlin. [59, 87, 90, 95, 115, 213, 214]

Briggs, G. A. D. (1984). Scanning electron acoustic microscopy and scanning acoustic microscopy: a favourable comparison. *Scanning Electron Microsc.* **3,** 1041–52. [17]

Briggs, G. A. D. (1985). *An introduction to scanning acoustic microscopy,* Royal Microscopical Society Handbook, no. 12. Oxford University Press, Oxford. [49]

Briggs, G. A. D. (1988). What can you see with a scanning acoustic microscope? (Vas kann man mit dem akustischen Rastermikroskop erfassen?) *Vortragsveranstaltung des Arbeitskreises Rastermikroskopie in der Materialprüfung* **13,** 105–12. [12]

Briggs, G. A. D. (1990*a*). How sensitive is acoustic microscopy? In *Advanced materials and processes* (ed. H. E. Exner and V. Schumacher), pp. 1409–14. DGM Informationsgesellschaft mbH, Oberursel. [12]

Briggs, G. A. D. (1990*b*). *Ultrasound—its chemical, physical and biological effects* (ed. K. S. Suslik). *Interdisciplinary Sci. Rev.* **15,** 190–1. [178]

Briggs, G. A. D. and Hoppe, M. (1991). Acoustic microscopy. In *Images of materials* (ed. D. B. Williams, A. R. Pelton, and R. Gronsky). Oxford University Press, NY. [12, 204]

Briggs, G. A. D., Somekh, M. G., and Ilett, C. (1982). Acoustic microscopy in materials science. In *Microscopy–techniques and capabilities* (ed. L. R. Baker). *SPIE* **368,** 74–80. [131]

Briggs, G. A. D., Rowe, J. M., Sinton, A. M., and Spencer, D. S. (1988). Quantitative methods in acoustic microscopy. *IEEE 1988 Ultrasonics Symposium,* pp. 743–9. IEEE, New York. [146, 147]

Briggs, G. A. D., Daft, C. M. W., Fagan, A. F., Field, T. A., Lawrence, C. W., Montoto, M., Peck, S. D., Rodriguez-Rey, A., and Scruby, C. B. (1989). Acoustic microscopy of old and new materials. In *Acoustical imaging,* Vol. 17 (ed. H. Shimizu, N. Chubachi, and J. Kushibiki), pp. 1–16. Plenum Press, New York. [12, 293]

Briggs, G. A. D., Jenkins P. J., and Hoppe, M. (1990). How fine a surface crack can you see in a scanning acoustic microscope? *J. Microsc.* **159,** 15–32. [277, 280, 282, 292]

Bukhny, M. A., Chernosatonskii, L. A., and Maev, R. G. (1990). Acoustic imaging of high-temperature superconducting materials. *J. Microsc.* **160,** 299–313. [113]

Bulstrode, C. J. K. (1987). Keeping up with orthopaedic epidemics. *Br. Med. J.* **295,** 514. [9]

Burnett, P. J. and Briggs, G. A. D. (1986). The elastic properties of ion-implanted silicon. *J. Mater. Sci.* **21,** 1828–36. [225]

Burnett, P. M., Briggs, G. A. D., Al-Shukri, S. M., Duffey, J. F., and De La Rue, R. M. (1986). Acoustic properties of proton-exchanged $LiNbO_3$ studied using the acoustic microscope $V(z)$ technique. *J. Appl. Phys.* **60,** 2517–22. [249]

Burton, N. J., Thaker, D. M., and Tsukamoto, S. (1985). Recent developments in the practical and industrial applications of scanning acoustic microscopy. *Ultrasonics Int. 85*, 334–8. [204]

Cantrell, J. H. and Qian, M. (1991). Microstress contrast in scanning electron acoustic microscopy of ceramics. In *Review of progress in quantitative nondestructive evaluation,* Vol. 10 (ed. D. O. Thompson and D. E. Chimenti). Plenum Press, New York. [18]

Cantrell, J. H., Qian, M., Ravichandran, M. V., and Knowles, K. M. (1990). Scanning electron acoustic microscopy of indention-induced cracks and residual stress in ceramics. *Appl. Phys. Lett.* **57,** 1870–2. [18]

Cargill, G. S. (1980). Ultrasonic imaging in scanning electron microscopy. *Nature* **286,** 691–3. [18]

Cargill, G. S. (1988). Electron beam-acoustic imaging. In *Physical acoustics XVIII* (ed. W. P. Mason and R. N. Thurston) pp. 125–65. Academic Press, San Diego. [17]

Cauchy, A. L. (1828). Sur l'equilibre et le mouvement d'une plaque solide. *Exercises de Mathématique* **3,** 328. [203]

Cauchy, A. L. (1829). Sur l'equilibre et le mouvement d'une plaque elastique dont l'elasticité n'est pas la même dans tous les sens. *Exercises de Mathématique* **4,** 1. [203]

Chan, K. H. and Bertoni, H. L. (1991). Ray representation of longitudinal lateral waves in acoustic microscopy. *IEEE Trans.* **UFFC 38,** 27–34. [122, 123]

Chimenti, D. E., Nayfeh, A. A., and Butler, D. L. (1982). Leaky Rayleigh waves on a layered half space. *J. Appl. Phys.* **53,** 170–6. [215]

Chiznik, D. (1991). Quantitative acoustic microscopy for measurement of material properties. *Ph.D. thesis,* Polytechnic University of New York. [125]

Chou, C.-H. and Khuri-Yakub, B. T. (1989). Design and implementation of mixed-mode transducers. *IEEE Trans.* **UFFC 36,** 337–41. [58, 72, 153]

Chou, C.-H. and Kino, G. S. (1987). The evaluation of $V(z)$ in a type II reflection microscope. *IEEE Trans.* **UFFC 34,** 341–5. [112]

Chou, C.-H., Khuri-Yakub, B. T., and Liang, K. K. (1987). Acoustic microscopy with shear wave transducers. *IEEE 1987 Ultrasonics Symposium,* pp. 813–16. IEEE, New York. [58, 72]

Chou, C.-H., Khuri-Yakub, B. T., and Kino, G. S. (1988). Lens design for acoustic microscopy. *IEEE Trans.* **UFFC 35,** 464–9. [57]

Christie, S. and Wyatt, A. F. G. (1982). Considerations of contrast in the helium acoustic microscope. In *Acoustical imaging,* Vol. 12 (ed. E. A. Ash and C. R. Hill), pp. 1–11. Plenum Press, New York. [39]

Chubachi, N. (1985). Ultrasonic microspectroscopy via Rayleigh waves. In *Rayleigh-wave theory and applications* (ed. E. A. Ash and E. G. S. Paige), pp. 291–7. Springer-Verlag, Berlin. [154]

Clarke, L. R., Chou, C.-H., Khuri-Yakub, B. T., and Marshall, D. B. (1985). Acoustic evaluation of grinding damage in ceramic materials. *IEEE 1985 Ultrasonics Symposium,* pp. 979–82. IEEE, New York. [276]

Cox, B. N. and Addison, R. C. (1984). Modelling the acoustic material signature in the presence of a surface-breaking crack. In *Review of progress in quantitative nondestructive evaluation* (ed. D. O. Thompson and D. E. Chimenti), pp. 1173–84. Plenum Press, New York. [276]

Crean, G. M., Somekh, M. G., Golanski, A., and Oberlin, J. C. (1987). The influence of thin film microstructure on surface acoustic wave velocity. *IEEE 1987 Ultrasonics Symposium,* pp. 843–7. IEEE, New York. [226]

Crostack, H.-A., Beller, U., Steffens, H.-D., and Reichel, K. (1990). Use of scanning acoustic microscopy to characterize cathodic arc deposited CrN films. In *Advanced materials and processes* (ed. H. E. Exner and V. Schumacher), pp. 1427–30. DGM Informationsgesellschaft mbH, Oberursel. [224]

Daft, C. M. W. and Briggs, G. A. D. (1989*a*). Wideband acoustic microscopy of tissue. *IEEE Trans.* **UFFC 36,** 258–63. [183]

Daft, C. M. W. and Briggs, G. A. D. (1989*b*). The elastic microstructure of various tissues. *J. Acoust. Soc. Am.* **85,** 416–22. [182–5]

Daft, C. M. W., Briggs, G. A. D., and O'Brien, W. D. (1989). Frequency dependence of tissue attenuation measured by acoustic microscopy. *J. Acoust. Soc. Am.* **85,** 2194–201. [178, 186]

Darling, A. I. (1956). Studies of the early lesion of enamel caries with transmitted light, polarised light and microradiography. *Br. Dent. J.* **101,** 289–97, 329–41 (cf. (1958) **105,** 119–35). [194]

Davids, D. A., Chizhik, D., and Bertoni, H. L. (1988). Measured characteristics of an acoustic microscope having a bow-tie transducer. *IEEE 1988 Ultrasonics Symposium,* pp. 763–6. IEEE, New York. [58, 154]

Davies, D. G. (1983). Scanning electron acoustic microscopy. *Scanning Electron Microsc.* **3,** 1163–76. [17]

De Billy, M., Casakany, A., Adler, L., and Quentin, G. (1983). Excitation of backward ultrasonic leaky Rayleigh and Leaky Lamb waves. *IEEE 1983 Ultrasonics Symposium,* pp. 1112–15. IEEE, New York. [120]

Del Grosso, V. A. and Mader, C. W. (1972). Speed of sound in pure water. *J. Acoust. Soc. Am.* **52,** 1442–6. [34, 35]

Derby, B., Briggs, G. A. D., and Wallach, E. R. (1983). Nondestructive testing and acoustic microscopy of diffusion bonds. *J. Mater. Sci.* **18,** 2345–53. [22]

Dicke, R. H. (1946). The measurement of thermal radiation at microwave frequencies. *Rev. Sci. Instrum.* **17,** 268–75. [29]

Dong, R. and Adler, L. (1984). Measurements of reflection and transmission coefficients of Rayleigh waves from cracks. *J. Acoust. Soc. Am.* **76,** 1761–3. [269]

Dransfeld, K. and Salzmann, E. (1970). Excitation, detection and attenuation of high-frequency elastic surface waves. In *Physical acoustics VII* (ed. W. P. Mason and R. N. Thurston), pp. 260–83. Academic Press, New York. [121]

Duquesne, J. Y., Yamanaka, K., Neron, C., Jen, C. K., Piche, L., and Lessard, G. (1989). Study of spherulites in a semi-crystalline polymer using acoustic microscopy. *Mat. Res. Soc. Symp. Proc.* **142,** 253–9. [209]

Fagan, A. F. (1990). Acoustic microscopy of brittle materials. *D.Phil. thesis,* Oxford University. [284, 293]

Fagan, A. F., Bell, J. M., and Briggs, G. A. D. (1989). Acoustic microscopy of polymers and polymer composites. In *Fractography and failure mechanisms of polymers and composites* (ed. A. C. Roulin-Moloney), pp. 213–30. Elsevier Applied Science, London. [209]

Fagan, A. F., Briggs, G. A. D., Czernuszka, J. T., and Scruby, C. B. (1990). Micro-structural observations of a deformed PSZ ceramic using acoustic microscopy. *Trans. R. Microsc. Soc.* **1,** 81–4. [281]

Fagan, A. F., Briggs, G. A. D., Czernuszka, J. T., and Scruby, C. B. (1991). Microstructural observations of two deformed PSZ ceramics using acoustic microscopy. *J. Mater. Sci.* (In press.) [281]

Faridian, F. and Somekh, M. G. (1986). Frequency modulation techniques in acoustic microscopy. *IEEE 1986 Ultrasonics Symposium,* pp. 769–73. IEEE, New York. [23, 74]

Farnell, G. W. (1970). Properties of elastic surface waves. In *Physical acoustics VI* (ed. W. P. Mason and R. N. Thurston), pp. 109–66. Academic Press, New York. [240]

Farnell, G. W. (1978). Types and properties of surface waves. In *Acoustic surface waves* (ed. A. A. Oliner), pp. 13–60. Springer Verlag, Berlin. [240]

Farnell, G. W. and Adler, E. L. (1972). Elastic wave propagation in thin layers. In *Physical acoustics IX* (ed. W. P. Mason and R. N. Thurston), pp. 35–127. Academic Press, New York. [218]

Fatkin, D. G. P., Scruby, C. B., and Briggs, G. A. D. (1989). Acoustic microscopy of low ductility materials. *J. Mater. Sci.* **24,** 23–40. [12, 291]

Forsgren, P.-O. (1990). Visualization and coding in three-dimensional image processing. *J. Microsc.* **159,** 195–202. [204]

Fossheim, K., Bye, T., Sathish, S., and Heggum, G. (1988). Acoustic scanning microscopy of grain structure in isotropic solids: pure aluminium and Al–2.5%Mg alloy. *J. Mater. Sci.* **23,** 1748–51. [253]

Foster, J. S. (1984). High resolution acoustic microscope in superfluid helium.

Proceedings of the 17th International Conference on Low Temperature Physics. Physica **126B,** 199–205. [41]

Foster, J. S. and Putterman, S. (1985). Parametric self-enhancement of the spontaneous decay of sound in superfluid helium. *Phys. Rev. Lett.* **54,** 1810–13. [38]

Foster, J. S. and Rugar, D. (1983). High resolution acoustic microscopy in superfluid helium. *Appl. Phys. Lett.* **42,** 869–71. [74]

Foster, J. S. and Rugar, D. (1985). Low-temperature acoustic microscopy. *IEEE Trans.* **SU-32,** 139–51. [32, 36–8, Fig. 3.3, 42, 45]

Fright, W. R., Bates, R. H. T., Rowe, J. M., Spencer, D. S., Somekh, M. G., and Briggs, G. A. D. (1989). Reconstruction of the complex reflectance function in acoustic microscopy. *J. Microsc.* **153,** 103–17. [112, 132, 134]

Gerchberg, R. W. and Saxton, W. O. (1972). A practical algorithm for the determination of phase from image and diffraction plane pictures. *Optik* **35,** 237–46. [132]

Gilmore, R. S., Tam, K. C., Young, J. D. and Howard, D. R. (1986). Acoustic microscopy from 10 to 100 MHz for industrial applications. *Phil. Trans. R. Soc. Lond.* **A320,** 215–35. [204]

Granato, A. V. and Lücke, K. (1966). The vibrating string model of dislocation damping. In *Physical acoustics IV-A* (ed. W. P. Mason), pp. 225–76. Academic Press, London. [257]

Gremaud, G., Kulik, A., and Sathish, S. (1990). Mechanical properties of surfaces and layers measured with a continuous wave acoustic microscope. In *Advanced materials and processes* (ed. H. E. Exner and V. Schumacher), pp. 1421–6. DGM Informationsgesellschaft mbH, Oberursel. [23]

Hadimioglu, B. and Foster, J. S. (1984). Advances in superfluid helium acoustic microscopy. *J. Appl. Phys.* **56,** 1976–80. [38]

Hadimioglu, B. and Quate, C. F. (1983). Water acoustic microscopy at suboptical wavelengths. *Appl. Phys. Lett.* **43,** 1006–7. [31, 45, 46]

Hall, L. (1948). The origin of ultrasonic absorption in water. *Phys. Rev.* **73,** 775–81. [81]

Hammer, R. and Hollis, R. L. (1982). Enhancing micrographs obtained with a scanning acoustic microscope using false-colour encoding. *Appl. Phys. Lett.* **40,** 678–80. [41]

Hecht, E. (1987). *Optics.* Addison-Wesley, Reading, Massachusetts. [13, 15, 28, 51, 52, 56, 65]

Heiserman, J. (1980). Cryogenic acoustic microscopy. In *Scanned image microscopy* (ed. E. A. Ash), pp. 71–96. Academic Press, London. [32, 36]

Heiserman, J., Rugar, D., and Quate, C. F. (1980). Cryogenic acoustic microscopy. *J. Acoust. Soc. Am.* **67,** 1629–37. [36]

Henderson, B. C. (1990). Mixers in microwave systems. *Watkins-Johnson Company Tech-notes* **17**(1), 1–15; **17**(2), 1–13. [69]

Heygster, G., Block, H., Gadomski, A., and Boseck, S. (1990). Modeling of the optical transfer function (OTF) of the scanning acoustic microscope (SAM) and its relation to the other scanning microscopes. *Optik* **85,** 89–98. [29, 204]

Hildebrand, J. A. and Lam, L. (1983). Directional acoustic microscopy for observation of elastic anisotropy. *Appl. Phys. Lett.* **42,** 413–15. [253]

Hildebrand, J. A. and Rugar, D. (1984). Measurement of cellular elastic properties by acoustic microscopy. *J. Microsc.* **134,** 245–60. [170, 171, 176]

Hildebrand, J. A., Rugar, D., Johnston, R. N., and Quate, C. F. (1981). Acoustic microscopy of living cells. *Proc. Natl. Acad. Sci. USA* **78,** 1656–60. [166]

Hildebrand, J. A., Liang, K. K., and Bennett, S. D. (1983). Fourier transform approach to materials characterization with the acoustic microscope. *J. Appl. Phys.* **54,** 7016–19. [112]

Hollis, R. L., Hammer, R., and Al-Jaroudi, M. Y. (1984). Subsurface imaging of glass fibres in a polycarbonate composite by acoustic microscopy. *J. Mater. Sci.* **19,** 1897–903. [207]

Honda, T. and Ohashi, K. (1990). A new approach for corrosion monitoring by a scanning acoustic microscope. *Proceedings of the 11th Corrosion Conference,* Florence. (In press.) [212]

Hoppe, M. and Bereiter-Hahn, J. (1985). Applications of scanning acoustic microscopy—survey and new aspects. *IEEE Trans.* **SU-32,** 289–301. [12, 26, 207, 224]

Howard, A. M. (1990). Evaluation of large die attach using acoustic microscopy. *IEE Colloquium on 'NDT Evaluation of Electronic Components and Assemblies'*, p. 3. [204]

Howard, J. N. (1985). Some sketches of Rayleigh. In *Rayleigh-wave theory and applications* (ed. E. A. Ash and E. G. S. Paige), pp. 2–9. Springer-Verlag, Berlin. [127]

Howie, A. (1987). An introduction to scanning acoustic microscopy. *Proc. R. Microsc. Soc.* **22,** 79. [165, 187]

Hung, B.-N. and Goldstein, A. (1983). Acoustic parameters of commercial plastics. *IEEE Trans.* **SU–30,** 249–54. [103]

Husson, D. (1985). A perturbation theory for the acoustoelastic effect of surface waves. *J. Appl. Phys.* **57,** 1562–8. [154]

Husson, D. and Kino, G. S. (1982). A perturbation theory for acoustoelastic effects. *J. Appl. Phys.* **53,** 7250–8. [154]

Huxley, A. (1990). 150th commemorative meeting: opening by Sir Andrew Huxley. *Proc. R. Microsc. Soc.* **25,** 94–5. [165]

Ilett, C., Somekh, M. G., and Briggs, G. A. D. (1984). Acoustic microscopy of elastic discontinuities. *Proc. R. Soc. Lond.* **A393,** 171–83. [112, 224, 260, 261, 263, 274, 281]

Ingel, R. P. and Lewis, D. (1988*a*). Errors in elastic constant measurements in single crystals. *J. Am. Ceram. Soc.* **71,** 261–4. [238]

Ingel, R. P. and Lewis, D. (1988*b*). Elastic anisotropy in zirconia single crystals. *J. Am. Ceram. Soc.* **71,** 265–71. [238, 286]

Ishikawa, I., Semba, T., Kanda, H., Katakura, K., Tani, Y., and Sato, H. (1989). Experimental observation of plastic deformation areas, using an acoustic microscope. *IEEE Trans.* **UFFC 36,** 274–9. [258]

Ishikawa, I., Semba, T., Tani, Y., and Sato, H. (1990). Development of anisotropic acoustic lenses and applications to material evaluation. *Spring Convention Academic Lecture Meeting: Reports of 1990 Precision Engineering Association,* p. 1111. [59, 154, 253, 259]

Jen, C. K. (1989). The role of acoustic properties in designs of acoustic and optical fibres. *Mater. Sci. Engng.* **A122,** 1–8. [227, 249]

Jen, C. K., Screenivas, K., and Sayer, M. (1988). Ultrasonic transducers for simultaneous generation of longitudinal and shear waves. *J. Acoust. Soc. Am.* **84,** 26–9. [72]

Jen, C. K., Neron, C., Bussiere, J. F., Abe, K., Li, L., Lowe, R., and Kushibiki, J. (1989). Acoustic microscopy of cladded optical fibres. *IEEE 1989 Ultrasonics Symposium,* pp. 831–5. IEEE, New York. [226]

Jenkins, P. J. (1990) Scanning acoustic microscopy of persistent slip bands. In *EMAG-MICRO '89* (ed. H. Y. Elder and P. J. Goodhew), *Inst. Phys. Conf. Ser.* **98,** pp. 153–6. Institute of Physics, Bristol. [288]

Jenkins, F. A. and White, H. E. (1976). *Fundamentals of optics.* McGraw-Hill, New York. [13, 51]

Jipson, V. B. (1979). Acoustic microscopy of interior planes. *Appl. Phys. Lett.* **35,** 385–7. [33]

Jipson, V. B. and Quate, C. F. (1978). Acoustic microscopy at optical wavelengths. *Appl. Phys. Lett.* **32,** 789–91. [26]

Katz, J. L. (1971). Hard tissue as a composite material—1. Bounds on elastic behaviour. *J. Biomechanics* **4,** 455–73. [198]

Katz, J. L. and Ukraincik, K. (1971). On the anisotropic properties of hydroxyapatite. *J. Biomechanics* **4,** 221–7. [198]

Kaye, G. W. C. and Laby, T. H. (1986). *Tables of physical and chemical constants.* Longman, London. [32–5, 103]

Kessler, L. W. (1985). Acoustic microscopy commentary: SLAM and SAM. *IEEE Trans.* **SU–32,** 136–8. [18]

Kessler, L. W. (1988). Acoustic microscopy—an industrial view. *IEEE 1988 Ultrasonics Symposium,* pp. 725–8. IEEE, New York. [18]

Kessler, L. W. and Yuhas, D. E. (1979). Acoustic microscopy—1979. *Proc. IEEE* **67,** 526–36. [Reprinted in Lee, H, and Wade, G. (1986). *Modern acoustical imaging* pp. 169–79. IEEE, New York]. [18]

Keyes, R. W. (1982). Device implications of the electronic effect in the elastic constants of silicon. *IEEE Trans.* **SU-29,** 99–103. [226]

Khuri-Yakub, B. T. and Chou, C.-H. (1986). Acoustic microscope lenses with shear wave transducers. *IEEE 1986 Ultrasonics Symposium,* pp. 741–4. IEEE, New York. [72]

Khuri-Yakub, B. T., Reinholdtsen, P., and Chou, C.-H. (1985). Acoustic imaging of subsurface defects in composites and samples with rough surfaces. *IEEE 1985 Ultrasonics Symposium,* pp. 746–9. IEEE, New York. [210]

Kino, G. S. (1980). Fundamentals of scanning systems. In *Scanned image microscopy* (ed. E. A. Ash), pp. 1–21. Academic Press, London. [28]

Kino, G. S. (1987). *Acoustic waves: devices, imaging and analog signal processing.* Prentice-Hall, Englewood Cliffs, New Jersey. [28, 57, 87, 159]

Kojima, S. (1987). Interference fringes in reflection acoustic microscopy. *Jap. J. Appl. Phys.* **26** (Suppl. 26–1), 233–5. [260]

Kolodziejczyk, E., Saurel, J. M., Fernandez-Graf, M. R., Attal, J., Fryder, V., Saied, A., Dillmann, M. L., and Rey, V. (1988). Acoustic microstructure of mature soya bean (Maple Arrow). *J. Microsc.* **150,** 57–64. [181]

Kolosov, O. V., Levin, V. M., Maev, R. G., and Senjushkina, T. A. (1987). The use of acoustic microscopy for biological tissue characterization. *Ultrasound Med. Biol.* **13,** 477–83. [179]

Kulik, A., Gremaud, G., and Satish, S. (1989). Continuous wave reflection scanning acoustic microscope (SAMCRUW). In *Acoustical imaging,* Vol. 17, (ed. H. Shimizu, N. Chubachi, and J. Kushibiki), pp. 71–8. Plenum Press, New York. [23, 75]

Kulik, A., Gremaud, G., and Satish, S. (1990). Acoustic microscopy as a polyvalent tool in materials science. *Trans. R. Microsc. Soc.* **1,** 85–90. [75, 204]

Kushibiki, J. and Chubachi, N. (1985). Material characterization by line-focus-beam acoustic microscope. *IEEE Trans.* **SU-32,** 189–212. [71, 82, 138, 141, 143, 150, 151, 194, 216]

Kushibiki, J. and Chubachi, N. (1987). Application of LFB acoustic microscope to film thickness measurements. *Electron. Lett.* **23,** 652–4. [151, 196, 221, 222]

Kushibiki, J., Sannomiya, T., and Chubachi, N. (1980). Performance of sputtered SiO_2 film as an acoustic antireflection coating at sapphire/water interface. *Electron. Lett.* **16,** 737–8. [60]

Kushibiki, J., Maehara, H., and Chubachi, N. (1981*a*). Acoustic properties of evaporated chalcogenide glass films. *Electron.* Lett. **17,** 322–3. [60]

Kushibiki, J., Ohkubo, A., and Chubachi, N. (1981*b*). Linearly focused acoustic beams for acoustic microscopy. *Electron. Lett.* **17,** 520–2. [58, 137]

Kushibiki, J., Ohkubo, A., and Chubachi, N. (1982). Effect of leaky SAW parameters on $V(z)$ curves obtained by acoustic microscopy. *Electron. Lett.* **18,** 668–70. [136]

Kushibiki, J., Horii, K., and Chubachi, N. (1983). Velocity measurement of multiple leaky waves on germanium by line-focus-beam acoustic microscope using FFT. *Electron. Lett.* **19,** 404–5. [247]

Kushibiki, J., Matsumoto, Y., Satake, M., and Chubachi, N. (1985). Non-destructive evaluation of acoustic inhomogeneity on wafers by line-focus-beam acoustic microscope. *Ultrasonics Int. 85*, 809–14. [249]

Kushibiki, J., Asano, H., Ueda, T., and Chubachi, N. (1986). Application of line-focus-beam acoustic microscope to inhomogeneity detection on SAW device materials. *IEEE 1986 Ultrasonics Symposium,* pp. 749–53. IEEE, New York. [249]

Kushibiki, J., Ha, K. L., Kato, H., Chubachi, N., and Dunn, F. (1987). Application of acoustic microscopy to dental material characterization. *IEEE 1987 Ultrasonics Symposium,* pp. 837–42. IEEE, New York. [196]

Kushibiki, J., Chubachi, N., and Tejima, E. (1989). Quantitative evaluation of materials by directional acoustic microscope. *Ultrasonics Int. 89*, 736–43. [59, 154, 253]

Kushibiki, J., Ishikawa, T., and Chubachi, N. (1990). Cut-off characteristics of leaky Sezawa and pseudo-Sezawa wave modes for thin-film characterization. *Appl. Phys. Lett.* **57,** 1967–9. [223]

Kushibiki, J., Takahashi, H., Kobayashi, T., and Chubachi, N. (1991*a*). Quantitative evaluation of elastic properties of $LiTaO_3$ crystals by line-focus-beam acoustic microscopy. *Appl. Phys. Lett.* **58,** 893–5. [249]

Kushibiki, J., Takahashi, H., Kobayashi, T., and Chubachi, N. (1991*b*). Characterization of $LiNbO_3$ crystals by line-focus-beam acoustic microscopy. *Appl. Phys. Lett.* **58,** 2622–4. [249]

Landau, L. D. and Lifshitz, E. M. (1970). *Theory of elasticity*. Pergamon, Oxford. [82]

Landolt, H. and Börnstein, R. (1979, 1984, 1987). *Numerical data and functional relationships in science and technology* (ed. O. Madelung) III, Vols 11, 18, 22. Springer-Verlag, Berlin. [238]

Lawrence, C. W. (1990). Acoustic microscopy of ceramic fibre composites. *D.Phil. thesis,* Oxford University. [3, 4, 227–31, 288, 289]

Lawrence, C. W., Scruby, C. B., and Briggs, G. A. D. (1989). A study of ceramic fibre composites by acoustic microscopy. In *EMAG-MICRO '89* (ed. H. Y. Elder and P. J. Goodhew), *Inst. Phys. Conf. Ser.* **98,** pp. 139–42. Institute of Bristol. [1]

Lawrence, C. W., Scruby, C. B., Briggs, G. A. D., and Dunhill, A. (1990). Crack detection in silicon nitride by acoustic microscopy. *NDT Int.* **23,** 3–10. [1]

Lee, C. C., Tsai, C. S., and Cheng, X. (1985). Complete characterization of thin- and thick-film materials using wideband reflection acoustic microscopy. *IEEE Trans.* **SU–32,** 248–58. [212]

Lee, C. C., Lahham, M., and Martin, B. G. (1990). Experimental verification of the Kramers–Kronig relationship for acoustic waves. *IEEE Trans.* **UFFC 37,** 286–94. [82]

Lees, S. and Rollins, F. R. (1972). Anisotropy in hard tissues. *J. Biomechanics* **5,** 557–66. [198]

Lemons, R. A. and Quate, C. F. (1974). Acoustic microscope—scanning version. *Appl. Phys. Lett.* **24,** 163–5. [19, 23]

Lemons, R. A. and Quate, C. F. (1979). Acoustic microscopy. In *Physical acoustics XIV* (ed. W. P. Mason and R. N. Thurston), pp. 1–92. Academic Press, London. [16, 19, 21, 22, 32, 36, 165]

Levin, V. M., Maev, R. G., Kolosov, O. B., Senjushkina, T. A., and Bukhny, M. A. (1990). Theoretical fundamentals of quantitative acoustic microscopy. *Acta Phys. Slov.* **40,** 171–84. [113]

Liang, K. K., Bennett, S. D., Khuri-Yakub, B. T., and Kino, G. S. (1982). Precision measurement of Rayleigh wave velocity perturbation. *Appl. Phys. Lett.* **41,** 1124–6. [144]

Liang, K. K., Bennett, S. D., Khuri-Yakub, B. T., and Kino, G. S. (1985*a*). Precise phase measurements with the acoustic microscope. *IEEE Trans.* **SU–32,** 266–73. [68]

Liang, K. K., Kino, G. S., and Khuri-Yakub, B. (1985*b*). Material characterisation by the inversion of $V(z)$. *IEEE Trans.* **SU–32,** 213–24. [112, 129, 130, 182, 219, 221]

Liang, K. K., Benettt, S. D., and Kino, G. S. (1986). Precision phase measurements with short tone burst signals in acoustic microscopy. *Rev. Sci. Instrum.* **57,** 446–52. [68, 72]

Lin, Z. C., Lee, H., and Wade, G. (1985). Scanning tomographic acoustic microscope: a review. *IEEE Trans.* **SU–32,** 168–80. [18]

Lindsay, R. B. (1982). Relaxation processes in sound propagation in fluids: a historical survey. In *Physical acoustics XVI* (ed. W. P. Mason and R. N. Thurston), pp. 1–36. Academic Press, London. [80, 81]

Litniewski, J. and Bereiter-Hahn, J. (1990). Measurements of cells in culture by scanning acoustic microscopy. *J. Microsc.* **158,** 95–107. [172, 174, 176]

Maddox, J. (1985). New ways with microscopes. *Nature* **315,** 177. [13]

Maeda, K., Murao, F., Yoshiga, T., Yamauchi, C., and Tsuzaki, T. (1986). Experimental studies on the suppression of cultured cell growth curves after irradiation with CW and pulsed ultrasound. *IEEE Trans.* **UFFC 33,** 186–93. [178]

Maev, R. G. and Levin, V. M. (1990). Basic principles of output signal formation in transmission raster acoustic microscopy. *Trans. R. Microsc. Soc.* **1,** 75–80. [113]

Maev, R. G. and Maslov, K. I. (1991). Temperature effects in the focal region of the acoustic microscope. *IEEE Trans.* **UFFC 38,** 166–71. [179]

Marshall, D. B., Evans, A. G., Khuri-Yakub, B. T., Tien, J. W., and Kino, G. S. (1983). The nature of machining damage in brittle materials. *Proc. R. Soc. Lond.* **A385,** 461–75. [276]

Martin, S. J. and Frye, G. C. (1990). Surface acoustic wave response to changes in viscoelastic film properties. *Appl. Phys. Lett.* **57,** 1867–9. [223]

Matthaei, E., Vetters, H., and Mayr, P. (1990). Reflective scanning acoustic microscopy for imaging subsurface structures in solid state materials. In *Advanced materials and processes* (ed. H. E. Exner and V. Schumacher), pp. 1415–20. DGM Informationsgesellschaft mbH, Oberursel. [204, 224]

Meckel, A. H., Griebstein, W. J., and Neal, R. J. (1965). Structure of human dental enamel as observed by electron microscopy. *Arch. Oral Biol.* **10,** 775–83. [196]

Meeks, S. W., Peter, D., Horne, D., Young, K., and Novotny, V. (1989). Microscopic imaging of residual stress using a scanning phase-measuring acoustic microscope. *Appl. Phys. Lett.* **55,** 1835–7. [Cf. Residual stress mapping with a scanning phase-measuring microscope, *IEEE 1989 Ultrasonics Symposium,* pp. 809–12. IEEE, New York.] [73, 154, Fig. 8.9]

Meunier, A., Katz, J. L., Christel, P., and Sedel, L. (1988). A reflection scanning acoustic microscope for bone and bone-biomaterials interface studies. *J. Orthopaedic Res.* **6,** 770–5. [202]

Miller, A. J. (1983). Aspects of SAM imaging of semiconductor devices. *Inst. Phys. Conf. Ser.* **67,** 393–8. [225]

Miller, A. J. (1985). Scanning acoustic microscopy in electronics research. *IEEE Trans.* **SU–32,** 320–4. [225]

Morse, P. M. and Ingard, K. U. (1987). *Theoretical acoustics.* Princeton University Press, Princeton, New Jersey. [80, 283]

Muha, M. S., Moulthrop, A. A., Kozlowski, G. C., and Hadimioglu, B. (1990). Acoustic microscopy at 15.3 GHz in pressurized superfluid helium. *Appl. Phys. Lett.* **56,** 1019–21. [40, 41]

Nagy, P. B. and Adler, L. (1989). On the origin of increased backward radiation from a liquid–solid interface at the Rayleigh angle. *J. Acoust. Soc. Am.* **85,** 1355–7. [120]

Narita, T. Miura, K., Ishikawa, I., and Ishikawa, T. (1990). Measurement of residual thermal stress and its distribution on silicon nitride ceramics joined to metals with scanning acoustic microscopy. *J. Japan. Inst. Metals* **54,** 1142–6. [153]

Negishi, K. and Ri, H. U. (1987). Propagation of multi-mode ultrasonic pulses in non-destructive material evaluation. In *Ultrasonic spectroscopy and its application to materials Science* (Ed. Y. Wada), pp. 70–4. Ministry of Education, Science, and Culture, Japan. [104, 106]

Nikoonahad, M. (1987). Differential amplitude contrast in acoustic microscopy. *Appl. Phys. Lett.* **51,** 1687–9. [72]

Nikoonahad, M. and Liu, D. C. (1990). Pulse-echo single frequency acoustic nonlinearity parameter (B/A) measurement. *IEEE Trans.* **UFFC 37,** 127–34. [45, 186]

Nikoonahad, M. and Sivers, E. A. (1989). Dual beam differential amplitude contrast scanning acoustic microscopy. In *Acoustical imaging,* Vol. 17 (ed. H. Shimizu, N. Chubachi, and J. Kushibiki), pp. 17–25. Plenum Press, New York. [72]

Nikoonahad, M., Yue, G.-Q., and Ash, E. A. (1983). Subsurface broadband acoustic microscopy of solids using reduced aperture lenses. In *Review of progress in quantitative nondestructive evaluation,* Vol. 2 (ed. D. O. Thompson and D. E. Chimenti), pp. 1611–23. Plenum Press, New York. [207]

Nikoonahad, M., Yue, G.-Q., and Ash, E. A. (1985). Pulse compression acoustic microscopy using SAW filters. *IEEE Trans.* **SU–32,** 152–63. [73, 74]

Obata, M., Shimada, H., and Mihara, T. (1990). Stress dependence of leaky surface wave on PMMA by line-focus-beam acoustic microscope. *Exp. Mechanics* **30,** 34–9. [153]

Okawai, H., Tanaka, M., Chubachi, N., and Kushibiki, J. (1987). Non-contact simultaneous measurement of thickness and acoustic properties of biological tissue using focused wave in a scanning acoustic microscope. Proc 7th Symp. Ultrasonic Electronics, Kyoto. *Jap. J. Appl. Phys.* **26** (Suppl. 26–1), 52–4. [169]

Oppenheim, A. V. and Schafer, R. W. (1975). *Digital signal processing.* Prentice-Hall, New Jersey. [160]

Pao, Y. H., Sachse, W., and Fukuoka, H. (1984). Acoustoelasticity and ultrasonic measurements of residual stress. In *Physical acoustics XVII* (ed. W. P. Mason and R. N. Thurston), pp. 61–143. Academic Press, New York. [152]

Papadakis, E. P. (1968). Ultrasonic attenuation caused by scattering in polycrystalline media. In *Physical acoustics IV-B* (ed. W. P. Mason), pp. 269–328. Academic Press, New York. [82, 203]

Parmon, W. and Bertoni, H. L. (1979). Ray interpretation of the material signature in the acoustic microscope. *Electron. Lett.* **15,** 684–6. [108]

Peck, S. D. and Briggs, G. A. D. (1986). A scanning acoustic microscope study of the small caries lesion in human enamel. *Caries Res.* **20,** 356–60. [187, 188]

Peck, S. D. and Briggs, G. A. D. (1987). The caries lesion under the scanning acoustic microscope. *Adv. Dent. Res.* **1,** 50–63. [49, 188–90]

Peck, S. D., Rowe, J. M., and Briggs, G. A. D. (1989). Studies on sound and carious enamel with the quantitative acoustic microscope. *J. Dent. Res.* **68,** 107–12. [192, 193, 195, 197]

Pinkerton, J. M. M. (1949). The absorption of ultrasonic waves in liquids and its relation to molecular constitution. *Proc. Phys. Soc.* **62,** 129–41. [81]

Pino, F., Sinclair, D. A., and Ash, E. A. (1981). New technique for sub-surface imaging using scanning acoustic microscopy. *Ultrasonics Int. 81*, 193–8. [54]

Poisson, S. P. (1829). Mémoire sur l'equilibre et le mouvement des corps elastiques. *Mém. Acad. Sci. Ser.* **28,** 357. [203]

Press, W. H., Flannery, B. P., Teukolsky, S. A., and Vetterling, W. T. (1986). *Numerical recipes.* Cambridge University Press, Cambridge. [159, 163]

Quate, C. F. (1980). Microwaves, acoustics and scanning microscopy. In *Scanned Image Microscopy.* (ed. E. A. Ash), pp. 23–55. Academic Press, London. [109]

Quate, C. F. (1985). Acoustic microscopy: recollections. *IEEE Trans.* **SU–32,** 132–5. [13]

Quate, C. F., Atalar, A., and Wickramasinghe, H. K. (1979). Acoustic microscopy with mechanical scanning: a review. *Proc. IEEE* **67,** 1092–114.

[Reprinted in Lee, H. and Wade, G. (1986). *Modern acoustical imaging* pp. 180–202. IEEE, New York. [112]
Quinten, A. and Arnold, W. (1989). Observation of stable crack growth in Al_2O_3 ceramics using a scanning acoustic microscope. *Mat. Sci. Engng.* **A122,** 15–19. [280]
Quinten, A., Sklarczyk, C., and Arnold, W. (1991). Observation of stable crack growth in Al_2O_3-ceramics by acoustic microscopy and acoustic emission. (In press.) [280]
Rayleigh, Lord (1879). Investigations in optics, with special reference to the spectroscope. *Phil. Mag.* **VIII,** 261–74. [Reprinted in *Scientific papers by Lord Rayleigh,* Vol. III Cambridge University Press, (1902) and Dover Publications, New York (1964).] [27]
Rayleigh, Lord (1885). On waves propagated along the plane surface of an elastic solid. *Proc. Lond. Math. Soc.* **17,** 4–11. [Reprinted in *Scientific papers by Lord Rayleigh,* Vol. IV. Cambridge University Press (1903) and Dover Publications, New York (1964).] [87, 104]
Rayleigh, Lord (1889). On the free vibrations of an infinite plate of homogeneous isotropic elastic matter. *Proc. Lond. Math. Soc.* **20,** 225–34. [Reprinted in *Scientific papers by Lord Rayleigh,* Vol. V. Cambridge University Press (1912) and Dover Publications, New York (1964).] [203]
Reinholdtsen, P. and Khuri-Yakub, B. T. (1986). Removing the effects of surface roughness in acoustic microscopy. *IEEE 1986 Ultrasonics Symposium,* pp. 759–63. IEEE, New York. [210]
Reinholdtsen, P. and Khuri-Yakub, B. T. (1991). Image processing for a scanning acoustic microscope that measures amplitude and phase. *IEEE Trans.* **UFFC 38,** 141–7. [210]
Revay, L., Lindblad, G., and Lind, L. (1990). IC package defects revealed by scanning acoustic microscopy (SAM). *CERT '90; Components engineering, Reliability and Test Conference* (*Electron. Components Inst.*), pp. 115–22. [204]
Robinson, F. N. H. (1974). *Noise and fluctuations in electronic devices and circuits.* Clarendon Press, Oxford. [29]
Rodriguez-Rey, A., Briggs, G. A. D., Field, T. A., and Montoto, M. (1990). Acoustic microscopy of rocks. *J. Microsc.* **160,** 21–9. [5, 6, Fig. 1.4]
Routh, H. F., Pusateri, T. L., and Nikoonahad, M. (1989). Differential phase contrast acoustic microscopy. *IEEE 1989 Ultrasonics Symposium,* pp. 817–20. IEEE, New York. [72]
Rowe, J. M. (1988). Quantitative acoustic microscopy of surfaces. *D.Phil. thesis,* Oxford University. [146]
Rowe, J. M., Kushibiki, J., Somekh, M. G., and Briggs, G. A. D. (1986). Acoustic microscopy of surface cracks. *Phil. Trans. R. Soc. Lond.* **A320,** 201–14. [269, 287]
Rugar, D. (1981). Cryogenic acoustic microscopy. *Ph.D. thesis,* Ginzton Lab. Report 3370, Stanford University. [60]
Rugar, D. (1984). Resolution beyond the diffraction limit in the acoustic microscope: a nonlinear effect. *J. Appl. Phys.* **56,** 1338–46. [43]
Rugar, D., Foster, J. S., and Heiserman, J. (1982). Acoustic microscopy at temperatures less than 0.2 K. In *Acoustical imaging,* Vol. 12 (ed. E. A. Ash and C. R. Hill), pp. 13–25. Plenum Press, New York. [37]
Sayers, C. M. (1985) Angular dependence of the Rayleigh wave velocity in

polycrystalline metals with small anisotropy. *Proc. R. Soc. Lond.* **A400,** 175–82. [250]

Schenk, E. A., Waag, R. W., Schenk, A. B., and Aubuchon, J. P. (1988). Acoustic microscopy of red blood cells. *J. Histochem. Cytochem.* **36,** 1341–51. [169]

Schmid, H. G. (1987). The mechanisms of fracture of WC–11wt%Co between 20°C and 1000°C. *Mater. Forum* **10,** 184–97. [294]

Schmid, H. G., Mari, D., Benoit, W., and Bonjour, C. (1987). Creep and fracture of WC–11wt%Co. In *Creep and fracture of engineering materials and structures* (ed. B. Wilshire and R. W. Evans), pp. 975–88. Institute of Metals, London. [294]

Scruby, C. B., Jones, K. R., and Antoniazzi, L. (1987). Diffraction of elastic waves by defects in plates. *J. NDE* **5,** 145–56. [90]

Scruby, C. B., Lawrence, C. W., Fatkin, D. G. P., Briggs, G. A. D., Dunhill, A., Gee, A. E., and Chao, C.-L. (1989). Non-destructive testing of ceramics by acoustic microscopy. *Br. Ceram. Trans. J.* **88,** 127–32. [1]

Selfridge, A. R. (1985). Approximate material properties in isotropic materials. *IEEE Trans.* **SU–32,** 381–94. [32, 103]

Sheppard, C. J. R. and Cogswell, C. J. (1990). 3-D image formation in confocal microscopy. *J. Microsc.* **159,** 179–94. [182, 204]

Sheppard, C. J. R. and Wilson, T. (1981). Effects of high angles of convergence on $V(z)$ in the scanning acoustic microscopy. *Appl. Phys. Lett.* **38,** 858–9. [57, 112, 204]

Sherar, M. D. and Foster, F. S. (1988). Ultrasound backscatter microscopy. *IEEE 1988 Ultrasonics Symposium,* pp. 959–66. IEEE, New York. [181]

Sherar, M. D., Noss, M. B., and Foster, F. S. (1987). Ultrasound backscatter microscopy images the internal structure of living tumour spheroids. *Nature* **330,** 493–5. [181]

Shimada, H. (1987). Propagation of multi-mode ultrasonic pulses in non-destructive material evaluation. In *Ultrasonic spectroscopy and its application to Materials science* (ed. Y. Wada), pp. 50–6. Ministry of Education, Science and Culture, Japan. [152]

Shotton, D. M. (1989). Confocal scanning optical microscopy and its applications for biological specimens. *J. Cell. Sci.* **94,** 175–206. [182, 204]

Silk, M. G. (1987). Changes in ultrasonic defect location and sizing. *NDT Int.* **20,** 9–14. [294]

Sinton, A. M., Briggs, G. A. D., and Tsukahara, Y. (1989). Time-resolved acoustic microscopy of polymer coatings. In *Acoustical imaging,* Vol. 17 (ed. H. Shimizu, N. Chubachi, and J. Kushibiki), pp. 87–95. Plenum Press, New York. [210–12, 219]

Smith, I. R. and Wickramasinghe, H. K. (1982). Dichromatic differential phase contrast microscopy. *IEEE Trans.* **SU–29,** 321–6. [Reprinted in Lee, H. and Wade, G. (1986). *Modern acoustical imaging,* pp. 209–14. IEEE, New York.] [72]

Smith, I. R., Wickramasinghe, H. K., Farnell, G. W., and Jen, C. K. (1983). Confocal surface acoustic wave microscopy. *Appl. Phys. Lett.* **42,** 411–13. [49]

Smith, I. R., Harvey, R. A., and Fathers, D. J. (1985). An acoustic microscope for industrial applications. *IEEE Trans.* **SU–32,** 274–88. [204]

Somekh, M. G. (1987). Consequences of resonant surface-wave excitation on contrast in reflection scanning acoustic microscope. *IEEE Proc.* **134,** 290–300. [126]

Somekh, M. G., Briggs, G. A. D., and Ilett, C. (1984). The effect of anisotropy on contrast in the scanning acoustic microscope. *Phil. Mag.* **A49,** 179–204. [243, 254, 256]

Somekh, M. G., Bertoni, H. L., Briggs, G. A. D., and Burton, N. J. (1985). A two-dimensional imaging theory of surface discontinuities with the scanning acoustic microscope. *Proc. R. Soc. Lond.* **A401,** 29–51. [264, 285]

Stanke, F. E. and Kino, G. S. (1984). Unified theory for elastic wave propagation in polycrystalline materials. *J. Acoust. Soc. Am.* **75,** 665–81. [82, 203]

Suslik, K. S. (ed.) (1988). *Ultrasound—its chemical, physical and biological effects.* VCH, New York. [178]

Tamir, T. (1972). Inhomogeneous wave types at planar structures: I. The lateral wave. *Optik* **36,** 209–32. [122]

Tan, M. R. T., Ransom, H. L., Cutler, C. C., and Chodorow, M. (1985). Oblique, off-specular, linear and nonlinear observations with a scanning micron wavelength acoustic microscope. *J. Appl. Phys.* **57,** 4931–35. [44]

Tanaka, M. (1989). Usefulness of ultrasonic imaging in the medical field. In *Acoustical imaging,* Vol. 17 (ed. H. Shimizu, N. Chubachi, and J. Kushibiki), pp. 453–66. Plenum Press, New York. [169]

Ter Haar, G. R., Daniels, S., and Morton, K. (1986). Evidence for acoustic cavitation *in vivo:* thresholds for bubble formation with 0.75-MHz continuous wave and pulsed beams. *IEEE Trans.* **UFFC 33,** 162–4. [178]

Tew, R. H. (1990). Ray theory of the diffraction of sound by an inhomogeneous membrane. *IMA J. Appl. Math.* **44,** 95–110. [268]

Tew, R. H., Ockendon, J. R., and Briggs, G. A. D. (1988). Acoustical scattering by a shallow surface-breaking crack in an elastic solid under light fluid loading. In *Recent developments in surface acoustic waves* (ed. D. F. Parker and G. A. Maugin), pp. 309–16. Springer-Verlag, Berlin. [107, 144, 262, 268]

Thompson, D. O. and Hsu, D. K. (1988). Technique for generation of unipolar ultrasonic pulses. *IEEE Trans.* **UFFC 35,** 450–6. [75]

Thompson, R. B., Skillings, B. J., Zachary, L. W., Schmerr, L. W., and Buck, O. (1983). Effects of crack closure on ultrasonic transmission. In *Review of progress in quantitative nondestructive evaluation,* Vol. 2 (ed. D. O. Thompson and D. E. Chimenti), pp. 325–41. Plenum Press, New York. [225, 281]

Thompson, R. B., Fiedler, C. J., and Buck, O. (1984). Inference of fatigue crack closure stresses from ultrasonic transmission measurements. In *Nondestructive methods for materials property determination* (ed. C. O. Ruud and R. B. Thompson), pp. 161–70. Plenum Press, New York. [281]

Thompson, R. B., Li. Y., Spitzig, W. A., Briggs, G. A. D., Fagan, A. F., and Kushibiki, J. (1990*a*). Characterization of the texture of heavily deformed metal–metal composites with acoustic microscopy. In *Review of progress in quantitative nondestructive evaluation,* Vol. 9 (ed. D. O. Thompson and D. E. Chimenti), pp. 1433–40. Plenum Press, New York. [250]

Thompson, R. B., Smith, J. F., and Lee, S. S. (1990*b*). Effects of plastic deformation on the inference of stress and texture from the velocities of ultrasonic plate modes. In *Review of progress in quantitative nondestructive evaluation,* Vol. 9 (ed. D. O. Thompson and D. E. Chimenti), pp. 1773–80. Plenum Press, New York. [258]

Tien, J. W., Khuri-Yakub, B. T., Kino, G. S., Evans, A. G., and Marshall, D. B. (1982). Long wavelength measurements of surface cracks in silicon nitride. In *Review of progress in quantitative nondestructive evaluation* Vol. 1 (ed. D. O. Thompson and D. E. Chimenti), pp. 569–71. Plenum Press, New York. [276]

Tsai, C. S. and Lee, C. C. (1987). Nondestructive imaging and characterization of electronic materials and devices using scanning acoustic microscopy. In *Pattern recognition and acoustical imaging* (ed. L. A. Ferrari). *SPIE* **768,** 260–6. [114, 207]

Tsukahara, Y. and Ohira, K. (1989). Attenuation measurements in polymer films and coatings by ultrasonic spectroscopy. *Ultrasonics Int. 89*, 924–9. [208]

Tsukahara, Y., Takeuchi, E., Hayashi, E., and Tani, Y. (1984). A new method of measuring surface layer-thickness using dips in angular dependence of reflection coefficients. *IEEE 1984 Ultrasonics Symposium,* pp. 992–6. IEEE, New York. [219]

Tsukahara, Y., Nakaso, N., Kushibiki, J., and Chubachi, N. (1989*a*). An acoustic micrometer and its application to layer thickness measurements. *IEEE Trans.* **UFFC 36,** 326–31. [218–20]

Tsukahara, Y., Ohira, K., Saito, M., and Briggs, G. A. D. (1989*b*). Evaluation of polymer coatings by ultrasonic spectroscopy. In *Acoustical imaging,* Vol. 17 (ed. H. Shimizu, N. Chubachi, and J. Kushibiki), pp. 257–64. Plenum Press, New York. [219]

Tsukahara, Y., Ohira, K., and Nakaso, N. (1990). An ultrasonic microspectrometer for the evaluation of elastic properties with microscopic resolution. *IEEE 1990 Ultrasonics Symposium,* pp. 925–30 [154]

Tsukahara, Y., Nakaso, N., and Ohira, K. (1991). Angular spectral approach to reflection of focussed beams with oblique incidence in spherical-planar-pair lenses. *IEEE Trans.* **UFFC 38,** 468–80. [154]

Vetters, H., Matthaei, A., Schulz, A., and Mayr, P. (1989). Scanning acoustic microprobe analysis for testing solid state materials. *Mater. Sci. Engng.* **A122,** 9–14. [204, 212, 224]

Wade, G. and Meyyappan, A. (1987). Scanning tomographic acoustic microscopy: principles and recent developments. In *Pattern recognition and acoustical imaging* (ed. L. A. Ferrari). *SPIE* **768,** 267–74. [18]

Wang, J., Gundle, R., and Briggs, G. A. D. (1990). The measurement of acoustic properties of living human cells. *Trans. R. Microsc. Soc.* **1,** 91–4. [156, 174]

Wang, J. K. and Tsai, C. S. (1984). Reflection acoustic microscopy for thick specimens. *J. Appl. Phys.* **55,** 80–8. [212, 215]

Wang, J. K. and Tsai, C. S. (1985). Acoustic transmission and image contrast of tilted plate specimens in transmission acoustic microscopy. *IEEE Trans.* **SU–32,** 241–7. [114]

Wang, J. K., Tsai, C. S. and Lee, C. C. (1980). Spectroscopic study of defects in thick specimen using transmission scanning acoustic microscopy. In *Scanned image microscopy* (ed. E. A. Ash), pp. 137–47. Academic Press, London. [114]

Weaver, J. M. R. (1986). The ultrasonic imaging of plastic deformation. *D.Phil. thesis,* Oxford University. [61, 62, 66, 253]

Weaver, J. M. R. (1991). An optimal R.F. system for quantitative acoustic microscopy. *IEEE Trans.* **UFFC.** (In press.) [69, 70]

Weaver, J. M. R. and Briggs, G. A. D. (1985). Acoustic microscopy techniques for observing dislocation damping. *J. Physique* **12** (C10), 743–50. [258]

Weaver, J. M. R., Briggs, G. A. D., and Somekh, M. G. (1983). Acoustic microscopy of ultrasonic attenuation. *J. Physique* **12** (C9), 371–6. [257]

Weaver, J. M. R., Daft, C. M. W., and Briggs, G. A. D. (1989). A quantitative acoustic microscope with multiple detection modes. *IEEE Trans.* **UFFC 36,** 554–60. [75, 144, 145, 290, 291]

Weglein, R. D. (1979*a*). A model for predicting acoustic materials signatures. *Appl. Phys. Lett.* **34,** 179–81. [108]

Weglein, R. D. (1979*b*). SAW dispersion and film-thickness measurement by acoustic microscopy. *Appl. Phys. Lett.* **35,** 215–17. [215]

Weglein, R. D. (1983). Integrated circuit inspection via acoustic microscopy. *IEEE Trans.* **SU–30,** 40–2. [207]

Weglein, R. D. and Wilson, R. G. (1978). Characteristic materials signatures by acoustic microscopy. *Electron. Lett.* **14,** 352–4. [104]

Wey, A. C. and Kessler, L. W. (1989). Image enhancement for scanning laser acoustic microscopy. *Ultrasonics Int. 89*, 756–61. [18]

Wickramasinghe, H. K. (1978). Contrast in reflection acoustic microscopy. *Electron. Lett.* **14,** 305–6. [112]

Wickramasinghe, H. K. (1979). Contrast and imaging performance in the scanning acoustic microscope. *J. Appl. Phys.* **50,** 664–72. [112]

Wickramasinghe, H. K. (1983). Scanning acoustic microscopy: a review. *J. Microsc.* **129,** 63–73. [11]

Wickramasinghe, H. K. and Petts C. R. (1980). Gas medium acoustic microscopy. In *Scanned image microscopy* (ed. E. A. Ash), pp. 57–70. Academic Press, London. [31, 32, 34]

Wilson, T. and Sheppard, C. J. R. (1984). *Theory and practice of scanning optical microscopy.* Academic Press, London. [28, 39, 47, 48, 182, 204]

Yajima, S., Hasegawa, Y., Hayashi, J., and Iimura, M. (1978). Synthesis of continuous silicon carbide fibre with high tensile strength and high Young's modulus. *J. Mater. Sci.* **13,** 2569–76. [3, 227]

Yamanaka, K. (1982). Analysis of SAW attenuation measurement using acoustic microscopy. *Electron Lett.* **18,** 587–9. [143]

Yamanaka, K. (1983). Surface acoustic wave measurements using an impulsive converging beam. *J. Appl. Phys,* **54,** 4323–9. [76, 144]

Yamanaka, K. and Enomoto, Y. (1982). Observation of surface cracks with scanning acoustic microscope. *J. Appl. Phys.* **53,** 846–50. [262]

Yamanaka, K., Kushibiki, J., and Chubachi, N. (1984). Anisotropy detection in hot-pressed silicon nitride by acoustic microscopy using the line-focus beam. *Electron. Lett.* **21,** 165–7. [291]

Yamanaka, K., Enomoto, Y., and Tsuya, Y. (1985). Acoustic microscopy of ceramic surfaces. *IEEE Trans.* **SU–32,** 313–19. [281]

Yamanaka, K., Nagata, Y., and Koda, T. (1989). Low temperature acoustic microscopy with continuous temperature control. *Ultrasonics. Int. 89*, 744–9. [209]

Yin, Q. R., Ilett, C., and Briggs, G. A. D. (1982). Acoustic microscopy of ferroelectric ceramics. *J. Mater. Sci.* **17,** 2449–52. [224]

Yoon, H. S. and Newnham, R. E. (1969). Elastic properties of fluorapatite. *Am. Mineralogist* **54,** 1193–7. [198]

Yue, G. Q., Nikoonahad, M., and Ash, E. A. (1982). Subsurface acoustic microscopy using pulse compression techniques. *IEEE 1982 Ultrasonics Symposium,* pp. 935–8. IEEE, New York. [73]

Zemanek (1971). Beam behaviour within the nearfield of a vibrating piston. *J. Acoust. Soc. Am.* **49,** 181–91. [56]

Zhai, T.-G., Lin, S., and Xiao, J.-M. (1990). Influence of non-geometric effect of PSB on crack initiation in aluminium single crystal. *Acta Metallurgia et Materialia* **38,** 1687–92. [288]

Zimmerman, M. C., Meunier, A., Katz, J. L., and Christel, P. (1990). The evaluation of cortical bone remodeling with a new ultrasonic technique. *IEEE Trans Biomedical Engineering* **37,** 433–41. [200]

Index

Page numbers indicate pages on which the term may be found, or relevant sections. Additional help may be found from the list of symbols (xiii–xv), where the range of equations where each symbol is used is indicated, and from the list of references (297–315), where the pages on which each reference is cited are given.